Colloids and Interface Science Series
Volume 5

Colloids in Agrochemicals

Tharwat F. Tadros

Colloids and Interface Science Series
edited by Tharwat F. Tadros

Other Volumes

Volume 1: Colloid Stability – Part I

ISBN 978-3-527-31462-1

Volume 2: Colloid Stability – Part II

ISBN 978-3-527-31503-1

Volume 3: Colloid Stability and Application in Pharmacy

ISBN 978-3-527-31463-8

Volume 4: Colloids in Cosmetics and Personal Care

ISBN 978-3-527-31464-5

Forthcoming Volume

Volume 6: Colloids in Paints

ISBN 978-3-527-31466-9

Complete Set

Colloids and Interface Science Series
Volumes 1–6

ISBN 978-3-527-31461-4

Colloids and Interface Science Series
Volume 5

Colloids in Agrochemicals

Colloids and Interface Science

Tharwat F. Tadros

WILEY-VCH

WILEY-VCH Verlag GmbH & Co. KGaA

The Author

Prof. Dr. Tharwat F. Tadros
89 Nash Grove Lane
Wokingham, Berkshire RG40 4HE
United Kingdom

Library of Congress Card No.: applied for

British Library Cataloguing-in-Publication Data
A catalogue record for this book is available from the British Library.

Bibliographic information published by the Deutsche Nationalbibliothek
The Deutsche Nationalbibliothek lists this publication in the Deutsche Nationalbibliografie; detailed bibliographic data are available on the Internet at http://dnb.d-nb.de

Printed in the Federal Republic of Germany
Printed on acid-free paper

Cover design Adam Design, Weinheim
Typesetting Asco Typesetters, Hong Kong
Printing Strauss GmbH, Mörlenbach
Bookbinding Litges & Dopf GmbH, Heppenheim

ISBN 978-3-527-31465-2

This book is dedicated to my grandchildren:
Nadia, Dominic, Theodore, Bruno and Viola

Contents

Colloids and Interface Science Series, Vol. 5
Colloids in Agrochemicals. Tharwat F. Tadros

ISBN: 978-3-527-31465-2

Preface

Most agrochemical formulations consist of disperse systems such as solid–liquid (suspensions), liquid–liquid (emulsions) and mixtures of these (suspoemulsions). In addition, other types of formulations include microemulsions, capsules, granules (for controlled release), multiple emulsions and oil-based suspensions. The formulation may also contain other ingredients such as rheology modifiers (thickeners) and adjuvants for optimization of biological control. Hence in most cases the formulation contains several components which may interact with each other and the final system can be complex.

During the last few decades, most agrochemical formulations were prepared by a simple trial and error procedure. Apart from being tedious and time consuming, this procedure suffers from the disadvantage of lack of understanding of the complex interactions in these disperse systems. This also results in a lack of prediction of the long-term physical stability of the formulations. This book has been written with the objective of understanding the interactions between the various components of the formulations and procedures that can be applied to control and predict the long-term physical stability of the formulation. For this purpose, the fundamental principles of colloid and interface science are applied to the various formulations in an attempt to throw some light on the possible interactions between the particles or droplets in the disperse system. Such an approach also leads to the optimum methods of selection of surfactants and dispersants that are used for formulation of the product.

The main factors that affect biological control are also discussed at a fundamental level. After a short introduction (Chapter 1) describing the different types of agrochemical formulations, four chapters are devoted to the basic principles of colloid and interface science. Chapter 2 describes the various surfactant classes used in the formulation. This is followed by a chapter on the physical chemistry of surfactant solutions and the process of micellization, which is important in biological control. The various interfacial aspects that are involved at the air/liquid, liquid/liquid and solid/liquid interfaces are described in Chapter 4. The interaction forces between particles or droplets in a disperse system and their effect on the colloid stability are described in Chapter 5.

The following chapters deal with the various agrochemical formulations. Chapter 6 describes emulsion concentrates (EWs), which are now commonly used to

Colloids and Interface Science Series, Vol. 5
Colloids in Agrochemicals. Tharwat F. Tadros

ISBN: 978-3-527-31465-2

replace emulsifiable concentrates (ECs). The latter suffer from the disadvantage of having to dissolve the active ingredient in an oil and addition of high surfactant concentrations for self-emulsification when added to water in the spray tank. Chapter 7 describes suspension concentrates (SCs), which are by far the most attractive formulation of insoluble active ingredients. By dispersing the powder in water and grinding the particles to reduce their size, one can prepare a colloidally stable suspension whose sedimentation can be controlled by addition of thickeners. Chapter 8 deals with the formulation of oil-based suspensions, which may be required for actives that are chemically unstable in water. The oil-based suspension is self-emulsified in water by proper choice of surfactants. Chapter 9 describes formulations made of mixtures of suspensions and emulsions (suspoemulsions), which allow one to formulate two active ingredients, one being a solid and the other a liquid. Chapter 10 deals with the formulation of agrochemicals as microemulsions which are thermodynamically stable and can also enhance the biological efficacy. Chapter 11 deals with a specific formulation of multiple emulsions which allows the formulation chemist to prepare a system of various active ingredients and also can be applied for controlled release. The latter topic is described in detail in Chapter 12, which gives the various forms of controlled-release formulations including microencapsulation and the use of granules. The final chapter deals with the basic principles of the selection of adjuvants, which are now commonly used in agrochemical formulations to control and enhance biological efficacy.

This text deals with the fundamentals of colloid and interface science as applied to agrochemical formulations. It also gives basic guidelines for formulating a product. This will enable the formulation chemist to shorten the time required for formulating an agrochemical. It also provides valuable information on how to control and predict the long-term physical stability of the system.

Wokingham, February 2009 *Tharwat Tadros*

1
General Introduction

The formulations of agrochemicals cover a wide range of systems that are prepared to suit a specific application. Agrochemicals are usually effective at levels from several grams to hundreds of grams of active ingredient per 1000 m^2. It is therefore difficult to apply such a small amount uniformly to the crop. In all cases, the active ingredient is first formulated in a suitable diluent such as water or an organic solvent, and when the formulation is applied it is further diluted in the spray tank to ensure uniform deposition on spraying. In some cases, an agrochemical is a water-soluble compound, of which paraquat and glyphosate (both are herbicides) are probably the most familiar. Paraquat is a 4,4′-bypyridinium salt and the counterions are normally chloride. It is formulated as a 20% aqueous solution, which on application is simply diluted into water at various ratios (1 : 50 up to 1 : 200 depending on the application). To such an aqueous solution, surface-active agents (also called surfactants, and sometimes referred to as wetters) are added, which are essential for a number of reasons. The most obvious reason for adding surfactants is to enable the spray solution to adhere to the target surface and spread over it to cover a large area. However, such a picture is an oversimplification since the surfactant plays a more subtle role in the optimization of the biological efficacy. Thus, the choice of the surfactant system in an agrochemical formulation is crucial since it has to perform a number of functions. To date, such a choice is made by a trial-and-error procedure, due to the complex nature of application and a lack of understanding of the mode of action of the chemical. It is the objective of this book to apply the basic principle of colloid and interface science in agrochemical formulations, subsequent application and optimization of biological efficacy.

The main purpose of any agrochemical formulation is to make handling and application of the active ingredient as easy as possible. An important function of the formulation is to optimize the biological efficacy. This is achieved in most cases by controlling the physical characteristics of the formulation and the use of adjuvants. An important criterion for any agrochemical is its safety, both to the crop and to the agrochemical worker. This requires adequate control of the spray droplet spectrum, reduction of any drift and removal of any toxic effect on contact with the individual. These stringent requirements can be achieved by careful analysis of all the interfacial phenomena that are involved on application. The concept of a ‘pesticide delivery system’ (PDS) must be applied, whereby the

Colloids and Interface Science Series, Vol. 5
Colloids in Agrochemicals. Tharwat F. Tadros

ISBN: 978-3-527-31465-2

active ingredients are made available to a specified target at a concentration and duration designed to accomplish an intended effect, i.e. obtain the fullest biological efficacy while minimizing the various harmful effects.

Most agrochemicals are water-insoluble compounds with various physical properties, which have first to be determined in order to decide on the type of formulation. One of the earliest types of formulations were wettable powders (WPs), which are suitable for formulating solid water-insoluble compounds that can be produced in a powder form. The chemical (which may be micronized) is mixed with a filler such as china clay and a solid surfactant, such as sodium alkyl or alkyl aryl sulfate or sulfonate, is added. When the powder is added to water, the particles are spontaneously wetted by the medium and on agitation dispersion of the particles takes place. It is clear that the particles should remain suspended in the continuous medium for a period of time depending on the application. Some physical testing methods are available to evaluate the suspensibility of the WP. Clearly, the surfactant system plays a crucial role in wettable powders. First, it allows spontaneous wetting and dispersion of the particles. Second, by adsorption on the particle surface, it provides a repulsive force that prevents aggregation of the particles. Such a process of aggregation will enhance the settling of the particles and may also cause problems on application, such as nozzle blockage.

The second and most familiar type of agrochemical formulations is emulsifiable concentrates (ECs). These are produced by mixing an agrochemical oil with another one such as xylene or trimethylbenzene or a mixture of various hydrocarbon solvents. Alternatively, a solid pesticide could be dissolved in a specific oil to produce a concentrated solution. In some cases, the pesticide oil may be used without any extra addition of oils. In all cases, a surfactant system (usually a mixture of two or three components) is added, for a number of purposes. First, the surfactant permits self-emulsification of the oil on addition to water. This occurs by a complex mechanism that involves a number of physical changes, such as lowering of the interfacial tension at the oil/water interface or enhancement of turbulence at that interface, with the result of spontaneous production of droplets. Second, the surfactant film that adsorbs at the oil/water interface stabilizes the emulsion produced against flocculation and/or coalescence. As we will see in later chapters, emulsion breakdown must be prevented, otherwise excessive creaming or sedimentation or oil separation might take place during application. This results in an inhomogeneous application of the agrochemical on the one hand and possible losses on the other. The third role of the surfactant system in agrochemicals is in enhancement of biological efficacy. As we will see in subsequent chapters, it is essential to arrive at optimum conditions for effective use of the agrochemicals. In this case, the surfactant system will help in spreading of the pesticide on the target surface and may enhance its penetration.

In recent years, there has been a great demand to replace ECs with concentrated aqueous oil-in-water (O/W) emulsions. Several advantages may be envisaged for such replacements. First, one is able to replace the added oil with water, which is of course much cheaper and environmentally acceptable. Second, removal of the oil could help in reducing undesirable effects such as phytotoxicity and skin irritation. Third, by formulating the pesticide as an O/W emulsion one is able to control the droplet size to an optimum value which may be crucial for biological efficacy. Fourth, water-soluble surfactants, which may be desirable for biological optimization, can be added to the aqueous continuous phase. As we will see later, the choice of a surfactant or a mixed surfactant system is crucial for the preparation of a stable O/W emulsion. In recent years, macromolecular surfactants have been designed to produce very stable O/W emulsions which could be easily diluted into water and applied without any detrimental effects to the emulsion droplets.

A similar concept has been applied to replace wettable powders, namely with aqueous suspension concentrates (SCs). These systems are more familiar than ECs and they have been in use for several decades. Indeed, SCs are probably the most widely used systems in agrochemical formulations. Again, SCs are much more convenient to apply than WPs. Dust hazards are absent and the formulation can be simply diluted in the spray tanks, without the need for any vigorous agitation. As we will see later, SCs are produced by a two- or three-stage process. The pesticide powder is first dispersed in an aqueous solution of a surfactant or a macromolecule (usually referred to as the dispersing agent) using a high-speed mixer. The resulting suspension is then subjected to a wet milling process (usually bead milling) to break any remaining aggregates or agglomerates and reduce the particle size. One usually aims at a particle size distribution ranging from 0.1 to 5 μm, with an average of 1–2 μm. The surfactant or polymer added adsorbs on the particle surface, resulting in colloidal stability. The particles need to be maintained stable over a long period of time, since any strong aggregation in the system may cause various problems. First, the aggregates, being larger than the primary particles, tend to settle faster. Second, any gross aggregation may result in a lack of dispersion on dilution. The large aggregates can block the spray nozzles and may reduce biological efficacy as a result of the inhomogeneous distribution of the particles on the target surface. Apart from their role in ensuring the colloidal stability of the suspension, surfactants are added to many SCs to enhance their biological efficacy. This is usually achieved by solubilization of the insoluble compound in the surfactant micelles. This will be discussed in later chapters. Another role that a surfactant may play in SCs is in the reduction of crystal growth (Ostwald ripening). This process may occur when the solubility of the agrochemical is appreciable (say greater than 100 ppm) and when the SC is polydisperse. The smaller particles will have a higher solubility than the larger particles. With time, the small particles dissolve and become deposited on the larger particles. Surfactants may reduce this Ostwald ripening by adsorption on the crystal surfaces, thus preventing deposition of the molecules on the surface. This will be described in detail in Chapter 7 on SCs.

Recently, microemulsions have been considered as potential systems for formulating agrochemicals. Microemulsions are isotropic, thermodynamically stable systems consisting of oil, water and surfactant(s), whereby the free energy of formation of the system is zero or negative. It is obvious why such systems, if they can be formulated, are very attractive, since they will have an indefinite shelf-life (within a certain temperature range). Since the droplet size of microemulsions is very small (usually less than 50 nm), they appear transparent. As we will see in later chapters, the microemulsion droplets may be considered as swollen micelles and hence they will solubilize the agrochemical. This may result in considerable enhancement of the biological efficacy. Hence microemulsions may offer several advantages over the commonly used macroemulsions. Unfortunately, formulating the agrochemical as a microemulsion is not straightforward since one usually uses two or more surfactants, an oil and the agrochemical. These tertiary systems produce various complex phases and it is essential to investigate the phase diagram before arriving at the optimum composition of microemulsion formation. As we will see in Chapter 10 on microemulsions, a high concentration of surfactant (10–20%) is needed to produce such a formulation. This makes such systems relatively more expensive to produce compared with macroemulsions. However, the extra cost incurred could be offset by an enhancement of biological efficacy, which means that a lower agrochemical application rate could be achieved.

An important application in agrochemicals is that of controlled-release formulations. Several methods are used for controlled release, of which microcapsules are probably the most widely used. These are small particles with a size range of 1–1000 μm consisting of a core material and an outer wall. The latter isolates the core material from the environment and protects it from degradation and interaction with other materials. The core active ingredient is designed to be released in a controlled manner as required. Microencapsulation of agrochemicals is usually carried out by interfacial condensation, *in situ* polymerization or coacervation. Interfacial condensation whereby two monomers, one oil soluble (placed, say, in an emulsion droplet) and one water soluble placed in the continuous medium, undergo interfacial polycondensation, producing a capsule wall of polyurea or polyurethane. The polymer wall must have appropriate molecular weight, glass transition temperature and thickness to achieve the desirable controlled release. The polymer wall should not interact with the agrochemical. This polymer wall must not cause any environmental damage on degradation after application and hence a biodegradable polymer is preferred. The polymer wall, which must be easily manufactured, should also be stable on storage and usage. The main advantages of microcapsule formulations are controlled or slow release of the core active ingredients, thus improving residual activity, reduction of application dosage, stabilization of the core active ingredient against environmental degradation, reduction of mammalian and fish toxicity, reduction of phytotoxicity and reduction of environmental pollution.

It can be seen from the above short discussion that agrochemical formulations are complex multi-phase systems and their preparation, stabilization and subsequent application require the application of the basic principles of colloid and interface science, and this is the objective of the present book. Chapter 2 gives a brief description of surfactants and dispersants that are commonly used in agrochemical formulations. This is followed by Chapter 3 on the solution properties of surfactants, in particular describing the phenomenon of micellization, Krafft temperature and phase behavior of surfactants. Chapter 4 deals with the interfacial aspects of agrochemical formulations. In this respect, the adsorption of surfactants at the air/water interface that is essential for spray application is described, with particular reference to the dynamics of adsorption. This is then followed by a description of the adsorption isotherms at the oil/water interface. A section is devoted to the adsorption of surfactants and polymeric surfactants at the solid/liquid interface. Chapter 5 considers the interaction forces between particles or droplets in disperse systems. The main interaction forces, namely van der Waals attraction and electrostatic and steric repulsion, are described at a fundamental level. Combination of these interaction forces gives the general trends of the energy–distance curves between particles or droplets in disperse systems. These curves can describe the stability/instability of dispersions. Chapter 6 deals with liquid/liquid dispersions (emulsions) that are referred to as EWs. The various breakdown processes that occur on storage are described and the methods that can be applied to overcome them are discussed. Chapter 7 is concerned with suspension concentrates (SCs) that are used to formulate solid agrochemicals with high melting points. The problem of sedimentation and its prevention are discussed. In particular, the use of rheology modifiers for the prevention of sedimentation is described. Chapter 8 deals with oil-based suspensions, which have recently been introduced, allowing one to formulate agrochemical solids that are chemically unstable in aqueous media. In addition, such oil-based suspensions allow one to incorporate water-insoluble adjuvants, thus enhancing the biological efficacy. Chapter 9 considers suspoemulsions (mixtures of suspensions and emulsions). The various interactions that may occur between the particles and droplets are analyzed at a fundamental level. Chapter 10 describes the use of microemulsions for the formulation of agrochemicals and their main advantages. Chapter 11 gives a short summary of the potential application of multiple emulsions in agrochemical formulations. Controlled-release formulations are discussed in Chapter 12. Finally, Chapter 13 describes the basic principles for the selection of adjuvants in agrochemical formulations.

Further Reading

1 Th.F. Tadros, *Surfactants in Agrochemicals*, Marcel Dekker, New York (1994).
2 Th.F. Tadros, *Applied Surfactants*, Wiley-VCH, Weinheim (2005).

2
General Classification of Surface-active Agents Used in Agrochemical Formulations

A simple classification of surfactants based on the nature of the hydrophilic group is commonly used. Three main classes may be distinguished, namely anionic, cationic and amphoteric [1, 2]. A useful technical reference is McCutcheon [3], which is produced annually to update the list of available surfactants. A list of the physicochemical properties of selected anionic, cationic and nonionic surfactants is available in a book by van Os *et al.* [4]. Another useful text is the *Handbook of Surfactants* by Porter [5]. It should be mentioned also that a fourth class of surfactants, usually referred to as polymeric surfactants, has been used for many years for the preparation of emulsion concentrates (EWs) and suspension concentrates (SCs) and their stabilization.

2.1
Anionic Surfactants

These are the most widely used class of surfactants in agrochemical applications [6, 7]. This is due to their relatively low cost of manufacture and they are used in practically every type of formulation. Linear chains are preferred since they are more effective and more degradable than the branched chains. The most commonly used hydrophilic groups are carboxylates, sulfates, sulfonates and phosphates. General formulae can be assigned to anionic surfactants as follows:

Carboxylates: $C_nH_{2n+1}COO^- \; X^+$
Sulfates: $C_nH_{2n+1}OSO_3^- \; X^+$
Sulfonates: $C_nH_{2n+1}SO_3^- \; X^+$
Phosphates: $C_nH_{2n+1}OPO(OH)O^- \; X^+$

with n being in the range 8–16 atoms and the counterion X^+ is usually Na^+.

Several other types of anionic surfactants are commercially available, such as sulfosuccinates, isothionates and taurates, and these are sometimes used for special applications. Below a brief description of the above anionic classes is given with some of their applications.

Colloids and Interface Science Series, Vol. 5
Colloids in Agrochemicals. Tharwat F. Tadros

ISBN: 978-3-527-31465-2

2.1.1 Carboxylates

These are perhaps the earliest known surfactants, since they constitute the earliest soaps, e.g. sodium or potassium stearate, $C_{17}H_{35}COONa$, and sodium myristate, $C_{14}H_{29}COONa$. The alkyl group may contain unsaturated portions, e.g. sodium oleate, which contains one double bond in the C_{17} alkyl chain. Most commercial soaps will be a mixture of fatty acids obtained from tallow, coconut oil, palm oil, etc. The main attraction of these simple soaps is their low cost, ready biodegradability and low toxicity. Their main disadvantage is their ready precipitation in water containing bivalent ions such as Ca^{2+} and Mg^{2+}. To avoid their precipitation in hard water, the carboxylates are modified by introducing some hydrophilic chains, e.g. ethoxycarboxylates with the general structure $RO(CH_2CH_2O)_nCH_2COO^-$, ester carboxylates containing hydroxyl or multiple COOH groups, sarcosinates which contain an amide group with the general structure $RCON(R')COO^-$.

The addition of the ethoxylated groups results in increased water solubility and enhanced chemical stability (no hydrolysis). The modified ether carboxylates are also more compatible with electrolytes. They are also compatible with other nonionic, amphoteric and sometimes even cationic surfactants. The ester carboxylates are very soluble in water, but they suffer from the problem of hydrolysis. The sarcosinates are not very soluble in acidic or neutral solutions but they are fairly soluble in alkaline media. They are compatible with other anionics, nonionics and cationics. The phosphate esters have very interesting properties, being intermediate between ethoxylated nonionics and sulfated derivatives. They have good compatibility with inorganic builders and they can be good emulsifiers. A specific salt of a fatty acid is lithium 12-hydroxystearate, which forms the major constituent of greases.

2.1.2 Sulfates

These are the largest and most important class of synthetic surfactants, which were produced by reaction of an alcohol with sulfuric acid, i.e. they are esters of sulfuric acid. In practice, sulfuric acid is seldom used and chlorosulfonic or sulfur dioxide–air mixtures are the most common methods of sulfating the alcohol. However, due to their chemical instability (hydrolyzing to the alcohol, particularly in acidic solutions), they have now been overtaken by the sulfonates, which are chemically stable.

The properties of sulfate surfactants depend on the nature of the alkyl chain and the sulfate group. The alkali metal salts show good solubility in water, but they tend to be affected by the presence of electrolytes. The most common sulfate surfactant is sodium dodecyl sulfate (abbreviated SDS, and sometimes referred to as sodium lauryl sulfate) which is extensively used both for fundamental studies and in many applications in industry. At room temperature (~25 °C) this surfactant is fairly soluble and 30% aqueous solutions are fairly fluid (low viscosity). However, below 25 °C the surfactant may separate out as a soft paste as the temperature falls below its Krafft point (the temperature above which the surfactant shows a rapid increase in solubility with further increase in temperature). The latter depends on the distribution of chain lengths in the alkyl chain: the wider the distribution, the lower is the Krafft temperature. Hence, by controlling this distribution, one may achieve a Krafft temperature of ~10 °C. As the surfactant concentration is increased to 30–40% (depending on the distribution of the chain length in the alkyl group), the viscosity of the solution increases very rapidly and may produce a gel. The critical micelle concentration (cmc) of SDS (the concentration above which the properties of the solution show abrupt changes) is 8×10^{-3} mol dm^{-3} (0.24%).

As with the carboxylates, the sulfate surfactants are also chemically modified to change their properties. The most common modification is to introduce some ethylene oxide (EO) units into the chain, giving what are usually referred to as alcohol ether sulfates, for example sodium dodecyl 3-mole ether sulfate, which is essentially dodecyl alcohol reacted with 3 mol of EO, then sulfated and neutralized with NaOH. The presence of EO units confers improved solubility compared with the straight alcohol sulfates. In addition, the surfactant becomes more compatible with electrolytes in aqueous solution. The ether sulfates are also more chemically stable than the alcohol sulfates. The cmc of the ether sulfates is also lower than those of the corresponding surfactants without the EO units.

2.1.3 Sulfonates

With sulfonates, the sulfur atom is directly attached to the carbon atom of the alkyl group and this gives the molecule stability against hydrolysis, compared with the sulfates (whereby the sulfur atom is indirectly linked to the carbon of the hydrophobe via an oxygen atom). The alkyl aryl sulfonates are the most common type of these surfactants (e.g. sodium alkyl benzene sulfonate) and are usually prepared by reaction of sulfuric acid with alkyl aryl hydrocarbons, e.g. dodecylbenzene. A special class of sulfonate surfactants is the naphthalene and alkyl naphthalene sulfonates, which are commonly used as dispersants for SCs.

As with the sulfates, some chemical modification is used by introducing ethylene oxide units, e.g. sodium nonylphenol 2-mole ethoxylate ethane sulfonate, $C_9H_{19}C_6H_4(OCH_2CH_2)_2SO_3Na$.

The linear alkyl benzene sulfonates (LABS) are manufactured from alkylbenzenes; the alkyl chain length can vary from C_8 to C_{15} and their properties are mainly influenced by the average molecular weight and the spread of carbon number of the alkyl side chain. The cmc of sodium dodecyl benzene sulfonate is 5×10^{-3} mol dm^{-3} (0.18%).

Another class of sulfonates is the α-alkene sulfonates, which are prepared by reacting a linear α-alkene with sulfur trioxide, typically yielding a mixture of alkene sulfonates (60–70%), 3- and 4-hydroxyalkane sulfonates (~30%) and some disulfonates and other species. The two main α-alkene fractions used as starting material are C_{12}–C_{16} and C_{16}–C_{18}.

A special class of sulfonates are sulfosuccinates, which are esters of sulfosuccinic acid:

$$
\begin{array}{l}
\quad CH_2COOH \\
\quad\ | \\
HSO_3CHCOOH
\end{array}
$$

Both mono- and diesters are produced. A widely used diester in many formulations is sodium di(2-ethylhexyl)sulfosuccinate (which is sold commercially under the trade name Aerosol OT). This is a particularly excellent wetting agent for agrochemical powders in water, which is essential for the preparation of SCs. The diesters are soluble both in water and in many organic solvents. They are particularly useful for the preparation of water-in-oil (W/O) microemulsions.

2.1.4 Phosphate-containing Anionic Surfactants

Both alkyl phosphates and alkyl ether phosphates are made by treating the fatty alcohol or alcohol ethoxylates with a phosphorylating agent, usually phosphorus pentoxide, P_4O_{10}. The reaction yields a mixture of mono- and diesters of phosphoric acid. The ratio of the two esters is determined by the ratio of the reactants and the amount of water present in the reaction mixture. The physicochemical properties of the alkyl phosphate surfactants depend on the ratio of the esters. Phosphate surfactants are used in the metalworking industry due to their anticorrosive properties.

2.2 Cationic Surfactants

The most common cationic surfactants are the quaternary ammonium compounds [8, 9] with the general formula $R'R''R'''R''''N^+ X^-$, where X^- is usually chloride and the Rs represent alkyl groups. A common class of cationics is the alkyltrimethylammonium chloride, where R contains 8–18 C atoms, e.g. dodecyltrimethylammonium chloride, $C_{12}H_{25}(CH_3)_3NCl$. Another cationic surfactant class is that containing two long-chain alkyl groups, e.g. dialkyldimethylammonium chloride, with the alkyl groups having a chain length of 8–18 C atoms. These dialkyl surfactants are less soluble in water than the monoalkyl quaternary compounds, but they are sometimes used in agrochemical formulations as adjuvants and/or rheology modifiers. A special cationic surfactant is alkyldimethylbenzylammonium chloride (sometimes referred to as benzalkonium chloride), which may also be used in some formulations as an adjuvant. Imidazolines can also form quaternaries, the most common product being the ditallow derivative quaternized with dimethyl sulfate.

Cationic surfactants can also be modified by incorporating poly(ethylene oxide) chains, e.g. dodecylmethyl[poly(ethylene oxide)]ammonium chloride.

Cationic surfactants are generally water soluble when there is only one long alkyl group. They are generally compatible with most inorganic ions and hard water. Cationics are generally stable to pH changes, both acidic and alkaline. They are incompatible with most anionic surfactants, but they are compatible with nonionics. These cationic surfactants are insoluble in hydrocarbon oils. In contrast, cationics with two or more long alkyl chains are soluble in hydrocarbon solvents, but they become only dispersible in water (sometimes forming bilayer vesicle-type structures). They are generally chemically stable and can tolerate electrolytes. The cmc of cationic surfactants is close to that of anionics with the same alkyl chain length.

2.3 Amphoteric (Zwitterionic) Surfactants

These are surfactants containing both cationic and anionic groups [10]. The most common amphoterics are the *N*-alkylbetaines, which are derivatives of trimethylglycine, $(CH_3)_3NCH_2COOH$ (which was described as betaine). An example of a betaine surfactant is laurylamidopropyldimethylbetaine, $C_{12}H_{25}CON(CH_3)_2CH_2$-COOH. These alkylbetaines are sometimes described as alkyl dimethylglycinates.

The main characteristic of amphoteric surfactants is their dependence on the pH of the solution in which they are dissolved. In acidic solutions, the molecule acquires a positive charge and it behaves like a cationic surfactant, whereas in alkaline solutions, they become negatively charged and behave like an anionic surfactant. A specific pH can be defined at which both ionic groups show equal ionization (the isoelectric point of the molecule). This can be described as shown in Scheme 2.1.

$N^+\cdots COOH$	$\leftrightarrow$	$N^+\cdots COO^-$	$\leftrightarrow$	$NH\cdots COO^-$
acidic pH < 3		isoelectric		pH > 6 alkaline

Scheme 2.1

Amphoteric surfactants are sometimes referred to as zwitterionic molecules. They are soluble in water, but the solubility shows a minimum at the isoelectric point. Amphoterics show excellent compatibility with other surfactants, forming mixed micelles. They are chemically stable in both acids and alkalis. The surface activities of amphoterics vary widely and depend on the distance between the charged groups, and they show a maximum in surface activity at the isoelectric point.

Another class of amphoterics are the *N*-alkylaminopropionates having the structure $RNHCH_2CH_2COOH$. The NH group is reactive and can react with another acidic molecule (e.g. acrylic) to form an aminodipropionate, $RN(CH_2CH_2COOH)_2$. Alkylimidazoline-based products can also be produced by reacting an alkylimidozoline with a chloro acid. However, the imidazoline ring breaks down during the formation of the amphoteric.

The change in charge with pH of amphoteric surfactants affects their properties, such as wetting and foaming. At the isoelectric point, the properties of amphoterics resemble those of nonionics very closely. Below and above the isoelectric point, the properties shift towards those of cationic and anionic surfactants, respectively. Zwitterionic surfactants have excellent dermatological properties and they also exhibit low eye irritation.

2.4 Nonionic Surfactants

The most common nonionic surfactants are those based on ethylene oxide, referred to as ethoxylated surfactants [11–13]. Several classes can be distinguished: alcohol ethoxylates, alkylphenol ethoxylates, fatty acid ethoxylates, monoalkanolamide ethoxylates, sorbitan ester ethoxylates, fatty amine ethoxylates and ethylene oxide–propylene oxide copolymers (sometimes referred to as polymeric surfactants).

Another important class of nonionics is the multi-hydroxy products such as glycol esters, glycerol (and polyglycerol) esters, glucosides (and polyglucosides) and sucrose esters. Amine oxides and sulfinyl surfactants represent nonionics with a small head group.

2.4.1 Alcohol Ethoxylates

These are generally produced by ethoxylation of a fatty chain alcohol such as dodecanol. Several generic names are given to this class of surfactants, such as ethoxylated fatty alcohols, alkyl polyoxyethylene glycol and monoalkyl poly(ethylene oxide) glycol ethers. A typical example is dodecylhexaoxyethylene glycol monoether, $C_{12}H_{25}(OCH_2CH_2O)_6OH$ (sometimes abbreviated to $C_{12}E_6$). In practice, the starting alcohol will have a distribution of alkyl chain lengths and the resulting ethoxylate will have a distribution of ethylene oxide chain length. Hence the numbers given in the literature refer to average numbers.

The cmc of nonionic surfactants is about two orders of magnitude lower than those of the corresponding anionics with the same alkyl chain length. The solubility of the alcohol ethoxylates depends both on the alkyl chain length and the number of ethylene oxide units in the molecule. Molecules with an average alkyl chain length of 12 C atoms and containing more than five EO units are usually soluble in water at room temperature. However, as the temperature of the solution is gradually raised, the solution becomes cloudy (as a result of dehydration of the PEO chain) and the temperature at which this occurs is referred to as the cloud point (CP) of the surfactant. At a given alkyl chain length, the CP increases with increase in the EO chain length of the molecule. The CP changes with change in concentration of the surfactant solution and the trade literature usually quotes the CP of a 1% solution. The CP is also affected by the presence of electrolytes in the aqueous solution. Most electrolytes lower the CP of a nonionic surfactant solution. Nonionics tend to have maximum surface activity near the CP. The CP of most nonionics increases markedly on addition of small quantities of anionic surfactants. The surface tension of alcohol ethoxylate solutions decreases with decrease in the number of EO units in the chain. The viscosity of a nonionic surfactant solution increases gradually with increase in its concentration, but at a critical concentration (which depends on the alkyl and EO chain lengths) the viscosity shows a rapid increase and ultimately a gel-like structure appears. This results from the formation of a liquid crystalline structure of the hexagonal type. In many cases, the viscosity reaches a maximum, after which it shows a decrease due to the formation of other structures (e.g. lamellar phases) (see Chapter 3).

2.4.2
Alkylphenol Ethoxylates

These are prepared by reaction of ethylene oxide with the appropriate alkylphenol. The most common surfactants of this type are those based on nonylphenol. These surfactants are cheap to produce, but they suffer from the problem of biodegradability and potential toxicity (the byproduct of degradation is nonylphenol, which has considerable toxicity). In spite of these problems, nonylphenol ethoxylates are still used in many agrochemical formulations, due to their advantageous properties, such as their solubility in both aqueous and nonaqueous media and their good emulsification and dispersion properties.

2.4.3
Fatty Acid Ethoxylates

These are produced by reaction of ethylene oxide with a fatty acid or a polyglycol and they have the general formula $RCOO(CH_2CH_2O)_nH$. When a polyglycol is used, a mixture of mono- and diesters [$RCOO(CH_2CH_2O)_nOCOR$] is produced. These surfactants are generally soluble in water provided that there are enough EO units and the alkyl chain length of the acid is not too long. The monoesters are much more soluble in water than the diesters. In the latter case, a longer EO chain is required to render the molecule soluble. The surfactants are compatible with aqueous ions, provided that here is not much unreacted acid. However, these surfactants undergo hydrolysis in highly alkaline solutions.

2.4.4
Sorbitan Esters and Their Ethoxylated Derivatives (Spans and Tweens)

The fatty acid esters of sorbitan (generally referred to as Spans, an Atlas commercial trade name) and their ethoxylated derivatives (generally referred to as Tweens) are among the most commonly used nonionics. The sorbitan esters are produced by reaction of sorbitol with a fatty acid at a high temperature (>200 °C). The sorbitol dehydrates to 1,4-sorbitan and then esterification takes place. If 1 mol of fatty acid is reacted with 1 mol of sorbitol, one obtains a monoester (some diester is also produced as a byproduct). Thus, sorbitan monoester has the following general formula:

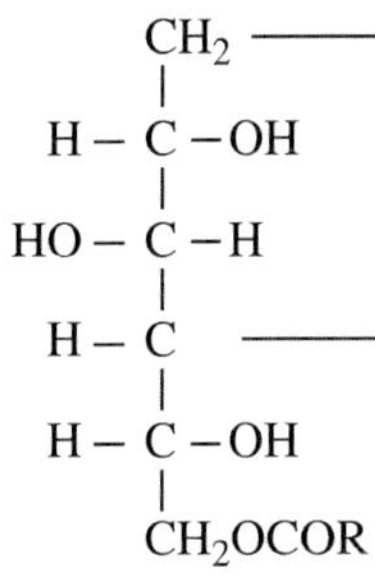

The free OH groups in the molecule can be esterified, producing di- and triesters. Several products are available depending on the nature of the alkyl group of the acid and whether the product is a mono-, di- or triester. Some examples are given below:

Sorbitan monolaurate – Span 20
Sorbitan monopalmitate – Span 40
Sorbitan monostearate – Span 60
Sorbitan mono-oleate – Span 80
Sorbitan tristearate – Span 65
Sorbitan trioleate – Span 85

The ethoxylated derivatives of Spans (Tweens) are produced by reaction of ethylene oxide with any hydroxyl group remaining on the sorbitan ester group. Alternatively, the sorbitol is first ethoxylated and then esterified. However, the final product has different surfactant properties to the Tweens. Some examples of Tween surfactants areas follows:

Polyoxyethylene (20) sorbitan monolaurate – Tween 20
Polyoxyethylene (20) sorbitan monopalmitate – Tween 40
Polyoxyethylene (20) sorbitan monostearate – Tween 60
Polyoxyethylene (20) sorbitan mono-oleate – Tween 80
Polyoxyethylene (20) sorbitan tristearate – Tween 65
Polyoxyethylene (20) sorbitan tri-oleate – Tween 85

The sorbitan esters are insoluble in water, but soluble in most organic solvents (low HLB (Hydrophilic–Lipophilic–Balance) number surfactants). The ethoxylated products are generally soluble in number and they have relatively high HLB numbers. One of the main advantages of the sorbitan esters and their ethoxylated derivatives is their approval as food additives. They are also widely used in cosmetics and some pharmaceutical preparations.

2.4.5 Ethoxylated Fats and Oils

A number of natural fats and oils have been ethoxylated, e.g. lanolin (wool fat) and castor oil ethoxylates. These products are useful for application in agrochemical formulations, e.g. as solubilizers.

2.4.6 Amine Ethoxylates

These are prepared by addition of ethylene oxide to primary or secondary fatty amines. With primary amines both hydrogen atoms on the amine group react with ethylene oxide and therefore the resulting surfactant has the following structure:

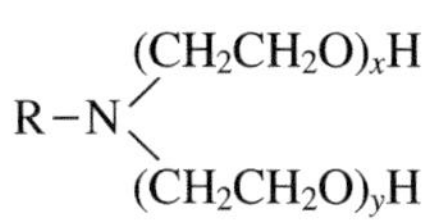

The above surfactants acquire a cationic character if the EO units are small in number and if the pH is low. However, at high EO levels and neutral pH they behave very similarly to nonionics. At low EO content, the surfactants are not soluble in water, but become soluble in an acidic solution. At high pH, the amine ethoxylates are water soluble provided that the alkyl chain length of the compound is not long (usually a C_{12} chain is adequate for reasonable solubility at sufficient EO content).

2.4.7 Surfactants Derived from Mono- and Polysaccharides

Several surfactants have been synthesized starting from mono- or oligosaccharides by reaction with the multifunctional hydroxy groups. The technical problem is one of joining a hydrophobic group to the multihydroxy structure. Several surfactants were made, e.g. by esterification of sucrose with fatty acids or fatty glycerides to produce sucrose esters.

The most interesting sugar surfactants are the alkyl polyglucosides (APGs). These molecules are produced from starch or glucose first by reaction with butanol in the presence of an acid catalyst to produce a butyl oligoglycoside intermediate which then reacts with a fatty alcohol such as dodecanol (acid catalyst) to produce dodecyl polyglucoside with a low degree of polymerization n (1.1–3).

The basic raw materials are glucose and fatty alcohols (which may be derived from vegetable oils) and hence these surfactants are sometimes referred to as 'environmentally friendly'. A product with $n = 2$ has two glucose residues with four OH groups on each molecule (i.e. a total of eight OH groups). The chemistry is more complex and commercial products are mixtures with $n = 1.1–3$. The properties of APG surfactants depend on the alkyl chain length and the average degree of polymerization. APG surfactants have good solubility in water and they have high cloud points (>100 °C). They are stable in neutral and alkaline solutions but are unstable in strongly acidic solutions. APG surfactants can tolerate high electrolyte concentrations and they are compatible with most types of surfactants.

2.5 Speciality Surfactants – Fluorocarbon and Silicone Surfactants

These surfactants can lower the surface tension of water to values below 20 mN m^{-1} (most surfactants described above lower the surface tension of water to values above 20 mN m^{-1}, typically in the region of 25–27 mN m^{-1}). The fluorocarbon and silicone surfactants are sometimes referred to as superwetters as they cause enhanced wetting and spreading of their aqueous solution. However, they are much more expensive than conventional surfactants and they are only applied for specific applications whereby the low surface tension is a desirable property.

Fluorocarbon surfactants have been prepared with various structures consisting of perfluoroalkyl chains and anionic, cationic, amphoteric and poly(ethylene oxide) polar groups. These surfactants have good thermal and chemical stability and they are excellent wetting agents for low-energy surfaces.

Silicone surfactants, sometimes referred to as organosilicones, are those with a polydimethylsiloxane backbone. The silicone surfactants are prepared by incorporation of a water-soluble or hydrophilic group into a siloxane backbone. The latter can also be modified by incorporation of a paraffinic hydrophobic chain at the end or along the polysiloxane backbone. The most common hydrophilic groups are EO/PO and the structures produced are rather complex and most manufacturers of silicone surfactants do not reveal the exact structure. The mechanism by which these molecules lower the surface tension of water to low values is far from being well understood. The surfactants are widely applied as spreading agents on many hydrophobic surfaces.

Incorporating organophilic groups into the backbone of the polydimethylsiloxane backbone can give products that exhibit surface-active properties in organic solvents.

2.6 Polymeric Surfactants

2.6.1 Ethylene Oxide–Propylene Oxide Copolymers (EO/PO)

These surfactants are sold under various trade names, e.g. Pluronics (Wyandotte), Synperonic PE (ICI) and Ploxomers. Two types may be distinguished: those prepared by reaction of polyoxypropylene glycol (difunctional) with EO or mixed EO/PO, giving block copolymers with the following structure:

$$HO(CH_2CH_2O)_n(CH_2\underset{\displaystyle CH_3}{\underset{|}{C}}HO)_m(CH_2CH_2)_nOH$$

abbreviated $(EO)_n(PO)_m(EO)_n$

abbreviated $(EO)_n(PO)_m(EO)_n$. Various molecules are available, where n and m are varied systematically.

The second type of EO/PO copolymers are prepared by reaction of poly(ethylene glycol) (difunctional) with PO or mixed EO/PO. These will have the structure $(PO)_n(EO)_m(PO)_n$ and they are referred to as reverse Pluronics.

Trifunctional products are also available where the starting material is glycerol. These have the following structure:

$$\begin{array}{l} CH_2\text{—}(PO)_m(EO)_n \\ | \\ CH\text{—}(PO)_n(EO)_n \\ | \\ CH_2\text{—}(PO)_m(EO)_n \end{array}$$

Tetrafunctional products are available where the starting material is ethylene diamine. These have the following structures:

$$\begin{array}{ccc} (EO)_n & & (EO)_n \\ & \diagdown \quad \diagup & \\ & NCH_2CH_2N & \\ & \diagup \quad \diagdown & \\ (EO)_n & & (EO)_n \end{array}$$

$$\begin{array}{ccc} (EO)_n(PO)_m & & (PO)_m(EO)_n \\ & \diagdown \quad \diagup & \\ & NCH_2CH_2N & \\ & \diagup \quad \diagdown & \\ (EO)_n(PO)_m & & (PO)_m(EO)_n \end{array}$$

The recent development of speciality polymeric surfactants of the graft type ('comb' structures) have enabled one to obtain specific applications in dispersions.

An example of such molecules is the graft copolymer of the poly(methyl methacrylate) backbone with several PEO side chains (sold under the trade name Hypermer CG6 by ICI), which has excellent dispersing and stabilizing properties for concentrated dispersions of hydrophobic particles in water. Using such a dispersant, one can obtain highly stable concentrated suspensions. These surfactants have been modified in several ways to produce molecules that are suitable as emulsifiers, dispersants in extreme conditions such as high or low pH values, high electrolyte concentrations, high temperatures, etc. Other polymeric surfactants that are suitable for dispersing agrochemical particles in nonaqueous media have also been prepared, where the side chains were made oil soluble, such as polyhydroxystearic acid.

References

1 Th.F. Tadros, *Applied Surfactants*, Wiley-VCH, Weinheim (2005).
2 K. Holmberg, B. Jonsson, B. Kronberg and B. Lindman, *Surfactants and Polymers in Solution*, 2nd edn, John Wiley & Sons, Chichester (2003).
3 McCutcheon, *Detergents and Emulsifiers*, Allied Publishing, Baltimore, MD (published annually).
4 N.M. van Os, J.R. Haak and L.A.M. Rupert, *Physico-chemical Properties of Selected Anionic, Cationic and Nonionic Surfactants*, Elsevier, Amsterdam (1993).
5 M.R. Porter, *Handbook of Surfactants*, Blackie Academic and Professional, London (1994).
6 W.M. Linfield, Ed., *Anionic Surfactants*, Marcel Dekker, New York (1967).
7 E.H. Lucasssen-Reynders, *Anionic Surfactants – Physical Chemistry of Surfactant Action*, Marcel Dekker, New York (1981).
8 E. Jungermana, *Cationic Surfactants*, Marcel Dekker, New York (1970).
9 N. Rubingh and P.M. Holland, Eds, *Cationic Surfactants – Physical Chemistry*, Marcel Dekker, New York (1991).
10 B.R. Buestein and C.L. Hiliton, *Amphoteric Surfactants*, Marcel Dekker, New York (1982).
11 M.J. Schick, Ed., *Nonionic Surfactants*, Marcel Dekker, New York (1966).
12 M.J. Schick, Ed., *Nonionic Surfactants: Physical Chemistry*, Marcel Dekker, New York (1987).
13 N. Schönfeldt, *Surface Active Ethylene Oxide Adducts*, Pergamon Press, Oxford (1970).

3
Physical Chemistry of Surfactant Solutions

3.1
Properties of Solutions of Surfactants

The physical properties of surface-active agents differ from those of smaller or non-amphipathic molecules in one major respect, namely the abrupt changes in their properties above a critical concentration [1]. This is illustrated in Figure 3.1, in which a number of physical properties (osmotic pressure, turbidity, solubilization, magnetic resonance, surface tension, equivalent conductivity and self-diffusion) are plotted as a function of concentration for an ionic surfactant [1].

At low concentrations, most properties are similar to those of a simple electrolyte. One notable exception is the surface tension, which decreases rapidly with increase in surfactant concentration. However, all the properties (interfacial and bulk) show an abrupt change at a particular concentration, which is consistent with the fact that at and above this concentration, surface-active ions or molecules in solution associate to form larger units. These associated units are called micelles (self-assembled structures) and the first-formed aggregates are generally approximately spherical in shape. The concentration at which this association phenomenon occurs is known as the critical micelle concentration (cmc). Each surfactant molecules has a characteristic cmc value at a given temperature and electrolyte concentration. The most common technique for measuring the cmc is surface tension, γ, which shows a break at the cmc, after which it remains virtually constant with further increase in concentration. However, other techniques, such as self-diffusion measurements and NMR and fluorescence spectroscopy, can be applied. A compilation of cmc values was given in 1971 by Mukerjee and Mysels [2], which although clearly not an up-to-date text is an extremely valuable reference. As an illustration, the cmc values of a number of surface-active agents are given in Table 3.1, to show some of the general trends [3]. Within any class of surface-active agent, the cmc decreases with increase in chain length of the hydrophobic portion (alkyl group). As a general rule, the cmc decreases by a factor of 2 for ionics (without added salt) and by a factor of 3 for nonionics on adding one methylene group to the alkyl chain. With nonionic surfactants, increasing the length of the hydrophilic group [poly(ethylene oxide)] causes an increase in cmc.

Colloids and Interface Science Series, Vol. 5
Colloids in Agrochemicals. Tharwat F. Tadros

ISBN: 978-3-527-31465-2

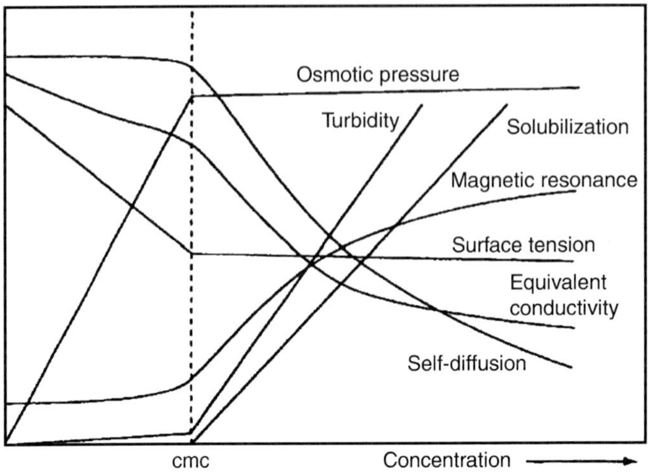

Figure 3.1 Variation of solution properties with concentration for surfactants.

In general, nonionic surfactants have lower cmc values than their corresponding ionic surfactants of the same alkyl chain length. Incorporation of a phenyl group in the alkyl group increases its hydrophobicity to a much smaller extent than increasing its chain length with the same number of carbon atoms. The

Table 3.1 Cmc values of surface-active agents.

Surface-active agent	cmc/mol dm^{-3}
Anionic	
Sodium octyl-l-sulfate	1.30×10^{-1}
Sodium decyl-l-sulfate	3.32×10^{-2}
Sodium dodecyl-l-sulfate	8.39×10^{-3}
Sodium tetradecyl-l-sulfate	2.05×10^{-3}
Cationic	
Octyltrimethylammonium bromide	1.30×10^{-1}
Decetryltrimethylammonium bromide	6.46×10^{-2}
Dodecyltrimethylammonium bromide	1.56×10^{-2}
Hexadecyltrimethylammonium bromide	9.20×10^{-4}
Nonionic	
Octyl hexaoxyethylene glycol monoether, C_8E_6	9.80×10^{-3}
Decyl hexaoxyethylene glycol monoether, $C_{10}E_6$	9.00×10^{-4}
Decyl nonaoxyethylene glycol monoether, $C_{10}E_9$	1.30×10^{-3}
Dodecyl hexaoxyethylene glycol monoether, $C_{12}E_6$	8.70×10^{-5}
Octylphenyl hexaoxyethylene glycol monoether, C_8E_6	2.05×10^{-4}

valency of the counterion in ionic surfactants has a significant effect on the cmc. For example, increasing the valency of the counter ion from 1 to 2 causes a reduction in the cmc by a factor of roughly 4.

The cmc for ionic surfactants is, to a first approximation, independent of temperature. The cmc varies in a non-monotonic way by ~10–20% over a wide temperature range. However, nonionic surfactants of the ethoxylate type show a monotonic decrease [4, 5] in cmc with increase in temperature.

The effect of addition of co-solutes, e.g. electrolytes and non-electrolytes, on the cmc can be very striking. For example, addition of 1 : 1 electrolyte to a solution of anionic surfactant gives a dramatic lowering of the cmc, which may amount to an order of magnitude. The effect is moderate for short-chain surfactants, but is much larger for long-chain surfactants. At high electrolyte concentrations, the reduction in cmc with increase in the number of carbon atoms in the alkyl chain is much stronger than without added electrolyte. This rate of decrease at high electrolyte concentrations is comparable to that of nonionics. The effect of added electrolyte also depends on the valency of the added counterions. In contrast, for nonionics, addition of electrolytes causes only small variations in the cmc.

Non-electrolytes such as alcohols can also cause a decrease in the cmc. The alcohols are less polar than water and are distributed between the bulk solution and the micelles. The more preference they have for the micelles, the more they stabilize them. A longer alkyl chain leads to a less favorable location in water and a more favorable location in the micelles.

The presence of micelles can account for many of the unusual properties of solutions of surface-active agents. For example, it can account for the nearly constant surface tension value above the cmc (see Figure 3.1). It also accounts for the reduction in the molar conductance of the surface-active agent solution above the cmc, which is consistent with the reduction in mobility of the micelles as a result of the counterion. The presence of micelles also accounts for the rapid increase in light-scattering or turbidity above the cmc.

The presence of micelles was originally proposed by McBain [6], who suggested that below the cmc most of the surfactant molecules are unassociated, whereas in the isotropic solutions immediately above the cmc, micelles and surfactant ions (molecules) are thought to coexist, the concentration of the latter changing very slightly as more surfactant is dissolved. However, the self-association of an amphiphile occurs in a stepwise manner with one monomer added to the aggregate at a time. For a long-chain amphiphile, the association is strongly cooperative up to a certain micelle size where counteracting factors become increasingly important. Typically the micelles have a closely spherical shape over a rather wide concentration range above the cmc Originally, it was suggested by Adam [7] and Hartley [8] that micelles are spherical in shape and have the following properties: (1) the association unit is spherical with a radius approximately equal to the length of the hydrocarbon chain; (2) the micelle contains about 50–100 monomeric units – the aggregation number generally increases with increase in alkyl chain length; (3) with ionic surfactants, most counterions are bound to the micelle surface, thus significantly reducing the mobility from the value to be expected from a micelle with non-counterion bonding; (4) micellization occurs over a narrow concentration range as a result of the high association number of surfactant micelles; and (5) the interior of the surfactant micelle has essentially the properties of a liquid hydrocarbon. This is confirmed by the high mobility of the alkyl chains and the ability of the micelles to solubilize many water-insoluble organic molecules, e.g. dyes and agrochemicals.

To a first approximation, micelles can, over a wide concentration range above the cmc, be viewed as microscopic liquid hydrocarbon droplets covered with polar head groups, which interact strongly with water molecules. It appears that the radius of the micelle core constituted of the alkyl chains is close to the extended length of the alkyl chain, i.e. in the range 1.5–3.0 nm. As we will see later, the driving force for micelle formation is the elimination of the contact between the alkyl chains and water. The larger a spherical micelle, the more efficient this is, since the volume-to-area ratio increases. It should be noted that the surfactant molecules in the micelles are not all extended. Only one molecule needs to be extended to satisfy the criterion that the radius of the micelle core is close to the extended length of the alkyl chain. The majority of surfactant molecules are in a disordered state. In other words, the interior of the micelle is close to that of the corresponding alkane in a neat liquid oil. This explains the large solubilization capacity of the micelle towards a broad range of nonpolar and weakly polar substances.

At the surface of the micelle, associated counterions (in the region of 50–80% of the surfactant ions) are present. However, simple inorganic counterions are very loosely associated with the micelle. The counterions are very mobile (see below) and there is no specific complex formed with a definite counterion–head group distance. In other words, the counterions are associated by long-range electrostatic interactions.

A useful concept for characterizing micelle geometry is the critical packing parameter (CPP). The aggregation number N is the ratio between the micellar core volume, V_{mic}, and the volume of one chain, v:

$$N = \frac{V_{mic}}{v} = \frac{\frac{4}{3}\pi R_{mic}^3}{v} \tag{1}$$

where R_{mic} is the radius of the micelle.

The aggregation number, N, is also equal to the ratio of the area of a micelle, A_{mic}, to the cross-sectional area, a, of one surfactant molecule:

$$N = \frac{A_{mic}}{a} = \frac{4\pi R_{mic}^2}{a} \tag{2}$$

Combining equations (1) and (2):

$$\frac{v}{R_{mic} a} = \frac{1}{3} \tag{3}$$

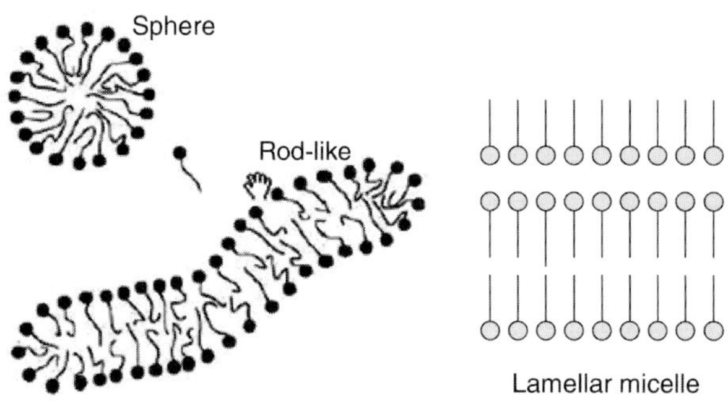

Figure 3.2 Shape of micelles.

Since R_{mic} cannot exceed the extended length of a surfactant alkyl chain, $l_{\max}$,

$$l_{\max} = 1.5 + 1.265 n_c \tag{4}$$

This means that for a spherical micelle,

$$\frac{v}{l_{\max} a} \leq \frac{1}{3} \tag{5}$$

The ratio $v/(l_{\max} a)$ is denoted the critical packing parameter (CPP).

Although the spherical micelle model accounts for many of the physical properties of solutions of surfactants, a number of phenomena remain unexplained, without considering other shapes. For example, McBain [9] suggested the presence of two types of micelles, spherical and lamellar, in order to account for the decrease in molar conductance of surfactant solutions. The lamellar micelles are neutral and hence they account for the reduction in the conductance. Later, Harkins *et al.* [10] used McBain's model of lamellar micelles to interpret his X-ray results in soap solutions. Moreover, many modern techniques such as light scattering and neutron scattering indicate that in many systems the micelles are not spherical. For example, Debye and Anacker [11] proposed a cylindrical micelle to explain that light scattering results on hexadecyltrimethylammonium bromide in water. Evidence for disc-shaped micelles has also been obtained under certain conditions. A schematic representation of the spherical, lamellar and rod-shaped micelles suggested by McBain, Hartley and Debye is given in Figure 3.2.

3.2
Solubility–Temperature Relationship for Surfactants

Many ionic surfactants show dramatic temperature-dependent solubility. The solubility may be very low at low temperatures and then increase by orders of

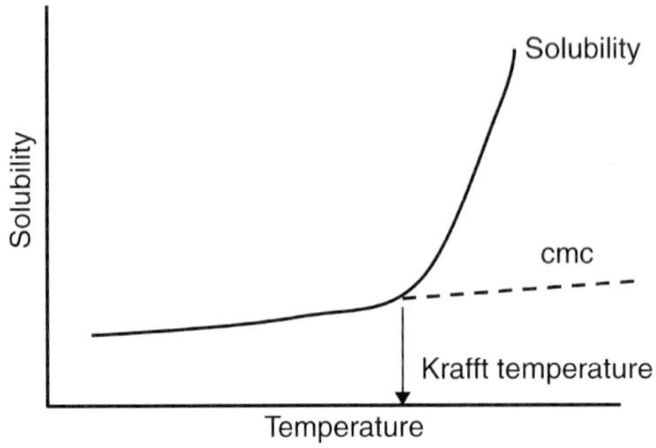

Figure 3.3 Variation of solubility and critical micelle concentration (cmc) with temperature.

magnitude in a relatively narrow temperature range. This is illustrated in Figure 3.3, which shows the change in solubility and cmc of sodium decyl sulfonate with temperature. This phenomenon is generally denoted the Krafft phenomenon, with the temperature of the onset of increasing solubility being known as the Krafft temperature. The cmc increases slowly with temperature and at the Krafft temperature the solubility is equal to the cmc. At this temperature, there is an equilibrium between hydrated surfactant solid, micelles and monomers (triple point). The Krafft temperature may vary dramatically with subtle changes in the surfactant chemical structure. In general, the Krafft temperature increases rapidly as the alkyl chain length of the surfactant increases. It decreases with increase in the alkyl chain distribution of the surfactant. It also depends on the head group and counterion. Addition of electrolytes causes an increase in the Krafft temperature.

With nonionic surfactants of the ethoxylate type, an increase in temperature for a solution at a given concentration causes dehydration of the poly(ethylene oxide) (PEO) chains and at a critical temperature the solution become cloudy.

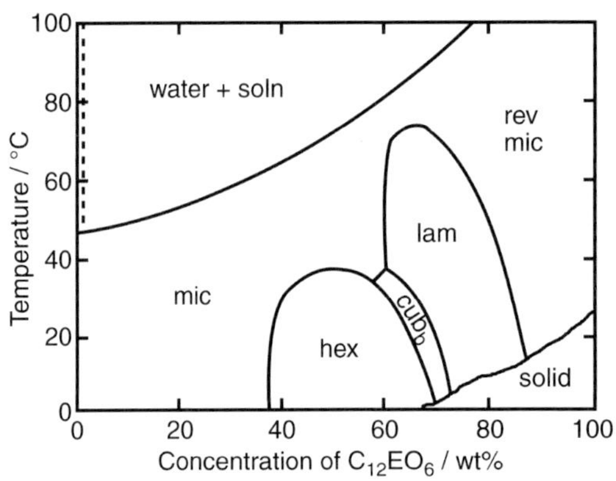

Figure 3.4 Phase diagram of nonionic surfactants.

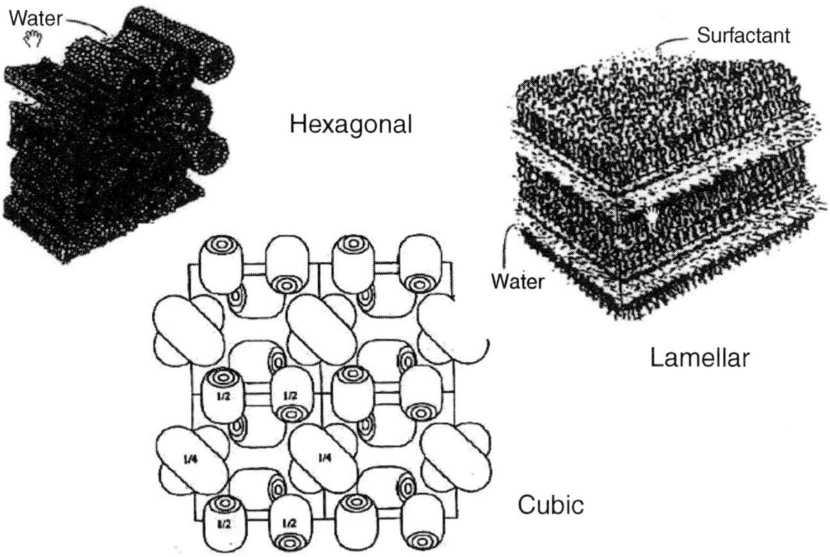

Figure 3.5 Schematic diagram of liquid crystalline phases.

This is illustrated in Figure 3.4, which shows the phase diagram of $C_{12}E_6$. Below the cloud point (CP) curve one can identify different liquid crystalline phases, hexagonal, cubic and lamellar, which are shown schematically in Figure 3.5.

3.3 Thermodynamics of Micellization

As mentioned above, the process of micellization is one of the most important characteristics of surfactant solutions, hence it is essential to understand its mechanism (the driving force for micelle formation). This requires analysis of the dynamics of the process (i.e. the kinetic aspects) and also the equilibrium aspects whereby the laws of thermodynamics may be applied to obtain the free energy, enthalpy and entropy of micellization. Below, a brief description of both aspects is given and this is followed by a discussion of the driving force for micelle formation.

3.3.1 Kinetic Aspects

Micellization is a dynamic phenomenon in which n monomeric surfactant molecules associate to form a micelle S_n, i.e.

$$nS \rightleftharpoons S_n \tag{6}$$

Hartley [8] envisaged a dynamic equilibrium whereby surface-active agent molecules are constantly leaving the micelles while other molecules from solution enter the micelles. The same applies to the counterions with ionic surfactants, which can exchange between the micelle surface and bulk solution.

Experimental investigation using fast kinetic methods such as stop flow, temperature and pressure jumps and ultrasonic relaxation measurements have shown that there are two relaxation processes for micellar equilibrium [12–14] characterized by relaxation times τ_1 and τ_2. The first relaxation time, τ_1, is of the order of 10^{-7} s (10^{-8}–10^{-3} s) and represents the lifetime of a surface-active molecule in a micelle, i.e. it represents the association and dissociation rate (represented by association and dissociation rate constants k^+ and k^-, respectively) for a single molecule entering and leaving the micelle, which may be represented by the equation

$$S + S_{n-1} \leftrightarrow S_n \tag{7}$$

The slower relaxation time τ_2 corresponds to a relatively slow process, namely the micellizationdissolution process represented by equation 1. The value of τ_2 is of the order of milliseconds (10^{-3}–1 s) and hence can be conveniently measured by stopped flow methods. The fast relaxation time τ_1 can be measured using various techniques depending on its range. For example, τ_1 values in the range 10^{-8}–10^{-7} s are accessible to ultrasonic absorption methods, whereas τ_1 in the range 10^{-5}–10^{-3} s can be measured by pressure jump methods. The value of τ_1 depends on surfactant concentration, chain length and temperature; τ_1 increases with increase in the chain length of surfactants, i.e. the residence time increases with increase in chain length.

The above discussion emphasizes the dynamic nature of micelles and it is important to realize that these molecules are in continuous motion and that there is a constant interchange between micelles and solution. The dynamic nature also applies to the counterions which exchange rapidly with lifetimes in the range 10^{-9}–10^{-8} s. Furthermore, the counterions appear to be laterally mobile and not to be associated with (single) specific groups on the micelle surfaces [3].

3.3.2
Equilibrium Aspects: Thermodynamics of Micellization

Two general approaches has been employed in tackling the problem of micelle formation. The first and simplest approach treats micelles as a single phase and this is referred to as the phase separation model. In this model, micelle formation is considered as a phase separation phenomenon and the cmc is then the saturation concentration of the amphiphile in the monomeric state, whereas the micelles constitute the separated pseudophase. Above the cmc, a phase equilibrium exists with a constant activity of the surfactant in the micellar phase. The Krafft point is viewed as the temperature at which solid hydrated surfactant, micelles and a solution saturated with undissociated surfactant molecules are in equilibrium at a given pressure.

In the second approach, micelles and single surfactant molecules or ions are considered to be in association–dissociation equilibrium. In its simplest form, a single equilibrium constant is used to treat the process represented by equation 1. The cmc is merely a concentration range above which any added surfactant appears in solution in a micellar form. Since the solubility of the associated surfactant is much greater than that of the monomeric surfactant, the solubility of the surfactant as a whole will not increase markedly with temperature until it reaches the cmc region. Thus, in the mass action approach, the Krafft point represents the temperature at which the surfactant solubility equals the cmc.

3.3.3 Phase Separation Model

Consider an anionic surfactant, in which n surfactant anions S^- and n counterions M^+ associate to form a micelle, i.e.

$$nS^- + nM^+ \leftrightarrow S_n \tag{8}$$

The micelle is simply a charged aggregate of surfactant ions plus an equivalent number of counterions in the surrounding atmosphere and is treated as a separate phase.

The chemical potential of the surfactant in the micellar state is assumed to be constant, at any given temperature and this may be adopted as the standard chemical potential, $\mu_m{}^\circ$, by analogy with a pure liquid or a pure solid. Considering the equilibrium between micelles and monomer, then

$$\mu^\circ{}_m = \mu^\circ{}_1 + RT \ln a \tag{9}$$

where m_1 is the standard chemical potential of the surfactant monomer and a_1 is its activity, which is equal to $f_1 x_1$, where f_1 is the activity coefficient and x_1 the mole fraction. Therefore, the standard free energy of micellization per mole of monomer, $\Delta G^\circ{}_m$, is given by,

$$\Delta G^\circ{}_m = \mu^\circ{}_m - \mu^\circ{}_m = RT \ln a_1 \approx RT \ln x_1 \tag{10}$$

where f_1 is taken as unity (a reasonable value in very dilute solution). The cmc may be identified with x_1 so that

$$\Delta G^\circ{}_m = RT \ln[\text{cmc}] \tag{11}$$

In equation 11, the cmc is expressed as a mole fraction, which is equal to $C/(55.5 + C)$, where C is the concentration of surfactant in mol dm^{-3}, i.e.

$$\Delta G^\circ{}_m = RT \ln C - RT \ln(55.5 + C) \tag{12}$$

It should be stated that ΔG° should be calculated using the cmc expressed as a mole fraction as indicated by equation 13. However, most cmc values quoted in the literature are given in mol dm^{-3} and in many cases ΔG° values have been quoted when the cmc was simply expressed in mol dm^{-3}. Strictly, this is incorrect, since ΔG° should be based on x_1 rather than on C. The value of ΔG° when the cmc is expressed in mol dm^{-3} is substantially different from the ΔG° value when the cmc is expressed in mole fraction. For example, for dodecyl hexaoxyethylene glycol, the quoted cmc value is 8.7×10^{-5} mol dm^{-3} at 25 °C. Therefore,

$$\Delta G^\circ = RT \ln \frac{8.7 \times 10^{-5}}{55.5 + 8.7 \times 10^{-5}} = -33.1\,\text{kJ mol}^{-1} \tag{13}$$

when the mole fraction scale is used, but

$$\Delta G^\circ = RT \ln 8.7 \times 10^{-5} = -23.2\,\text{kJ mol}^{-1} \tag{14}$$

when the molarity scale is used.

The phase separation model has been questioned for two main reasons [15, 16]. First, according to this model a clear discontinuity in the physical property of a surfactant solution, such as surface tension, turbidity, etc., should be observed at the cmc This is not always found experimentally and the cmc is not a sharp break point. Second, if two phases actually exist at the cmc, then equating the chemical potential of the surfactant molecule in the two phases would imply that the activity of the surfactant in the aqueous phase would be constant above the cmc If this was the case, the surface tension of a surfactant solution should remain constant above the cmc. However, careful measurements have shown that the surface tension of a surfactant solution decreases slowly above the cmc, particularly when using purified surfactants.

3.3.4 Mass Action Model

This model assumes a dissociation–association equilibrium between surfactant monomers and micelles and an equilibrium constant can be calculated. For a nonionic surfactant, where charge effects are absent, this equilibrium is simply represented by equation 1, which assumes a single equilibrium. In this case, the equilibrium constant K_m is given by the equation

$$K_m = \frac{[\text{S}_n]}{[\text{S}]^n} \tag{15}$$

The standard free energy per monomer is then given by

$$-\Delta G^\circ{}_\text{m} = -\frac{\Delta G}{n} = \frac{RT}{n} \ln K_\text{m} = \frac{RT}{n} \ln[\text{S}_n] - RT \ln[\text{S}] \tag{16}$$

For many micellar systems, $n > 50$ and, therefore, the first term on the right-hand side of equation 16 may be neglected, resulting in the following expression for ΔG°_m:

$$\Delta G^\circ_m = RT \ln[S] = RT \ln[\text{cmc}] \tag{17}$$

which is identical with the equation derived using the phase separation model.

The mass action model allows a simple extension to be made to the case of ionic surfactants, in which micelles attract a substantial proportion of counterions, into an attached layer. For a micelle made of n surfactant ions, where $n - p$ charges are associated with counterions, i.e. having a net charge of p units and degree of dissociation p/n, the following equilibrium may be established (for an anionic surfactant with Na^+ counterions):

$$nS^- + (n-p)Na^+ \leftrightarrow S_n{}^{p-} \tag{18}$$

$$K_m = \frac{[S_n{}^{p-}]}{[S^-]^n[Na^+]^{(n-p)}} \tag{19}$$

A convenient solution for relating ΔG_m to [cmc] was given by Phillips [17], who arrived at the following expression:

$$\Delta G^\circ_m = [2 - (p/n)]RT \ln[\text{cmc}] \tag{20}$$

For many ionic surfactants, the degree of dissociation (p/n) is ~0.2, so that

$$\Delta G^\circ_m = 1.8RT \ln[\text{cmc}] \tag{21}$$

Comparison with equation 20 clearly shows that for similar ΔG_m, the [cmc] is about two orders of magnitude higher for ionic surfactants than nonionic surfactants of the same alkyl chain length (see Table 3.1).

In the presence of excess added electrolyte, with mole fraction x, the free energy of micellization is given by the expression

$$\Delta G^\circ_m = RT \ln[\text{cmc}] + [1 - (p/n)] \ln x \tag{22}$$

which shows that as x increases, [cmc] decreases.

It is clear from equation 21 that as $p \rightarrow 0$, i.e. when most charges are associated with counterions,

$$\Delta G^\circ_m = 2RT \ln[\text{cmc}] \tag{23}$$

whereas when $p \sim n$, i.e. the counterions are bound to micelles,

$$\Delta G^{\circ}{}_{\mathrm{m}} = RT \ln[\mathrm{cmc}] \tag{24}$$

which is the same equation as for nonionic surfactants.

Although the mass action approach could account for a number of experimental results, such as the small change in the properties around the cmc, it has not escaped criticism. For example, the assumption that surfactants exist in solution in only two forms, namely single ions and micelles of uniform size, is debatable. Analysis of various experimental results has shown that micelles have a size distribution which is narrow and concentration dependent. Hence the assumption of a single aggregation number is an oversimplification and in reality there is a micellar size distribution [18].

3.3.5
Enthalpy and Entropy of Micellization

The enthalpy of micellization can be calculated from the variation of cmc with temperature. This follows from

$$-\Delta H^{\circ} = RT^2 \frac{\mathrm{d}\ln[\mathrm{cmc}]}{\mathrm{d}T} \tag{25}$$

The entropy of micellization can then be calculated from the relationship between ΔG° and ΔH°, i.e.

$$\Delta G^{\circ} = \Delta H^{\circ} - T\Delta S^{\circ} \tag{26}$$

Therefore, ΔH° may be calculated from the surface tensionlog C plots at various temperatures. Unfortunately, the errors in locating the cmc (which in many cases is not a sharp point) leads to a large error in the value of ΔH°. A more accurate and direct method of obtaining ΔH° is microcalorimetry. As an illustration, the thermodynamic parameters ΔG°, ΔH° and $T\Delta S^{\circ}$ for octylhexaoxyethylene glycol monoether (C_8E_6) are given in Table 3.2.

Table 3.2 Thermodynamic quantities for micellization of octylhexaoxyethylene glycol monoether.

Temperature/°C	ΔG°/kJ mol^{-1}	ΔH°/kJ mol^{-1} (from cmc)	ΔH°/kJ mol l^{-1} (from calorimetry)	$T\Delta S^{\circ}$/kJ mol^{-1}
25	-21.3 ± 2.1	8.0 ± 4.2	20.1 ± 0.8	41.8 ± 1.0
40	-23.4 ± 2.1		14.6 ± 0.8	38.0 ± 1.0

Table 3.3 Change of thermodynamic parameters of micellization of alkyl sulfoxide with increasing chain length of the alkyl group.

Surfactant	ΔG°/kJ mol^{-1}	ΔH°/kJ mol^{-1}	$T\Delta S^\circ$/kJ mol^{-1}
$C_6H_{13}S(CH_3)O$	−12.0	10.6	22.6
$C_7H_{15}S(CH_3)O$	−15.9	9.2	25.1
$C_8H_{17}S(CH_3)O$	−18.8	7.8	26.4
$C_9H_{19}S(CH_3)O$	−22.0	7.1	29.1
$C_{10}H_{21}S(CH_3)O$	−5.5	5.4	30.9
$C_{11}H_{23}S(CH_3)O$	−8.7	3.0	31.7

It can be seen from Table 3.2 that ΔG° is large and negative. However, ΔH° is positive, indicating that the process is endothermic. In addition, $T\Delta S^\circ$ is large and positive, which implies that in the micellization process there is a net increase in entropy. As we will see in the next section, this positive enthalpy and entropy points to a different driving force for micellization from that encountered in many aggregation processes.

The influence of the alkyl chain length of the surfactant on the free energy, enthalpy and entropy of micellization was demonstrated by Rosen [19], who listed these parameters as a function of alkyl chain length for sulfoxide surfactants. The results are given in Table 3.3, where it can be seen that the standard free energy of micellization becomes increasingly negative as the chain length increases. This is to be expected since the cmc decreases with increase in alkyl chain length. However, ΔH° becomes less positive and $T\Delta S$ becomes more positive with increase in chain length of the surfactant. Thus, the large negative free energy of micellization is made up of a small positive enthalpy, which decreases slightly with increase in the chain length of the surfactant, and a large positive entropy term $T\Delta S^\circ$, which becomes more positive as the chain is lengthened. As we will see in the next section, these results can be accounted for in terms of the hydrophobic effect, which will be described in some detail.

3.3.6 Driving Force for Micelle Formation

Until recently, the formation of micelles was regarded primarily as an interfacial energy process, analogous to the process of coalescence of oil droplets in an aqueous medium. If this was the case, micelle formation would be a highly exothermic process, as the interfacial free energy has a large enthalpy component. As mentioned above, experimental results have clearly shown that micelle formation involves only a small enthalpy change and is often endothermic. The negative free energy of micellization is the result of a large positive entropy. This led to the conclusion that micelle formation must be a predominantly entropy-driven

Table 3.4 Thermodynamic parameters for transfer of hydrocarbons from water to liquid hydrocarbon at 25 °C.

Hydrocarbon	$\Delta G°$/kJ mol^{-1}	$\Delta H°$/kJ mol^{-1}	$\Delta S°$/kJ mol^{-1}	$\Delta C_p°$/kJ mol^{-1}	$C_p^{\circ,\text{gas}}$/kJ mol^{-1}
C_2H_6	−16.4	10.5	88.2	–	–
C_3H_8	−20.4	7.1	92.4	–	–
C_4H_{10}	−24.8	3.4	96.6	−273	−143
C_6H_{14}	−32.5	0	109.2	−441	−197
C_6H_6	−19.3	−2.1	58.8	−227	−134
$C_6H_5CH_3$	−22.7	−1.7	71.4	−265	−155
$C_6H_5C_2H_5$	−26.0	−2.0	79.8	−319	−185
$C_6H_5C_3H_8$	−29.0	−2.3	88.2	−395	–

process. Two main sources of entropy have been suggested. The first is related to the so-called 'hydrophobic effect'. The latter effect was first established from a consideration of the free energy enthalpy and entropy of transfer of hydrocarbon from water to a liquid hydrocarbon. Some results are given in Table 3.4. This table also lists the heat capacity change ΔC_p on transfer from water to a hydrocarbon, in addition to $C_p^{\circ,\text{gas}}$, i.e. the heat capacity in the gas phase [3]. It can be seen from Table 3.4 that the principal contribution to the value of $\Delta G°$ is the large positive value of $\Delta S°$, which increases with increase in hydrocarbon chain length, whereas $\Delta H°$ is positive or small and negative.

To account for this large positive entropy of transfer, several authors [19, 20] have suggested that the water molecules around a hydrocarbon chain are ordered, forming 'clusters' or 'icebergs'. On transfer of an alkane from water to a liquid hydrocarbon, these clusters are broken, thus releasing water molecules, which now have a higher entropy. This accounts for the large entropy of transfer of an alkane from water to a hydrocarbon medium. This effect is also reflected in the much higher heat capacity change on transfer, $\Delta C_p°$, when compared with the heat capacity in the gas phase, $C_p^{\circ,\text{gas}}$.

The above effect is also operative on transfer of surfactant monomer to a micelle, during the micellization process. The surfactant monomers will also contain 'structured' water around their hydrocarbon chain. On transfer of such monomers to a micelle, these water molecules are released and they have a higher entropy.

The second source of entropy increase on micellization may arise from the increase in flexibility of the hydrocarbon chains on their transfer from an aqueous to a hydrocarbon medium [21, 22]. The orientations and bendings of an organic chain are likely to be more restricted in an aqueous phase than an organic phase.

It should be mentioned that with ionic and zwitterionic surfactants, an additional entropy contribution, associated with the ionic head groups, must be considered. Upon partial neutralization of the ionic charge by the counterions when aggregation occurs, water molecules are released. This will be associated with an entropy increase, which should be added to the entropy increase resulting from the hydrophobic effect mentioned above. However, the relative contribution of the two effects is difficult to assess in a quantitative manner.

3.4 Micellization in Surfactant Mixtures (Mixed Micelles)

In most agrochemical formulations, more than one surfactant molecule is used in the system. It is necessary, therefore, to predict the type of possible interactions and whether this leads to some synergistic effects. Two general cases may be considered: surfactant molecules with no net interaction (with similar head groups) and systems with net interaction. This is discussed separately below [3].

3.4.1 Surfactant Mixtures with No Net Interaction

This is the case when mixing two surfactants with the same head group but with different chain lengths. In analogy with the hydrophilic-lipophilic balance (HLB) for surfactant mixtures, one can also assume that the cmc of a surfactant mixture (with no net interaction) is an average of the two cmcs of the single components [3]:

$$\mathrm{cmc} = x_1 \mathrm{cmc}_1 + x_2 \mathrm{cmc}_2 \tag{27}$$

where x_1 and x_2 are the mole fractions of the respective surfactants in the system. However, the mole fractions should not be those in the whole system, but those inside the micelle. This means that equation 27 should be modified:

$$\mathrm{cmc} = x_1^{\mathrm{m}} \mathrm{cmc}_1 + x_2^{\mathrm{m}} \mathrm{cmc}_2 \tag{28}$$

where the superscript m indicates that the values are inside the micelle. If x_1 and x_2 are the solution composition, then

$$\frac{1}{\mathrm{cmc}} = \frac{x_1}{\mathrm{cmc}_1} + \frac{x_2}{\mathrm{cmc}_2} \tag{29}$$

The molar composition of the mixed micelle is given by

$$x_1^{\mathrm{m}} = \frac{x_1 \mathrm{cmc}_2}{x_1 \mathrm{cmc}_2 + x_2 \mathrm{cmc}_1} \tag{30}$$

Figure 3.6 shows the cmc as a function of molar composition of the solution and in the micelles for a mixture of SDS and nonylphenol with 10 mol of ethylene oxide (NP-E_{10}). If the molar composition of the micelles is used as the

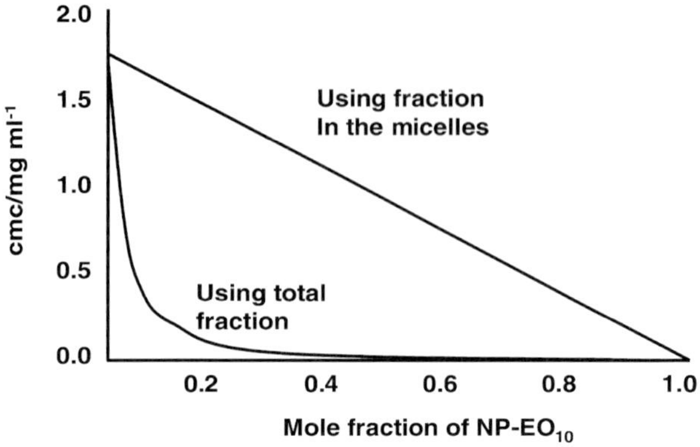

Figure 3.6 cmc as a function of surfactant composition, x_1, or micellar surfactant composition, x_1^{m}, for the system SDS + NP-E_{10}.

x-axis, the cmc is more or less the arithmetic mean of the cmcs of the two surfactants. If, on the other hand, the molar composition in the solution is used as the x-axis (which at the cmc is equal to the total molar concentration), then the cmc of the mixture shows a dramatic decrease at low fractions of NP-E_{10}. This decrease is due to the preferential absorption of NP-E_{10} in the micelle. This higher absorption is due to the higher hydrophobicity of the NP-E_{10} surfactant compared with SDS.

3.4.2
Surfactant Mixtures with a Net Interaction

With many agrochemical formulations, surfactants of different kinds are mixed together, for example anionics and nonionics. The nonionic surfactant molecules shield the repulsion between the negative head groups in the micelle and hence there will be a net interaction between the two types of molecules. Another example is the case when anionic and cationic surfactants are mixed, whereby a very strong interaction will take place between the oppositely charged surfactant molecules. To account for this interaction, equation 28 has to be modified by introducing activity coefficients of the surfactants, f_1^{m}, and f_2^{m}, in the micelle:

$$\mathrm{cmc} = x_1^{\mathrm{m}} f_1^{\mathrm{m}} \mathrm{cmc}_1 + x_2^{\mathrm{m}} f_2^{\mathrm{m}} \mathrm{cmc}_2 \tag{31}$$

An expression for the activity coefficients can be obtained using the regular solutions theory:

$$\ln f_1^{\mathrm{m}} = (x_1^{\mathrm{m}})^2 \beta \tag{32}$$

$$\ln f_2^{\mathrm{m}} = (x_2^{\mathrm{m}})^2 \beta \tag{33}$$

where β is an interaction parameter between the surfactant molecules in the micelle. A positive β value means that there is a net repulsion between the surfactant molecules in the micelle, whereas a negative β value means a net attraction.

The cmc of the surfactant mixture and the composition x_1 are given by the following equations:

$$\frac{1}{\text{cmc}} = \frac{x_1}{f_1^{\text{m}}\text{cmc}_1} + \frac{x_2}{f_2^{\text{m}}\text{cmc}_2} \tag{34}$$

$$x_1^{\text{m}} = \frac{x_1 f_2^{\text{m}}\text{cmc}_2}{x_1 f_2^{\text{m}}\text{cmc}_2 + x_2 f_2^{\text{m}}\text{cmc}_1} \tag{35}$$

As β becomes more negative, the cmc of the mixture decreases. β values in the region of -2 are typical for anionic–nonionic mixtures, whereas values in the region of -10 to -20 are typical of anionic–cationic mixtures. With increasing negative value of β, the mixed micelles tend towards a mixing ratio of 50 : 50, which reflects the mutual electrostatic attraction between the surfactant molecules.

The predicted cmc and micellar composition depend both on the ratio of the cmcs and the value of β. When the cmcs of the single surfactants are similar, the predicted value of the cmc is very sensitive to small variations in β. On the other hand, when the ratio of the cmcs is large, the predicted value of the mixed cmc and the micellar composition are insensitive to variations of β. For mixtures of nonionic and ionic surfactants, β decreases with increase in electrolyte concentration. This is due to the screening of the electrostatic repulsion on the addition of electrolyte. With some surfactant mixtures, β decreases with increase in temperature, i.e. the net attraction decreases with increase in temperature.

3.5 Surfactant–Polymer Interaction

Mixtures of surfactants and polymers are very common in many agrochemical formulations. With many suspension and emulsion systems stabilized with surfactants, polymers are added for a number of reasons. For example, polymers are added as suspending agents ('thickeners') to prevent sedimentation or creaming of these systems. The interaction between surfactants and water-soluble polymers results in some synergistic effects, e.g. enhancing the surface activity, stabilizing suspensions and emulsions, etc. It is important, therefore, to study the interaction between surfactants and water-soluble polymers in a systematic way.

One of the earliest studies of surfactant–polymer interactions came from surface tension measurements. Figure 3.7 shows some typical results for the effect of addition of polyvinylpyrrolidone (PVP) on the γ–log C curves for SDS [23].

In a system of fixed polymer concentration and varying surfactant concentrations, two critical concentrations appear, denoted T_1 and T_2. T_1 represents the con-

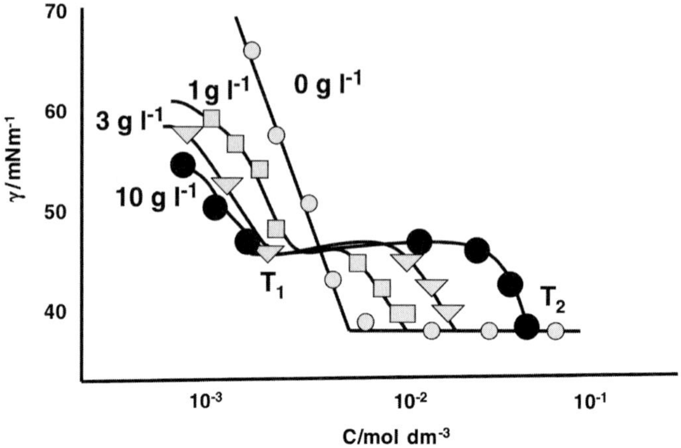

Figure 3.7 γ–log C curves for SDS solutions in the presence of different concentrations of PVP.

centration at which interaction between the surfactant and polymer first occurs. This is sometimes termed the critical aggregation concentration (CAC), i.e. the onset of association of surfactant with the polymer. Because of this, there is no further increase in surface activity and thus no lowering of surface tension. T_2 represents the concentration at which the polymer becomes saturated with surfactant. Since T_1 is generally lower than the cmc of the surfactant in the absence of polymer, then 'adsorption' or 'aggregation' of SDS on or with the polymer is more favorable than normal micellization. As the polymer is saturated with surfactant (i.e. beyond T_2), the surfactant monomer concentration and the activity start to increase again and there is lowering of γ until the monomer concentration reaches the cmc, after which γ remains virtually constant and normal surfactant micelles begin to form.

The above picture is confirmed if the association of surfactant is directly monitored (e.g. by using surfactant-selective electrodes, by equilibrium dialysis or by some spectroscopic techniques). At low surfactant concentration, there is no significant interaction. At the CAC, a strongly cooperative binding is indicated and at higher concentrations a plateau is reached. Further increase in surfactant concentration produces 'free' surfactant molecules until the surfactant activity or concentration joins the curve obtained in the absence of polymer. The binding isotherms show the strong analogy with micelle formation and the interpretation of these isotherms in terms of a depression of the cmc.

3.5.1
Factors Influencing the Association Between Surfactant and Polymer

Several factors influence the interaction between surfactant and polymer and these are summarized as follows. (1) *Temperature:* an increase in temperature generally increases the CAC, i.e. the interaction becomes less favorable. (2) *Addition of electrolyte:* this generally decreases the CAC, i.e. it increases the binding. (3) *Surfactant chain length:* an increase in the alkyl chain length decreases the CAC, i.e. it increases association. A plot of log CAC versus the number of carbon atoms, *n*, is linear (similar to the log cmc*n* relationship obtained for surfactants alone). (4) *Surfactant structure:* alkyl benzene sulfonates are similar to SDS, but introduction of EO groups in the chain weakens the interaction. (5) *Surfactant classes:* weaker interaction is generally observed with cationics compared with anionics. However, the interaction can be promoted by using a strongly interacting counterion for the cationic (e.g. CNS^-). The interaction between ethoxylated surfactants and nonionic polymers is weak. the interaction is stronger with alkyl phenol ethoxylates. (6) *Polymer molecular weight:* a minimum molecular weight of ~4000 for PEO and PVP is required for 'complete' interaction. (7) *Amount of polymer:* the CAC seems to be insensitive to (or slightly lower) with increase in polymer concentration. T_2 increases linearly with increase in polymer concentration. (8) *Polymer structure and hydrophobicity:* several uncharged polymers such as PEO, PVP and poly(vinyl alcohol) (PVOH) interact with charged surfactants. Many other uncharged polymers interact weakly with charged surfactants. e.g. hydroxyethylcellulose (HEC), dextran and polyacrylamide (PAAm). For anionic surfactants, the following order of increasing interaction has been listed: PVOH < PEO < MEC (methylcellulose) < PVAc [partially hydrolyzed poly(vinyl acetate)] < PPO ~ PVP. For cationic surfactants, the following order was listed: PVP < PEO < PVOH < MEC < PVAc < PPO. The position of PVP can be explained by the slight positive charge on the chain, which causes repulsion with cations and attraction with anionics.

3.5.2
Driving Force for Surfactant–Polymer Interaction

The driving force for polymer–surfactant interaction is the same as that for the process of micellization (see above). As with micelles, the main driving force is the reduction of the hydrocarbon–water contact area of the alkyl chain of the dissolved surfactant. A delicate balance between several forces is responsible for the surfactant–polymer association. For example, aggregation is resisted by the crowding of the ionic head groups at the surface of the micelle. Packing constraints also resist association. Molecules that screen the repulsion between the head groups, e.g. electrolytes and alcohol, promotes association. A polymer molecule with hydrophobic and hydrophilic segments (which is also flexible) can enhance association by ion–dipole association between the dipole of the hydrophilic groups and the ionic head groups of the surfactant. In addition, contact between the hydrophobic segments of the polymer and the exposed hydrocarbon areas of the micelles can enhance association. With SDS–PEO and SDS–PVP, the association complexes are approximately three monomer units per molecule of aggregated surfactant [24–26].

3.5.3 Structure of Surfactant–Polymer Complexes

Generally, there are two alternative pictures of mixed surfactant–polymer solutions, one describing the interaction in terms of a strongly cooperative association or binding of the surfactant to the polymer chain and one in terms of a micellization of surfactant on or in the vicinity of the polymer chain. For polymers with hydrophobic groups the binding approach is preferred, whereas for hydrophilic homopolymers the micelle formation picture is more likely. The latter picture was suggested by Cabane [27], who proposed a structure in which the aggregated SDS is surrounded by macromolecules in a loop configuration.

The consequences of the above model are (1) a more favorable free energy of association (CAC < cmc) and increased ionic dissociation of the aggregates and (2) an altered environment of the CH_2 groups of the surfactant near the head group. The micelle sizes are similar with polymer present and without and the aggregation numbers are typically similar to or slightly lower than those of the micelles forming in the absence of a polymer. In the presence of a polymer, the surfactant chemical potential is lowered with respect to the situation without a polymer [28].

3.5.4 Surfactant–Hydrophobically Modified Polymer Interaction

Water-soluble polymers are modified by grafting a low amount of hydrophobic groups (of the order of 1% of the monomers reacted in a typical molecule), resulting in the formation of 'associative structures'. These molecules are referred to as associative thickeners and are used as rheology modifiers in many agrochemical formulations. An added surfactant will interact strongly with the hydrophobic

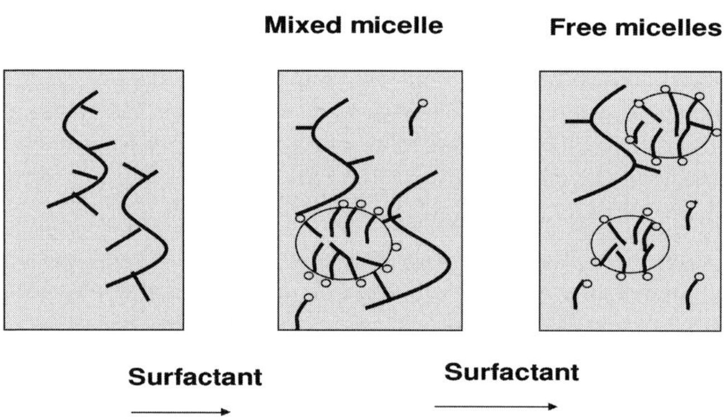

Figure 3.8 Schematic representation of the interaction between surfactant and HM polymer.

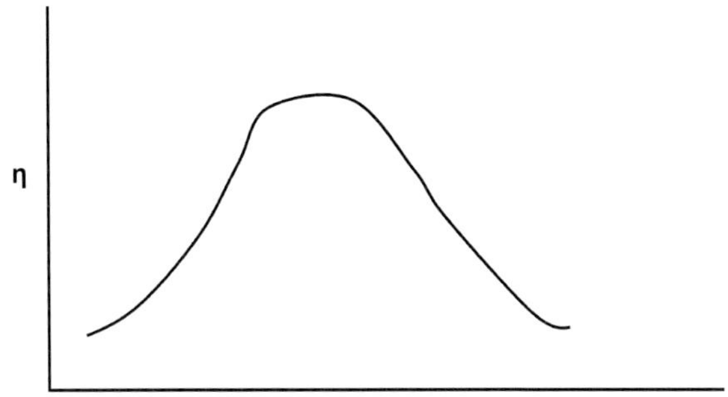

Figure 3.9 Viscosity–surfactant concentration relationship for HM-modified polymer solutions.

groups of the polymer, leading to a strengthened association between the surfactant molecules and the polymer chain. A schematic picture for the interaction between SDS and hydrophobically modified hydroxyethylcellulose (HM-HEC) is shown in Figure 3.8, which shows the interaction at various surfactant concentrations [29].

Initially the surfactant monomers interact with the hydrophobic groups of the HM polymer and at some surfactant concentration (CAC), the micelles can cross-link the polymer chains. At higher surfactant concentrations, the micelles which are now abundant will no longer be shared between the polymer chains, i.e. the cross-links are broken. These effects are reflected in the variation of viscosity with surfactant concentration for HM polymer, as illustrated in Figure 3.9. The viscosity of the polymer increases with increase in surfactant concentration, reaching a maximum at an optimum concentration (maximum cross-links), and then decreases with further increase in surfactant concentration. For the unmodified polymer, the changes in viscosity are relatively small.

References

1 B. Lindman, in *Surfactants*, ed. Th.F. Tadros, Academic Press, London (1984); K. Holmberg, B. Jonsson, B. Kronberg and B. Lindman, *Surfactants and Polymers in Aqueous Solution*, 2nd edn, John Wiley & Sons, New York (2003).

2 P. Mukerjee and K.J. Mysels, *Critical Micelle Concentrations of Aqueous Surfactant Systems*, National Bureau of Standards, Washington, DC (1971).

3 J. Israelachvili, *Intermolecular and Surface Forces, with Special Applications to Colloidal and Biological Systems*, Academic Press, London, p. 251 (1985).

4 P.H. Elworthy, A.T. Florence and C.B. Macfarlane, *Solubilization by Surface Active Agents*, Chapman and Hall, London (1968).

5 K. Shinoda, T. Nagakawa, B.I. Tamamushi and T. Isemura, *Colloidal Surfactants, Some Physicochemical Properties*, Academic Press, London (1963).

6 J.W. McBain, *Trans. Faraday Soc.*, *9*, 99 (1913).

7 N.K. Adam, *J. Phys. Chem.*, *29*, 87 (1925).

8 G.S. Hartley, *Aqueous Solutions of Paraffin Chain Salts*, Hermann, Paris (1936).

9 J.W. McBain, *Colloid Science*, Heath, Boston (1950).
10 W.D. Harkins, W.D. Mattoon and M.L. Corrin, *J. Am. Chem. Soc.*, *68*, 220 (1946); W.D. Harkins, W.D. Mattoon and M.L. Corrin, *J. Colloid Sci.*, *1*, 105 (1946).
11 P. Debye and E.W. Anaker, *J. Phys. Colloid Chem.*, *55*, 644 (1951).
12 E.A.G. Anainsson and S.N. Wall, *J. Phys. Chem.*, *78*, 1024 (1974); E.A.G. Anainsson and S.N. Wall, *J. Phys. Chem.*, *79*, 857 (1975).
13 E.A.G. Anainsson, S.N. Wall, M. Almagren, H. Hoffmann, W. Ulbricht, R. Zana, J. Lang and C. Tondre, *J. Phys. Chem.*, *80*, 905 (1976).
14 J. Rassing, P.J. Sams and E. Wyn-Jones, *J. Chem. Soc., Faraday Trans. 2*, *70*, 1247 (1974).
15 M.J. Jaycock and R.H. Ottewill, in *Proceedings of the Fourth International Congress on Surface Activity*, Vol. 2, p. 545 (1964).
16 T. Okubo, H. Kitano, T. Ishiwatari and N. Isem, *Proc. R. Soc. London, Ser. A*, *36*, 81 (1979).
17 J.N. Phillips, *Trans. Faraday Soc.*, *51*, 561 (1955).
18 M. Kahlweit and M. Teubner, *Adv. Colloid Interface Sci.*, *13*, 1 (1980).
19 M.L. Rosen, *Surfactants and Interfacial Phenomena*, Wiley-Interscience, New York (1978).
20 T. Tanford, *The Hydrophobic Effect*, 2nd edn, John Wiley & Sons, New York (1980).
21 G. Stainsby and A.E. Alexander, *Trans. Faraday Soc.*, *46*, 587 (1950).
22 R.H. Arnow and L. Witten, *J. Phys. Chem.*, *64*, 1643 (1960).
23 I.D. Robb, *Chem. Ind. (London)*, 530 (1972).
24 B. Cabane and R. Duplessix, *J. Phys. (Paris)*, *43*, 1529 (1982).
25 R. Nagarajan, *Colloids Surf.*, *13*, 1 (1985).
26 T. Gilyani and E. Wolfram, *Colloids Surf.*, *3*, 181 (1981).
27 B. Zabane, *J. Phys. Chem.*, *81*, 1639 (1977).
28 D.F. Evans and H. Winnerstrom, *The Colloidal Domain. Where Physics, Chemistry, Biology and Technology Meet*, John Wiley and Sons, New York, p. 312 (1994).
29 E.D. Goddard, *Colloids Surf.*, *19*, 301 (1986).

4
Interfacial Aspects of Agrochemical Formulations

4.1
Introduction

Several interfacial aspects must be considered when dealing with agrochemical formulations: (1) Adsorption of surfactants at the air/liquid interface, both equilibrium and dynamic aspects. These determine spray formation (spray droplet spectrum), impaction and adhesion of droplets on leaf surfaces and the various wetting and spreading phenomena. (2) Adsorption of surfactants at the oil/water interface, which determines emulsion formation and their stability. This subject is also important when dealing with microemulsions. (3) Adsorption of surfactants and polymers at the solid/liquid interface. This is important when dealing with dispersion of agrochemical powders in liquids, preparation of suspension concentrates and their stabilization.

This chapter deals with the above interfacial aspects, starting with the equilibrium aspects of surfactant adsorption at the air/water and oil/water interfaces. Due to the equilibrium aspects of adsorption (the rate of adsorption is equal to the rate of desorption) one can apply the second law of thermodynamics as analyzed by Gibbs (see below). This is followed by a section on dynamic aspects of surfactant adsorption, particularly the concept of dynamic surface tension and the techniques that can be applied in its measurement. The adsorption of surfactants both on hydrophobic surfaces (which represent the case with most agrochemical solids) and on hydrophilic surfaces (such as oxides) is analyzed using the Langmuir adsorption isotherms. The structure of surfactant layers on solid surfaces is described. The subject of polymeric surfactant adsorption is dealt with separately due to its complex nature, namely irreversibility of adsorption and conformation of the polymer at the solid/liquid interface.

4.2
Equilibrium Adsorption of Surfactants at the Air/Liquid and Liquid/Liquid Interfaces

Before describing surfactant adsorption at the air/liquid (A/L) and liquid/liquid (L/L) interface, it is essential to define the interface. The surface of a liquid is the

Colloids and Interface Science Series, Vol. 5
Colloids in Agrochemicals. Tharwat F. Tadros

ISBN: 978-3-527-31465-2

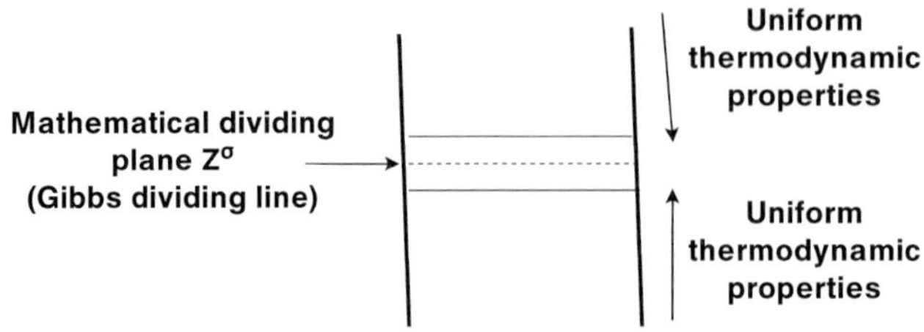

Figure 4.1 Gibbs convention for an interface.

boundary between two bulk phases, namely liquid and air (or the liquid vapor). Similarly, an interface between two immiscible liquids (oil and water) may be defined provided that a dividing line is introduced since the interfacial region is not a layer that is one molecule thick, but usually has a thickness δ with properties that are different from the two bulk phases α and β [1]. However, Gibbs [2] introduced the concept of a mathematical dividing plane Z_σ in the interfacial region (Figure 4.1). In this model, the two bulk phases α and β are assumed to have uniform thermodynamic properties up to Z_σ. This picture applies for both the A/L and L/L interfaces (with A/L interfaces, one of the phases is air saturated with the vapor of the liquid).

Using the Gibbs model, it is possible to obtain a definition of the surface or interfacial tension γ, starting from the Gibbs–Duhem equation:

$$dG^\sigma = -S^\sigma\,dT + A\,d\gamma + \sum n_i\,d\mu_i \qquad (1)$$

where G^σ is the surface free energy, S^σ is the entropy, A is the area of the interface, n_i is the number of moles of component i with chemical potential μ_i at the interface. At constant temperature and composition of the interface (i.e. in absence of any adsorption),

$$\gamma = \left(\frac{\partial G^\sigma}{\partial A}\right)_{T,n_i} \qquad (2)$$

It is obvious from equation 2 that for a stable interface γ should be positive. In other words, the free energy should increase if the area of the interface increases, otherwise the interface will become convoluted, increasing the interfacial area, until the liquid evaporates (for the A/L case) or the two 'immiscible' phases dissolved in each other (for the L/L case).

It is clear from equation 2 that surface or interfacial tension, i.e. the force per unit length tangentially to the surface measured in units of mN m^{-1}, is dimensionally equivalent to an energy per unit area measured in mJ m^{-2}. For this reason, it has been stated that the excess surface free energy is identical with the surface tension, but this is true only for a single-component system, i.e. a pure liquid (where the total adsorption is zero).

There are generally two approaches for treating surfactant adsorption at the A/L and L/L interfaces. The first approach, adopted by Gibbs, treats adsorption as an equilibrium phenomenon whereby the second law of thermodynamics may be applied using surface quantities. The second approach, referred to as the equation of state approach, treats the surfactant film as a two-dimensional layer with a surface pressure π that may be related the surface excess Γ (amount of surfactant adsorbed per unit area). Only the Gibbs approach will be described.

4.3 The Gibbs Adsorption Isotherm

Gibbs [2] derived a thermodynamic relationship between the surface or interfacial tension γ and the surface excess Γ (adsorption per unit area). The starting point of this equation is the Gibbs–Duhem equation given above (equation 1). At constant temperature, but in the presence of adsorption, equation 1 reduces to

$$d\gamma = -\sum \frac{n_i^\sigma}{A} d\mu_i = -\sum \Gamma_i \, d\mu_i \tag{3}$$

where $\Gamma_i = n_i^\sigma / A$ is the number of moles of component i and adsorbed per unit area.

Equation (2) is the general form for the Gibbs adsorption isotherm. The simplest case of this isotherm is a system of two components in which the solute (2) is the surface-active component, i.e. it is adsorbed at the surface of the solvent (1). For such a case, equation 3 may be written as

$$-d\gamma = \Gamma_1^\sigma \, d\mu_1 + \Gamma_2^\sigma \, d\mu_2 \tag{4}$$

and if the Gibbs dividing surface is used, $\Gamma_1 = 0$ and

$$-d\gamma = \Gamma_{2,1}^\sigma \, d\mu_2 \tag{5}$$

where $\Gamma_{2,1}^\sigma$ is the relative adsorption of (2) with respect to (1). Since

$$\mu_2 = \mu_2^\circ + RT \ln a_2^L \tag{6}$$

or

$$d\mu_2 = RT d \ln a_2^L \tag{7}$$

then

$$-d\gamma = \Gamma_{2,1}^\sigma RT d \ln a_2^L \tag{8}$$

or

$$\Gamma^{\sigma}_{2,1} = -\frac{1}{RT}\left(\frac{d\gamma}{d \ln a^{L}_{2}}\right) \tag{9}$$

where a^{L}_{2} is the activity of the surfactant in bulk solution that is equal to $C_2 f_2$ or $x_2 f_2$, where C_2 is the concentration of the surfactant in mol dm^{-3} and x_2 is its mole fraction.

Equation 9 allows one to obtain the surface excess (abbreviated Γ_2) from the variation of surface or interfacial tension with surfactant concentration. Note that $a_2 \sim C_2$ since in dilute solutions $f_2 \sim 1$. This approximation is valid since most surfactants have low critical micelle concentration (cmc) (usually less than 10^{-3} mol dm^{-3}) but adsorption is complete at or just below the cmc.

The surface excess Γ_2 can be calculated from the linear portion of the γ–log C_2 curves before the cmc Such γ–log C curves are illustrated in Figure 4.2 for the air/water (A/W) and oil/water (O/W) interfaces; [C_{SAA}] denotes the concentration of surface-active agent in bulk solution. It can be seen that for the A/W interface γ decreases from the value for water (72 mN m^{-1} at 20 °C) reaching about 25–30 mN m^{-1} near the cmc. This is clearly schematic since the actual values depend on the surfactant nature. For the O/W case, γ decreases from about 50 mN m^{-1} (for a pure hydrocarbon–water interface) to ~1–5 mN m^{-1} near the cmc (again depending on the nature of the surfactant).

As mentioned above, Γ_2 can be calculated from the slope of the linear position of the curves shown in Figure 4.2 just before the cmc is reached. From Γ_2, the area per surfactant ion or molecule can be calculated, since

$$\text{Area/molecule} = \frac{1}{\Gamma_2 N_A} \tag{10}$$

where N_A is Avogadro's number. Determining the area per surfactant molecule is very useful since it gives information on the surfactant orientation at the interface. For example, for ionic surfactants such as sodium dodecyl sulfate, the area

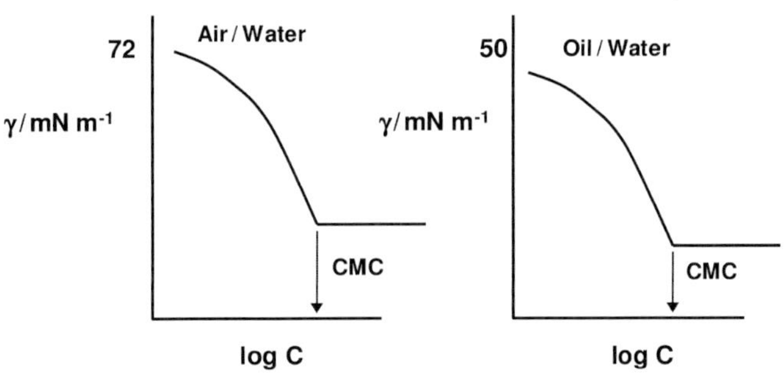

Figure 4.2 Surface or interfacial tension–log *C* curves.

per surfactant is determined by the area occupied by the alkyl chain and head group if these molecules lie flat at the interface, whereas for a vertical orientation, the area per surfactant ion is determined by that occupied by the charged head group, which at low electrolyte concentration will be in the region of $0.40\,\text{nm}^2$. Such an area is larger than the geometric area occupied by a sulfate group, as a result of the lateral repulsion between the head groups. On addition of electrolytes, this lateral repulsion is reduced and the area per surfactant ion for a vertical orientation will be lower than $0.4\,\text{nm}^2$ (reaching in some case $0.2\,\text{nm}^2$). On the other hand, if the molecules lie flat at the interface the area per surfactant ion will be considerably higher than $0.4\,\text{nm}^2$.

Another important point can be made from the γ–log C curves. At a concentration just before the break point, one has the condition of constant slope, which indicates that saturation adsorption has been reached. Just above the break point,

$$\left(\frac{\partial \gamma}{\partial \ln a_2}\right)_{p,T} = 0 \tag{11}$$

indicating the constancy of γ with log C above the cmc Integration of equation 11 gives

$$\gamma = \text{constant} \times \ln a_2 \tag{12}$$

Since γ is constant in this region, then a_2 must remain constant. This means that addition of surfactant molecules above the cmc must result in association to form units (micellar) with low activity.

As mentioned before, the hydrophilic head group may be un-ionized, e.g. alcohols or poly(ethylene oxide), alkane or alkylphenol compounds, weakly ionized such as carboxylic acids or strongly ionized such as sulfates, sulfonates and quaternary ammonium salts. The adsorption of these different surfactants at the air/water and oil/water interface depends on the nature of the head group. With nonionic surfactants, repulsion between the head groups is small and these surfactants are usually strongly adsorbed at the surface of water from very dilute solutions. As mentioned before, nonionic surfactants have much lower cmc values compared with ionic surfactants with the same alkyl chain length. Typically, the cmc is in the region of 10^{-5}–$10^{-4}\,\text{mol dm}^{-3}$. Such nonionic surfactants form closely packed adsorbed layers at concentrations lower than their cmc values. The activity coefficient of such surfactants is close to unity and is only slightly affected by addition of moderate amounts of electrolytes (or change in the pH of the solution). Thus, nonionic surfactant adsorption is the simplest case since the solutions can be represented by a two-component system and the adsorption can be accurately calculated using equation 9.

With ionic surfactants, on the other hand, the adsorption process is relatively more complicated since one has to consider the repulsion between the head groups and the effect of the presence of any indifferent electrolyte. Moreover, the Gibbs adsorption equation has to be solved taking into account the surfactant ions, the counterion and any indifferent electrolyte ions present. For a strong surfactant electrolyte such as an Na^+R^-

$$\Gamma_2 = \frac{1}{2RT}\frac{\partial \gamma}{\partial \ln a_{\pm}} \tag{13}$$

where the factor of 2 arises because both surfactant ion and counter ion must be adsorbed to maintain neutrally and $d\gamma/d\ln a_{\pm}$ is twice as large as for an un-ionized surfactant.

If a non-adsorbed electrolyte, such as NaCl, is present in large excess, then any increase in concentration of Na^+R^- produces a negligible increase in Na^+ concentration and therefore $d\mu_{Na}$ becomes negligible. Moreover, $d\mu_{Cl}$ is also negligible, so that the Gibbs adsorption equation reduces to

$$\Gamma_2 = -\frac{1}{RT}\left(\frac{\partial \gamma}{\partial \ln C_{NaR}}\right) \tag{14}$$

i.e. it becomes identical with that for a nonionic surfactant.

The above discussion clearly illustrates that for calculation of Γ_2 from the γ–log C curve, one has to consider the nature of the surfactant and the composition of the medium. For nonionic surfactants the Gibbs adsorption equation (equation 9) can be directly used. For ionic surfactants, in the absence of electrolytes, the right-hand side of the equation 9 should be divided by 2 to account for surfactant dissociation. This factor disappears in the presence of a high concentration of an indifferent electrolyte.

4.4 Dynamic Processes of Adsorption

Most spraying processes work under dynamic conditions and improvement of their efficiency requires the use of surfactants that lower the liquid surface tension γ_{LV} under these dynamic conditions. The interfaces involved (e.g. droplets formed in a spray or impacting on a surface) are freshly formed and have only a small effective age of some seconds or even less than a millisecond.

The most frequently used parameter to characterize the dynamic properties of liquid adsorption layers is the dynamic surface tension (which is a time-dependent quantity). Techniques should be available to measure γ_{LV} as a function of time (ranging from a fraction of a millisecond to minutes and hours or days) [3].

To optimize the use of surfactants, polymers and their mixtures, a specific knowledge of their dynamic adsorption behavior rather than equilibrium properties is of great interest [3]. It is necessary, therefore, to describe the dynamics of surfactant adsorption at a fundamental level.

4.4.1 General Theory of Adsorption Kinetics

The first physically sound model for adsorption kinetics was derived by Ward and Tordai [4]. It is based on the assumption that the time dependence of surface or interfacial tension, which is directly proportional to the surface excess Γ (mol m^{-2}), is caused by diffusion and transport of surfactant molecules to the interface. This is referred to as 'the diffusion-controlled adsorption kinetics model'. The interfacial surfactant concentration at any time t, $\Gamma(t)$, is given by the following expression:

$$\Gamma(t) = 2\left(\frac{D}{\pi}\right)^{\frac{1}{2}}\left[c_0 t^{\frac{1}{2}} - \int_0^{t^{\frac{1}{2}}} c(0, t-\tau)\, \mathrm{d}(\tau)^{\frac{1}{2}}\right] \tag{15}$$

where D is the diffusion coefficient, c_0 is the bulk concentration and τ is the thickness of the diffusion layer.

The above diffusion-controlled model assumes transport by diffusion of the surface-active molecules to be the rate-controlled step. The so-called 'kinetic-controlled model' is based on the transfer mechanism of molecules from solution to the adsorbed state and vice versa [3].

A schematic diagram of the interfacial region is given in Figure 4.3, which shows three main states: (1) adsorption when the surface concentration Γ is lower than the equilibrium value Γ_0; (2) equilibrium state when $\Gamma = \Gamma_0$ and (3) desorption when $\Gamma > \Gamma_0$.

The transport of surfactant molecules from the liquid layer adjacent to the interface (subsurface) is simply determined by molecular movements (in the absence of forced liquid flow). At equilibrium, i.e. when $\Gamma = \Gamma_0$, the flux of adsorption is

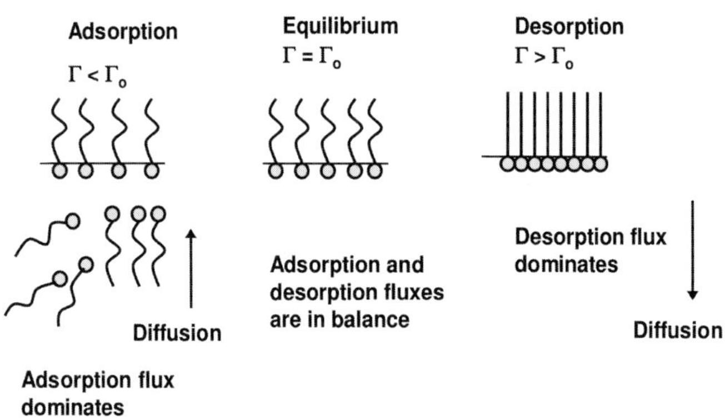

Figure 4.3 Representation of the fluxes of adsorbed surfactant molecules in the absence of liquid flow.

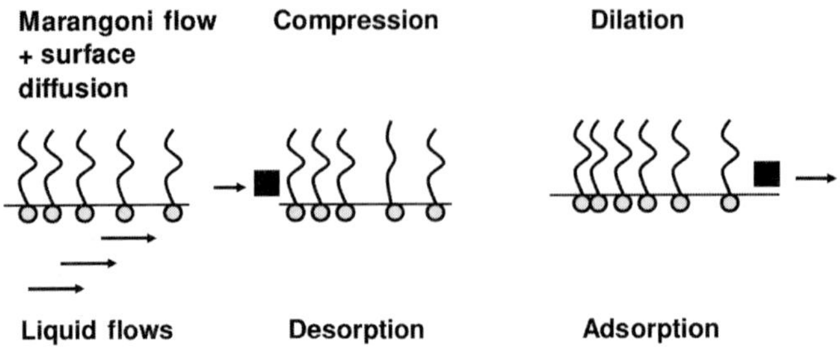

Figure 4.4 Representation of surfactant transport at the surface and in the bulk of a liquid.

equal to the flux of desorption. Clearly, when $\Gamma < \Gamma_0$, the flux of adsorption predominates, whereas when $\Gamma > \Gamma_0$, the flux of desorption predominates [3].

In the presence of liquid flow, the situation becomes more complicated due to the creation of surface concentration gradients [3]. These gradients, described by the Gibbs dilational elasticity [5], initiate a flow of mass along the interface in the direction of higher surface or interfacial tension (Marangoni effect). This situation can arise, for example, if an adsorption layer is compressed or stretched, as is illustrated in Figure 4.4.

A qualitative model that can describe adsorption kinetics is given by equation 16, which affords a rough estimate and results from equation 15 when the second term on the right-hand side is neglected:

$$\Gamma(t) = \left(\frac{Dt}{\pi}\right)^{\frac{1}{2}} \tag{16}$$

An equivalent equation to equation 16 was derived by Panaitov and Petrov [5]:

$$c(0,t) = c_0 - \frac{2}{(D\pi)^{\frac{1}{2}}} \int_0^{t^{\frac{1}{2}}} \frac{\mathrm{d}\Gamma(t-\tau)}{\mathrm{d}t} \mathrm{d}\tau^{\frac{1}{2}} \tag{17}$$

Hansen [6] and Miller and Lunkenheimer [7] gave numerical solutions to the integrals in equation 15 and 17 and obtained a simple expression using a Langmuir isotherm:

$$\Gamma(t) = \Gamma_\infty \frac{c(0,t)}{a_\mathrm{L} + c(0,t)} \tag{18}$$

where a_L is the constant in the Langmuir isotherm (mol m^{-3}). The corresponding equation from the variation of surface tension γ with time is as follows (Langmuir–Szyszowski equation):

$$\gamma = \gamma_0 + RT\Gamma_\infty \ln(1 - \Gamma_\infty) \tag{19}$$

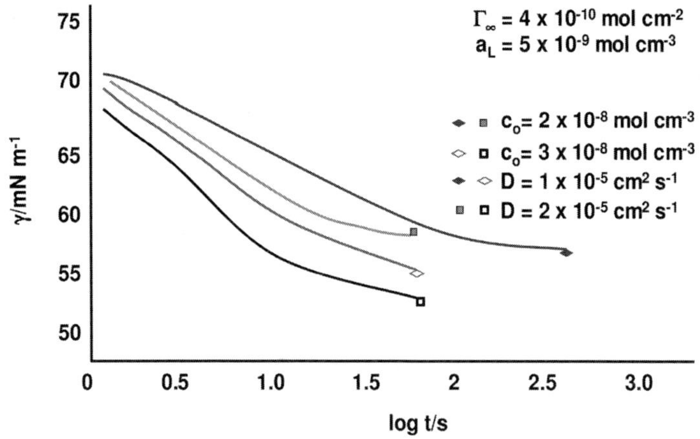

Figure 4.5 Surface tension γ–log t curves calculated on the basis of equations 15–19.

Calculations based on equations 15–19 are given in Figure 4.5, with different values of c_0/a_L [3].

4.4.2
Adsorption Kinetics from Micellar Solutions

As mentioned in Chapter 3, surfactants form micelles above the cmc of different sizes and shapes, depending on the nature of the molecule, temperature, electrolyte concentration, etc. The dynamic nature of micellization can be described by two main relaxation processes, τ_1 (the lifetime of a monomer in a micelle) and τ_2 (the lifetime of the micelle, i.e. complete dissolution into monomers).

The presence of micelles in equilibrium with monomers influences the adsorption kinetics remarkably. After a fresh surface has been formed surfactant monomers are adsorbed, resulting in a concentration gradient of these monomers. This gradient will be equalized by diffusion to re-establish a homogeneous distribution. Simultaneously, the micelles are no longer in equilibrium with monomers within the range of concentration gradient. This leads to a net process of micelle dissolution or rearrangement to re-establish the local equilibrium. As a consequence, a concentration gradient of micelles results, which is equalized by diffusion of micelles [3].

Based on the above concepts, one would expect that the ratio of monomers c_1 to micelles c_m, the aggregation number n, rate of micelle formation k_f and rate of micelle dissolution k_d will influence the rate of the adsorption process. A schematic diagram of the kinetic process in the presence of micelles is given in Figure 4.6.

The diagram shows that to describe the kinetics of adsorption, one must take into account the diffusion of monomers and micelles and also the kinetics of

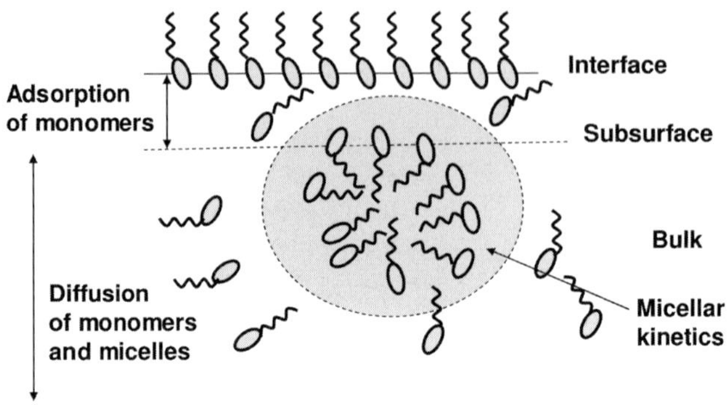

Figure 4.6 Representation of the adsorption process from a micellar solution.

micelle formation and dissolution. Several processes may take place and these are represented schematically in Figure 4.7. Three main mechanisms may be considered, namely (1) formation–dissolution, (2) rearrangement and (3) stepwise aggregation–dissolution. To describe the effect of micelles on adsorption kinetics, one should know several parameters, such as the micelle aggregation number and rate constants of micelle kinetics [8].

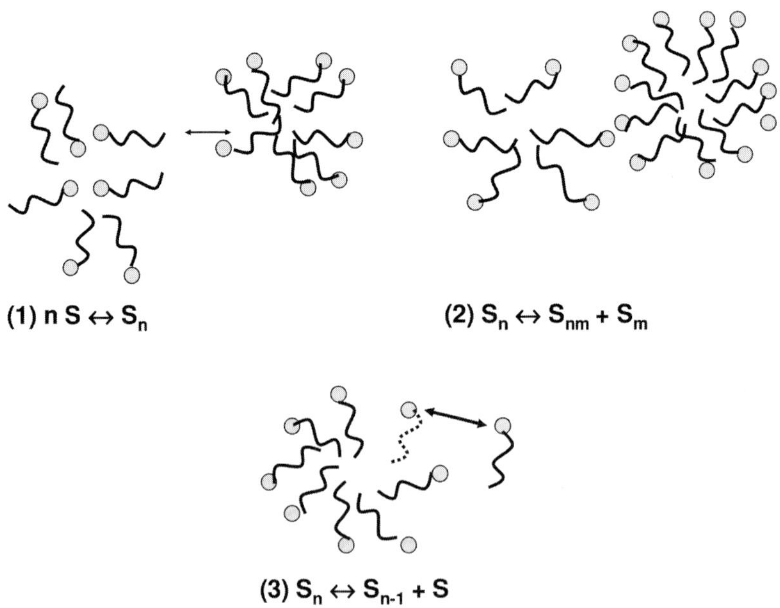

Figure 4.7 Schematic diagram of micelle kinetics.

4.4.3
Experimental Techniques for Studying Adsorption Kinetics

The two most suitable techniques for studying adsorption kinetics are the drop volume method and the maximum bubble pressure method. The first method can provide information on adsorption kinetics in the range from seconds to some minutes. However, it has the advantage of measurement both at the air/liquid and liquid/liquid interfaces. The maximum bubble pressure method allows one to make measurements in the millisecond range, but it is restricted to the air/liquid interface. A description of both techniques is given below.

4.4.3.1 The Drop Volume Technique

A schematic representation of the drop volume apparatus [9] is given in Figure 4.8. A metering system in the form of a motor-driven syringe allows the formation of the liquid drop at the tip of a capillary, which is positioned in a sealed cuvette. The cuvette is either filled with a small amount of the measuring liquid, to saturate the atmosphere, or with a second liquid in the case of interfacial studies. A light barrier arranged below the forming drop permits the detection of drop detachment from the capillary. Both the syringe and the light barriers are computer controlled and allow fully automatic operation of the setup. The syringe and the cuvette are temperature controlled by a water jacket, which makes interfacial tension measurements possible in the temperature range 10–90 °C.

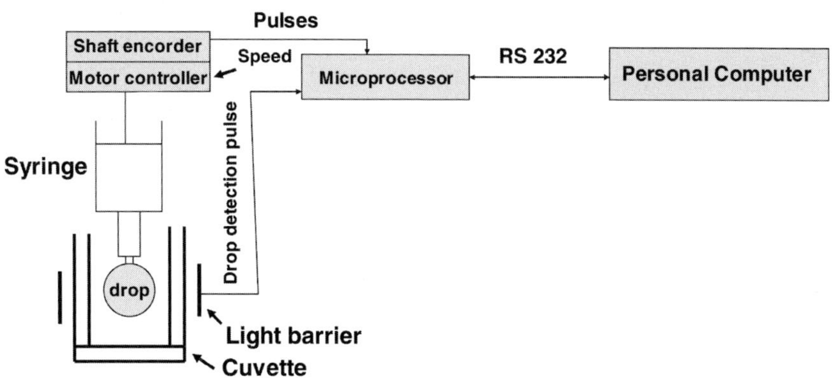

Figure 4.8 Representation of the drop volume apparatus.

As mentioned above, the drop volume method is of dynamic character and it can be used for adsorption processes in the time interval of seconds up to some minutes. At small drop times, the so-called hydrodynamic effect has to be considered [10]. This gives rise to apparently higher surface tension. Kloubek *et al.* [11] used an empirical equation to account for this effect:

$$V_e = V(t) - \frac{K_V}{t} \tag{20}$$

where V_e is the unaffected drop volume, $V(t)$ is the measured drop volume and K_v is a proportionality factor that depends on surface tension γ, density difference $\Delta\rho$ and tip radius r_{cap}.

Dukhin *et al.* [3] obtained the following equation for the variation of drop volume $V(t)$ with time:

$$V(t) = V_e + t_0 F = V_e \left(1 + \frac{t_0}{t - t_0}\right) \tag{21}$$

where F is the liquid flow per unit time, which is given by

$$F = \frac{V(t)}{t} = \frac{V_e}{t - t_0} \tag{22}$$

The drop volume technique is limited in its application. Under conditions of fast drop formation and larger tip radii, the drop formation shows irregular behavior.

4.4.3.2 Maximum Bubble Pressure Technique

This is the most useful technique for measuring adsorption kinetics at short times, particularly if a correction for the so-called 'dead time', τ_d, is made. The dead time is simply the time required to detach the bubble after it has reached its hemispherical shape. A schematic representation of the principle of maximum bubble pressure is presented in Figure 4.9, which shows the evolution of a bubble at the tip of a capillary. It also shows the variation of pressure p in the bubble with time.

At $t = 0$ (initial state), the pressure is low (note that the pressure is equal to $2\gamma/r$; since r of the bubble is large, p is small). At $t = \tau$ (smallest bubble radius that is equal to the tube radius) p reaches a maximum. At $t = \tau_b$ (detachment time) p decreases since the bubble radius increases. The design of a maximum

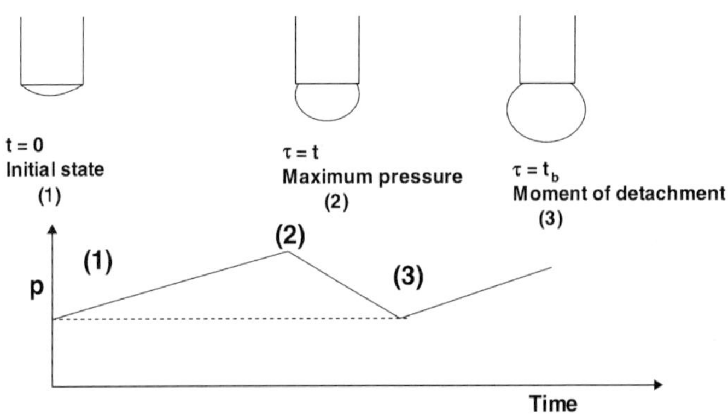

Figure 4.9 Scheme of bubble evolution and pressure change with time.

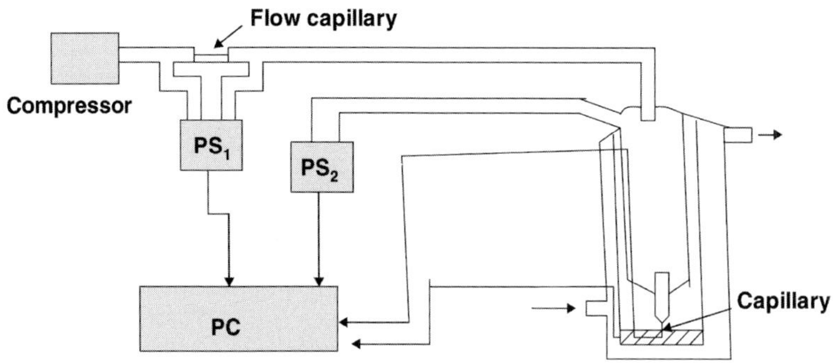

Figure 4.10 Maximum bubble pressure apparatus.

bubble pressure method for high bubble formation frequencies (short surface age) requires the following: (1) measurement of bubble pressure; (2) measurement of bubble formation frequency; and (3) estimation of surface lifetime and effective surface age. The first problem can easily be solved if the system volume (which is connected to the bubble) is large enough in comparison with the bubble separating from the capillary. In this case, the system pressure is equal to the maximum bubble pressure. The use of an electric pressure transducer for measuring bubble formation frequency presumes that pressure oscillations in the measuring system are distinct enough and this satisfies point (2). Estimation of the surface lifetime and effective surface age, i.e. point (3), requires estimation of the dead time τ_d. A schematic representation of the setup for measuring the maximum bubble pressure and surface age is shown in Figure 4.10. The air coming from a microcompressor flows first through the flow capillary. The air flow rate is determined by measuring the pressure difference at both ends of the flow capillary with the electric transducer PS_2. Thereafter, the air enters the measuring cell and the excess air pressure in the system is measured by a second electric sensor PS_2. A sensitive microphone is placed in the tube which leads the air to the measuring cell.

The measuring cell is equipped with a water jacket for temperature control, which simultaneously holds the measuring capillary and two platinum electrodes, one of which is immersed in the liquid under study and the second is situated exactly opposite the capillary and controls the size of the bubble. The electric signals from the gas flow sensor PS_1 and pressure transducer PS_2, the microphone and the electrodes, and also the compressor, are connected to a personal computer which operates the apparatus and acquires the data.

The value of τ_d, equivalent to the time interval necessary to form a bubble of radius R, can be calculated using Poiseuille's law:

$$\tau_d = \frac{\tau_b L}{Kp}\left(\frac{1 + 3r_{ca}}{2R}\right) \tag{23}$$

where K is given by Poiseuille's law:

$$K = \frac{\pi r^4}{8\eta l} \tag{24}$$

η is the gas viscosity, l is the length, L is the gas flow rate and r_{ca} is the radius of the capillary.

The calculation of dead time τ_d can be simplified by taking into account the existence of two gas flow regimes for the gas flow leaving the capillary: bubble flow regime when $\tau > 0$ and jet regime when $\tau = 0$ and hence $\tau_b = \tau_d$. A typical dependence of p on L is shown in Figure 4.11.

To the right of the critical point, the dependence of p on L is linear in accordance with Poiseuille's law. Under these conditions,

$$\tau_d = \tau_b \frac{L p_c}{L_c p} \tag{25}$$

where L_c and p_c are related to the critical point and L and p are the actual values of the dependence to the left of the critical point.

The surface lifetime can be calculated from

$$\tau = \tau_b - \tau_d = \tau_b \left(1 - \frac{L p_c}{L_c p}\right) \tag{26}$$

The critical point in the dependence of p and L can be easily located and is included in the soft ware of the computer program.

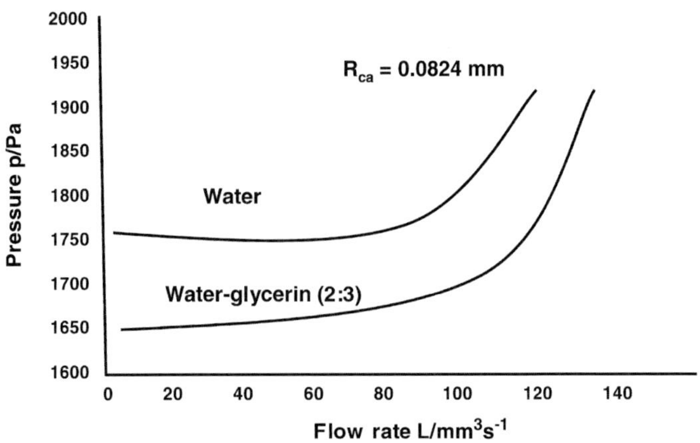

Figure 4.11 Dependence of p on gas flow rate L at 30 °C.

The surface tension value in the maximum bubble pressure method is calculated using the Laplace equation:

$$p = \frac{2\gamma}{r} + \rho hg + \Delta p \tag{27}$$

where ρ is the density of the liquid, g is the acceleration due to gravity, h is the depth to which the capillary is immersed in the liquid and Δp is a correction factor to allow for hydrodynamic effects.

4.5
Adsorption of Surfactants and Polymeric Surfactants at the Solid/Liquid Interface

The use of surfactants (ionic, nonionic and zwitterionic) and polymers to control the stability behavior of suspension concentrates in agrochemical formulations is of considerable importance. They are a particularly robust form of stabilization which is useful at high disperse volume fractions and high electrolyte concentrations, and also under extreme conditions of high temperature. In particular, surfactants and polymers are essential for the stabilization of suspensions in nonaqueous media, where electrostatic stabilization is less successful.

The key to understanding how surfactants and polymers (to be referred to as polymeric surfactants) function as stabilizers is to know their adsorption and conformation at the solid/liquid interface. This is the objective of the present section, which gives a survey of the general trends observed and some of the theoretical treatments.

Since surfactant and polymer adsorption processes are significantly different, the two subjects will be treated differently. Suffice to say at this stage that surfactant adsorption is relatively more simple than polymer adsorption. This stems from the fact that surfactants consist of a small number of units and they mostly are reversibly adsorbed, allowing one to apply thermodynamic treatments. In this case, it is possible to describe the adsorption in terms of the various interaction parameters, namely chain–surface, chain–solvent and surface–solvent. Moreover, the conformation of the surfactant molecules at the interface can be deduced from these simple interactions parameters. In contrast, the process of polymer adsorption is fairly complicated. In addition to the usual adsorption considerations described above, one of the principle problems to be resolved is the conformation of the polymer molecule at the surface. This can acquire various possible forms depending on the number of segments and chain flexibility. This requires the application of statistical thermodynamic methods.

4.6
Surfactant Adsorption

As mentioned above, surfactant adsorption may be described in terms of simple interaction parameters. However, in some cases these interaction parameters may involve ill-defined forces, such as hydrophobic bonding, solvation forces and chemisorption. In addition, the adsorption of ionic surfactants involves electrostatic forces, particularly with polar surfaces containing ionogenic groups. For that reason, the adsorption of ionic and nonionic surfactants will be treated separately. The surfaces (substrates) can also be hydrophobic or hydrophilic and these may be treated separately.

4.6.1 Adsorption of Ionic Surfactants on Hydrophobic Surfaces

The adsorption of ionic surfactants on hydrophobic surfaces such as agrochemical particles is governed by the hydrophobic interaction between the alkyl chain of the surfactant and the hydrophobic surface. In this case, electrostatic interaction will play a relatively smaller role. However, if the surfactant head group has the same sign of charge as that on the substrate surface, electrostatic repulsion may oppose adsorption. In contrast, if the head groups are of opposite charge sign to the surface, adsorption may be enhanced. Since the adsorption depends on the magnitude of the hydrophobic bonding free energy, the amount of surfactant adsorbed increases directly with increase in the alkyl chain length in accordance with Traube's rule.

The adsorption of ionic surfactants on hydrophobic surfaces may be represented by the Stern–Langmuir isotherm [12]. Consider a substrate containing N_s sites (mol m^{-2}) on which Γ mol m^{-2} of surfactant ions are adsorbed. The surface coverage θ is Γ/N_s and the fraction of uncovered surface is $1-\theta$. The Stern–Langmuir equation is written as

$$\frac{\theta}{1-\theta} = \frac{C}{55.5}\exp\left(-\frac{\Delta G^{\circ}_{\text{ads}}}{RT}\right) \qquad (28)$$

Equation (28) applies only at low surface coverage ($\theta < 0.1$), where lateral interaction between the surfactant ions can be neglected.

At high surface coverage ($\theta > 0.1$), one should take the lateral interaction between the chains into account, by introducing a constant A, e.g. using the Frumkin–Fowler–Guggenheim equation [12]:

$$\frac{\theta}{1-\theta}\exp(A\theta) = \frac{C}{55.5}\exp\left(-\frac{\Delta G^{\circ}_{\text{ads}}}{RT}\right) \qquad (29)$$

Various authors [13, 14] have used the Stern–Langmuir equation in a simple form to describe the adsorption of surfactant ions on mineral surfaces:

$$\Gamma = 2rC\exp\left(-\frac{\Delta G^{\circ}_{\text{ads}}}{RT}\right) \qquad (30)$$

Various contributions to the adsorption free energy may be envisaged. To a first approximation, these contributions may be considered to be additive. In the first instance, ΔG_{ads} maybe taken to consist of two main contributions:

$$\Delta G_{ads} = \Delta G_{elec} + \Delta G_{spec} \tag{31}$$

where ΔG_{elec} accounts for any electrical interactions and ΔG_{spec} is a specific adsorption term which contains all contributions to the adsorption free energy that are dependent on the 'specific' (non-electrical) nature of the system [15]. Several authors subdivided ΔG_{spec} into supposedly separate independent interactions [15, 16], e.g.

$$\Delta G_{spec} = \Delta G_{cc} + \Delta G_{cs} + \Delta G_{hs} + \cdots \tag{32}$$

where ΔG_{cc} is a term that accounts for the cohesive chain–chain interaction between the hydrophobic moieties of the adsorbed ions, ΔG_{cs} is the term for chain–substrate interaction and ΔG_{hs} is a term for the head group–substrate interaction. Several other contributions to ΔG_{spec} may be envisaged, e.g. ion–dipole, ion–induced dipole or dipole–induced dipole interactions.

Since there is no rigorous theory that can predict adsorption isotherms, the most suitable method to investigate adsorption of surfactants is to determine the adsorption isotherm. Measurement of surfactant adsorption is fairly straightforward. A known mass m (g) of the particles (substrate) with known specific surface area A_s ($m^2\,g^{-1}$) is equilibrated at constant temperature with surfactant solution with initial concentration C_1. The suspension is kept stirred for a sufficient time to reach equilibrium. The particles are then removed from the suspension by centrifugation and the equilibrium concentration C_2 is determined using a suitable analytical method. The amount of adsorption Γ (mol m^{-2}) is calculated as follows:

$$\Gamma = \frac{C_1 - C_2}{mA_s} \tag{33}$$

The adsorption isotherm is represented by plotting Γ versus C_2. A range of surfactant concentrations should be used to cover the whole adsorption process, i.e. from the initial values low to the plateau values. To obtain accurate results, the solid should have a high surface area (usually $>1\,m^2$).

Several examples may be quoted from the literature to illustrate the adsorption of surfactant ions on solid surfaces. For a model hydrophobic surface, carbon black has been chosen [17, 18]. Figure 4.12 shows typical results for the adsorption of sodium dodecyl sulfate (SDS) on two carbon black surfaces, namely Spheron 6 (untreated) and Graphon (graphitized), which also describes the effect of surface treatment. The adsorption of SDS on untreated Spheron 6 tends to show a maximum that is removed on washing. This suggests the removal of impurities from the carbon black which become extractable at high surfactant concentration. The plateau adsorption value is $\sim 2 \times 10^{-6}$ mol m^{-2} ($\sim 2\,\mu mol\,m^{-2}$). This plateau value is reached at ~ 8 mmol dm^{-3} SDS, i.e. close to the cmc of

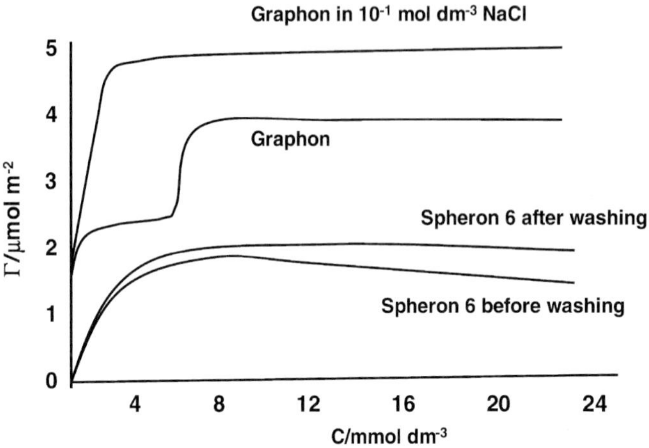

Figure 4.12 Adsorption isotherms for sodium dodecyl sulfate on carbon substrates.

the surfactant in the bulk solution. The area per surfactant ion in this case is ~0.7 nm^2.

Graphitization (Graphon) removes the hydrophilic ionizable groups (e.g. –C=O or –COOH), producing a surface that is more hydrophobic. The same occurs by heating Spheron 6 to 2700 °C. This leads to a different adsorption isotherm (Figure 4.12), showing a step (inflection point) at a surfactant concentration in the region of ~6 mmol dm^{-3}. The first plateau value is ~2.3 μmol m^{-2} whereas the second plateau value (which occurs at the cmc of the surfactant) is ~4 μmol m^{-2}. It is likely in this case that the surfactant ions adopt different orientations at the first and second plateaus. In the first plateau region, a more 'flat' orientation (alkyl chains adsorbing parallel to the surface) is obtained whereas at the second plateau a vertical orientation is more favorable, with the polar head groups being directed towards the solution phase. Addition of electrolyte (10^{-1} mol dm^{-3} NaCl) enhances the surfactant adsorption. This increase is due to the reduction of lateral repulsion between the sulfate head groups, and this enhances the adsorption.

The adsorption of ionic surfactants on other hydrophobic surfaces resembles that for carbon black [19, 20]. For example, Saleeb and Kitchener [19] found a similar limiting area for cetyltrimethylammonium bromide on Graphon and polystyrene (~0.4 nm^2). As with carbon black, the area per molecule depends on the nature and amount of added electrolyte. This can be accounted for in terms of reduction of head group repulsion and/or counterion binding.

Surfactant adsorption close to the cmc may appear Langmuirian, although this does not automatically imply a simple orientation. For example, rearrangement from a horizontal to a vertical orientation or electrostatic interaction and counterion binding may be masked by simple adsorption isotherms. It is essential, therefore, to combine the adsorption isotherms with other techniques such as microcalorimetry and various spectroscopic methods to obtain a full picture of surfactant adsorption.

4.6.2
Adsorption of Ionic Surfactants on Polar Surfaces

The adsorption of ionic surfactants on polar surfaces that contain ionizable groups may show characteristic features due to additional interaction between the head group and substrate and/or possible chain–chain interaction. This is best illustrated by the results of adsorption of sodium dodecyl sulfonate (SDSe) on alumina at pH 7.2 obtained by Fuerstenau [21] and shown in Figure 4.13. At the pH value, the alumina is positively charged (the isoelectric point of alumina is at pH ≈ 9) and the counter ions are Cl^- from the added supporting electrolyte. In Figure 4.13, the saturation adsorption Γ_1 is plotted versus equilibrium surfactant concentration C_1 using logarithmic scales. The figure also shows the results of zeta potential (ζ) measurements (which are a measure of the magnitude sign of charge on the surface). Both adsorption and zeta potential results show three distinct regions. The first region (I), which shows a gradual increase in adsorption with increase in concentration, with virtually no change in the value of the zeta potential, corresponds to an ion-exchange process [22]. In other words, the surfactant ions simply exchange with the counterions (Cl^-) of the supporting electrolyte in the electrical double layer. At a critical surfactant concentration, the desorption increases dramatically with further increase in surfactant concentration (region II). In this region, the positive zeta potential gradually decreases, reaching a zero value (charge neutralization), after which a negative value is obtained that increases rapidly with increase in surfactant concentration. The rapid increase in region II was explained in terms of 'hemi-micelle formation' that was originally postulated by Gaudin and Fuerstenau [23]. In other words, at a critical surfactant concentration (to be denoted the cmc of 'hemi-micelle formation' or, better, the critical aggregation concentration, CAC), the hydrophobic moieties of the adsorbed surfactant chains are 'squeezed out' from the aqueous solution by forming two-dimensional aggregates on the adsorbent surface. This is analogous

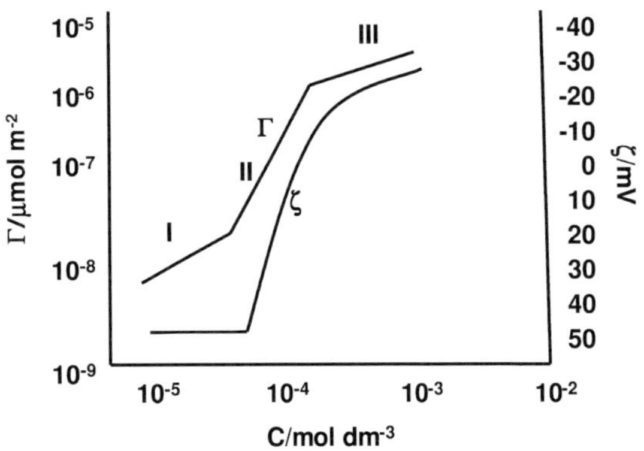

Figure 4.13 Adsorption isotherm for sodium dodecyl sulfonate on alumina and corresponding zeta potential (ζ).

to the process of micellization in bulk solution. However, the CAC is lower than the cmc, indicating that the substrate promotes surfactant aggregation. At a certain surfactant concentration in the hemi-micellization process, the isoelectric point is exceeded and, thereafter, the adsorption is hindered by the electrostatic repulsion between the hemi-micelles and hence the slope of the adsorption isotherm is reduced (region III).

4.6.3 Adsorption of Nonionic Surfactants

Several types of nonionic surfactants exist, depending on the nature of the polar (hydrophilic) group. The most common type is that based on a poly(oxyethylene) glycol group, i.e. $(CH_2CH_2O)_nOH$ (where n can vary from as little as 2 units to as high as 100 or more units) linked either to an alkyl (C_xH_{2x+1}) or alkylphenyl $(C_xH_{2x+1}–C_6H_4–)$ group. These surfactants may be abbreviated as C_xE_n or $C_x\varphi E_n$ (where C refers to the number of C atoms in the alkyl chain, φ denotes C_6H_4 and E denotes ethylene oxide). These ethoxylated surfactants are characterized by a relatively large head group compared with the alkyl chain (when $n > 4$). However, there are nonionic surfactants with a small head group such as amine oxides $(–N \rightarrow O)$, phosphate oxide $(–P \rightarrow O)$ and sulfinylalkanol $[–SO–(CH_2)_n–OH]$ [24]. Most adsorption isotherms in the literature are based on the ethoxylated-type surfactants.

The adsorption isotherms of nonionic surfactants are in many cases Langmuirian, like those of most other highly surface-active solutes adsorbing from dilute solutions, and adsorption is generally reversible. However, several other adsorption types are produced [23], and these are illustrated in Figure 4.14. The steps in the isotherm may be explained in terms of the various adsorbate–adsorbate, adsorbate–adsorbent and adsorbate–solvent interactions. These orientations are illustrated schematically in Figure 4.15.

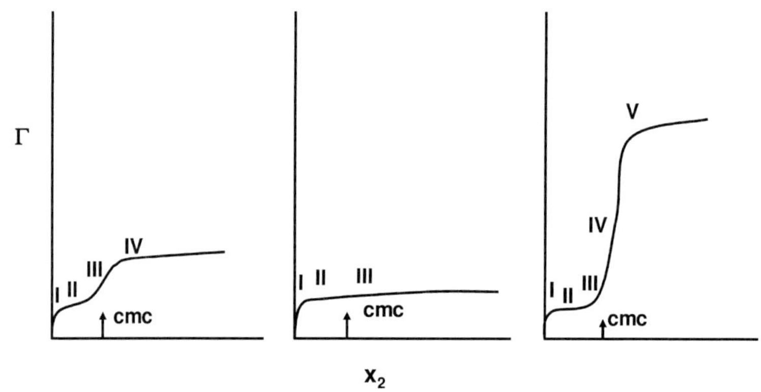

Figure 4.14 Adsorption isotherms corresponding to the three adsorption sequences shown in Figure 4.15.

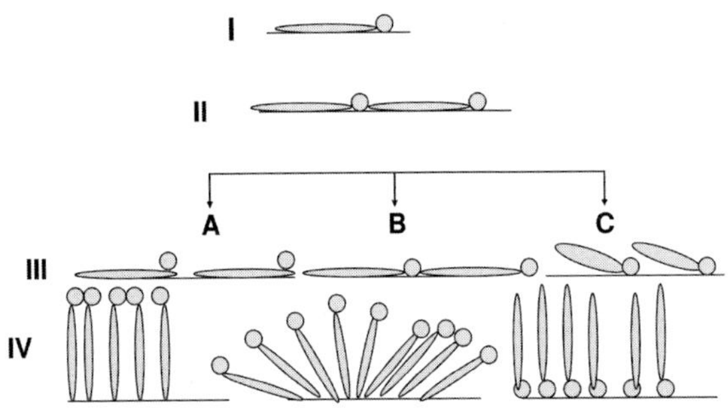

Figure 4.15 Model for adsorption of nonionic surfactants.

In the first stage of adsorption (denoted by I in Figures 4.14 and 4.15), surfactant–surfactant interaction is negligible (low coverage) and adsorption occurs mainly by van der Waals interaction. On a hydrophobic surface, the interaction is dominated by the hydrophobic portion of the surfactant molecule. This is mostly the case with agrochemicals which have hydrophobic surfaces. However, if the chemical is hydrophilic in nature, the interaction will be dominated by the EO chain. The approach to monolayer saturation with the molecules lying flat is accompanied by a gradual decrease in the slope of the adsorption isotherm (region II in Figure 4.14). An increase in the size of the surfactant molecule, e.g. increasing the length of the alkyl or EO chain, will decrease adsorption (when expressed in moles per unit area). On the other hand, an increase in temperature will increase adsorption as a result of desolvation of the EO chains, thus reducing their size. Moreover, an increase in temperature reduces the solubility of the nonionic surfactant and this enhances adsorption.

The subsequent stages of adsorption (regions III and IV) are determined by surfactant–surfactant interaction, although surfactant–surface interaction initially determines adsorption beyond stage II. This interaction depends on the nature of the surface and the hydrophilic–lipophilic balance (HLB) of the surfactant molecule. For a hydrophobic surface, adsorption occurs via the alkyl group of the surfactant. For a given EO chain, the adsorption will increase with increase in the alkyl chain length. On the other hand, for a given alkyl chain length, adsorption increases with decrease in the PEO chain length.

As the surfactant concentration approaches the cmc, there is a tendency for aggregation of the alkyl groups. This will cause a vertical orientation of the surfactant molecules (stage IV). The will compress the head group and for an EO chain it will result in a less coiled, more extended conformation. The larger the surfactant alkyl chain, the greater will be the cohesive forces and hence the smaller the cross-sectional area. This may explain why saturation adsorption increases with increasing alkyl chain length. The interaction occurring in the adsorption layer during the fourth and subsequent stages of adsorption are similar to those that occur in bulk solution. In this case aggregate units, as shown in Figure 4.15 V (hemi-micelles or micelles) may be formed.

4.7
Adsorption of Polymeric Surfactants at the Solid/Liquid Interface

The simplest type of polymeric surfactant is a homopolymer, which is formed from the same repeating units [25], e.g. poly(ethylene oxide) (PEO) and polyvinylpyrrolidone (PVP). Homopolymers have little surface activity at the oil/water (O/W) interface. However, homopolymers may adsorb significantly at the solid/liquid (S/L) interface Even if the adsorption energy per monomer segment is small (fraction of kT, where k is the Boltzmann constant and T is the absolute temperature), the total adsorption energy per molecule may be sufficient (several segments are adsorbed at the surface) to overcome the unfavorable entropy loss of the molecule at the S/L interface. Homopolymers may also adsorb at the solid surface by some specific interaction, e.g. hydrogen bonding (for example, adsorption of PEO or PVP on silica). In general, homopolymers are not the most suitable dispersants for suspension concentrates.

A small variant is to use polymers that contain specific groups that have a high affinity to the surface, e.g. partially hydrolyzed poly(vinyl acetate) (PVAc), technically referred to as poly(vinyl alcohol) (PVA). Commercially available PVA molecules contain 4–12% of acetate groups. The acetate groups give the molecule its amphipathic character. On a hydrophobic surface (such as agrochemical particles) the polymer adsorbs with preferential attachment of the acetate groups on the surface, leaving the more hydrophilic vinyl alcohol segments dangling in the aqueous medium.

The most convenient polymeric surfactants are those of the block and graft copolymer type. A block copolymer is a linear arrangement of blocks of varying composition [24]:

diblock: poly A–block poly B: ~~A~~~~~ ~~~~~~B~~
triblock: poly A–block poly B–poly A: ~~A~~~~~ ~~~~ B ~~~~
~~~~~A~~

and a graft copolymer is a non–linear array of one B block on which several A polymers are grafted:

```
~~~~~B~~~~~
 | | | | |
 A A A A A
```

Most block and graft copolymers have low cmc and in many cases it is not easy to measure the cmc for these block and graft copolymers. The aggregation process is also affected by temperature and solvency of the medium for the A chains. One of the most useful methods to follow the aggregation of block and graft copolymers is to use time-averaged light scattering. By measuring the intensity as a function of concentration, one can extrapolate the results to zero concentration and obtain the molecular weight of the micelle. This allows one to obtain the aggregation number from a knowledge of the molecular weight of the monomer.
```

Several examples of block and graft copolymers may be quoted. Triblock polymeric surfactants: Pluronics (BASF) or Synperonic PE (Croda): two poly A blocks of PEO and one block poly B of poly(propylene oxide) (PPO); several chain lengths of PEO and PPO are available. Triblocks of PPO–PEO–PEO (inverse Pluronics) are also available. Polymeric triblock surfactants can be applied as dispersants. The hydrophobic PPO chain resides at the hydrophobic surface, leaving the two PEO chains dangling in aqueous solution (providing steric stabilization).

The above triblocks are not the most efficient dispersants; the PPO chain is not sufficiently hydrophobic to provide a strong 'anchor' to a hydrophobic surface. Several other di- and triblock copolymers have been synthesized: diblocks of polystyrene–block poly(vinyl alcohol); triblocks of poly(methyl methacrylate)–block poly(ethylene oxide)–poly(methyl methacrylate); diblocks of polystyrene–poly(ethylene oxide); triblocks of poly(ethylene oxide)–polystyrene–poly(ethylene oxide).

An alternative (and perhaps more efficient) polymeric surfactant is the amphipathic graft copolymer consisting of a polymeric backbone B [polystyrene or poly(methyl methacrylate)] and several A chains ('teeth') such as poly(ethylene oxide). The graft copolymer is referred to as a 'comb' stabilizer; the polymer forms a 'brush' at the solid/liquid interface. Typical commercially available graft copolymers are Atlox 4913 and Hypermer CG-6 supplied by Croda. The 'grafting into' technique has also been used to synthesize polystyrene–poly(ethylene oxide) graft copolymers. These molecules are not commercially available.

Recently, a novel graft copolymer based on a naturally occurring polysaccharide, inulin (polyfructose), has been synthesized [26]. Inulin is a polydisperse polysaccharide consisting mainly, if not exclusively, of $\beta(2 \rightarrow 1)$-fructosylfructose unit ($F_m$) with normally, but not necessarily, one glucopyranose unit at the reducing end ($GF_n$) [27, 28]. To produce the amphipathic graft copolymer, the chains were modified by introduction of alkyl groups ($C_4$–$C_{18}$) on the polyfructose backbone through isocyanates. The alkyl groups represent the B chains (that are randomly distributed on the sugar backbone on primary hydroxyl functions and also on the secondary ones), which become strongly adsorbed on a hydrophobic solid such as an agrochemical particle. The sugar chain forms the stabilizing chain as this is highly water soluble. The graft copolymer will adsorb on hydrophobic surfaces with the alkyl groups strongly attached (multipoint anchoring), leaving the polyfructose chains dangling in solution and probably forming large loops. These graft copolymers can produce highly stable suspensions, in particular at high electrolyte concentrations [29].

Understanding the adsorption and conformation of polymeric surfactants at interfaces is key to knowing how these molecules act as stabilizers. Most basic ideas on adsorption and conformation of polymers have been developed for the solid/liquid interface [30]. The process of polymer adsorption involves a number of different interactions that must be considered separately. Three main interactions must be taken into account, namely the interaction of the solvent molecules with the surface which need to be displaced for the polymer segments to adsorb, the interaction between the chains and the solvent and the interaction between the polymer and the surface. Apart from knowing these interactions, one of the most fundamental considerations is the conformation of the polymer molecule at the interface. These molecules adopt various conformations, depending on their structure. The simplest case to consider is that of a homopolymer that consists of identical segments [e.g. poly(ethylene oxide)], which shows a sequence of loops, trains and tails, as illustrated in Figure 4.16a. It should be mentioned at this stage that for such a polymer to adsorb, the reduction in entropy of the chain as it approaches the interface must be compensated by an energy of adsorption between the segments and the surface. In other words, the chain segments must have a minimum adsorption energy, $\chi^s$, otherwise no adsorption occurs. With polymers that are highly water soluble, such as poly(ethylene oxide) (PEO), the interaction energy with the surface may be too small for adsorption to occur and if this takes place the whole molecule may not be strongly adsorbed on the surface.

For this reason, many commercially available polymers that are described as homopolymers, such as poly(vinyl alcohol) (PVA), contain some hydrophobic groups or short blocks (vinyl acetate in the case of PVA) that ensured their adsorption on hydrophobic surfaces. This is illustrated in Figure 4.16b. Clearly, if all the segments have a high affinity to the surface, the whole molecule may lie flat on the surface, as illustrated in Figure 4.16c. This situation is rarely the case, since the molecule will have very low solubility in the continuous medium.

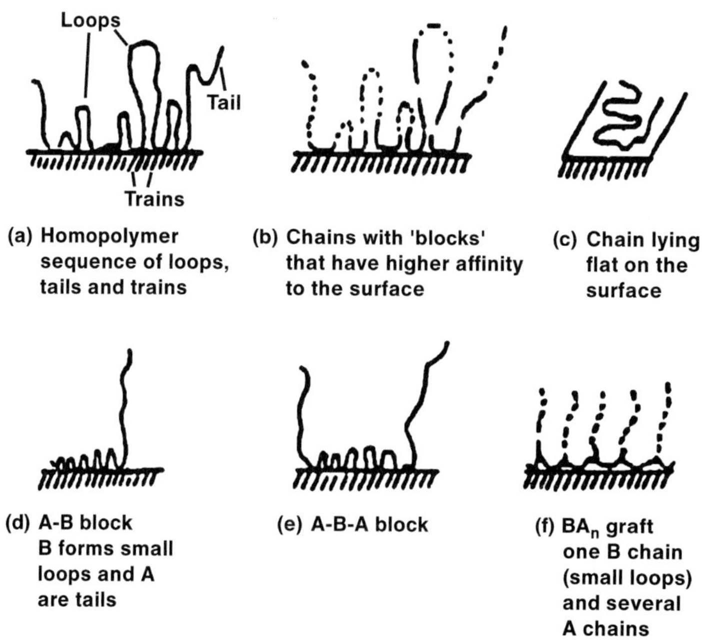


Figure 4.16 Various conformations of polymeric surfactants adsorbed on a plane surface.

The most favorable structures for polymeric surfactants are those represented in Figure 4.16d–f, referred to as block and graft copolymers. The molecule shown in Figure 4.16d is an A–B block, consisting of a B chain that has a high affinity to the surface (or soluble in the oil phase), referred to as the 'anchoring' chain, and an A chain that has a very low affinity to the surface and is strongly solvated by the medium. As will be discussed in Chapter 5, this is the most convenient structure, since the forces that ensure strong adsorption are opposite to those that ensure stability. A variant of the structure shown in Figure 4.16e is the A–B–A block copolymer. In this case, the anchor chain B contains two stabilizing chains (tails). Another variant is that shown in Figure 4.16f, described as a graft copolymer ('comb' type structure) with one B chain and several A chains (tails or 'teeth').

It is clear from the above description of polymer configurations that for full characterization of the process of adsorption, it is necessary to know the following parameters: the amount of polymer adsorbed per unit area of the surface, $\Gamma$ (mole $m^{-2}$ or mg $m^{-2}$), the fraction of segments in close contact with the surface, $p$, and the distribution of polymer segments, $\rho(z)$, from the surface towards the bulk solution. It is essential to know how far the segments extend into solution, i.e. the adsorbed layer thickness, $\delta$. It is important to know how these parameters change with polymer coverage (concentration), the structure of the polymer and its molecular weight. It is also essential to know how these parameters change with the environment, such as solvency of the medium for the chains and temperature.

Several theories exist that describe the process of polymer adsorption, which have been developed using either a statistical mechanical approach or quasi-lattice models. In the statistical mechanical approach, the polymer is considered to consist of three types of structures with different energy states, trains, loops and tails [31, 32]. The structures close to the surface (trains) are adsorbed with an internal partition function determined by short-range forces between the segment and surface (assigned an adsorption energy per segment $\chi^s$). The segments in loops and tails are considered to have an internal partition function equivalent to that of segments in bulk solution and these are assigned a segment–solvent interaction parameter $\chi$ (Flory–Huggins interaction parameter). By equating the chemical potential of the macromolecule in the adsorbed state and in bulk solution, the adsorption isotherm can be determined. In earlier theories, the case of an isolated chain on the surface (low coverage) was considered, but later the theories were modified to take into account the lateral interaction between the chains, i.e. at high coverage.

The quasi-lattice model was developed by Roe [33] and by Scheutjens and Fleer [34–36]. The basic procedure was to describe all chain conformations as step-weighted random walks on a quasi-crystalline lattice which extends in parallel layers away from the surface. This is illustrated in Figure 4.17, which shows a possible conformation of a polymer molecule at a surface.

The partition function was written in terms of number of chain configurations which were treated as connected sequences of segments. In each layer, random

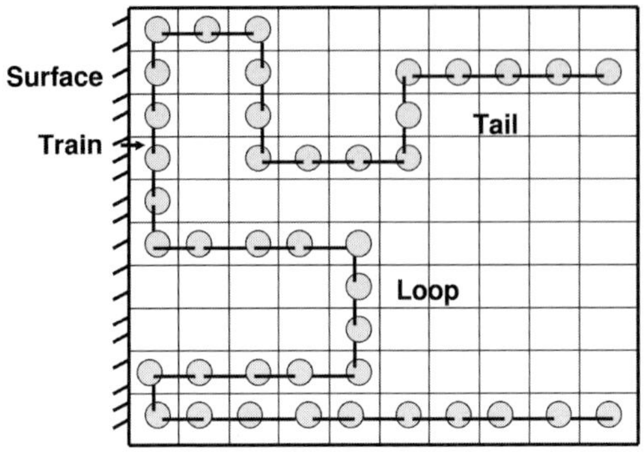


**Figure 4.17** Schematic representation of a polymer molecule adsorbing on a flat surface.

mixing (Bragg–William or mean field approximation) between segments and solvent molecules was assumed. Each step in the random walk was assigned a weighting factor $p_i$ that was considered to consist of three contributions, namely the adsorption energy $\chi^s$, the configurational entropy of mixing and the segment–solvent interaction parameter $\chi$.

The above theories gave a number of predictions for polymeric surfactant adsorption. Figure 4.18 shows typical adsorption isotherms plotted as surface coverage $\theta$ (in equivalent monolayers) versus polymer volume fraction $\phi_*$ in bulk solution $\phi_*$ was taken to vary between 0 and $10^{-3}$, which is the normal experimental range). The results in Figure 4.18 show the effect of increasing the chain length $r$ and effect of solvency (using two values for the Flory–Huggins interaction parameter, i.e. $\chi = 0$ (athermal solvent) and $\chi = 0.5$ (theta solvent). As the number of segments in the chain increases from low (with few segments) to high (many segments) values, the adsorption isotherm changes from a

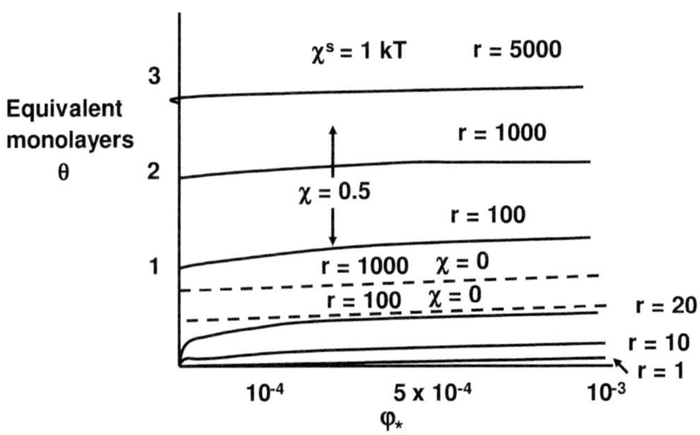


**Figure 4.18** Adsorption isotherms for oligomers and polymers in the dilute region based on the Scheutjens and Fleer theory.

Langmuirian type (characteristic of surfactant adsorption) to a high-affinity type. In the latter case, the first addition of polymer chains to the solution results in their virtually complete adsorption.

The adsorption isotherms for chains with $r = 100$ and above are typical of those obtained experimentally for most polymers that are not too polydisperse, i.e. showing a steep rise followed by a nearly horizontal pseudo-plateau (which increases by only a few percent per decade of $\phi_*$). Adsorption in this case is described as being 'irreversible', i.e. the equilibrium between adsorbed and free polymer is shifted towards the surface. This explains the strong anchoring of the polymer chains to the surface. As the solvency of the medium for the chains decreases, the amount of polymer adsorbed increases. This is clearly illustrated in Figure 4.18 when comparing the results obtained when $\chi = 0$ (very good solvent) with those obtained using a poor solvent with $\chi = 0.5$. In good solvents (dashed lines in Figure 4.18), $\theta$ is much smaller and levels off for long chains to attain an adsorption plateau which is essentially independent of molecular weight. This explains the relatively 'weaker' adsorption of homopolymers that are highly solvated by the medium. It is now clear from these theories why block and graft copolymers are preferred for the stabilization of dispersions. The poor solubility of the anchor chain B in the medium and its strong affinity to the surface ensure the strong adsorption of the molecule. In contrast, the high solubility of the stabilizing chain A ensures effective steric stabilization. Another prediction from the theories is that the higher the molecular weight of the polymer, the higher is the amount of adsorption, when the latter is expressed in mg $m^{-2}$.

Another point that emerges from Scheutjens and Fleer's theory [34–36] is the difference in shape between experimental and theoretical adsorption isotherms in the low-concentration region. The experimental isotherms are usually rounded, whereas those predicted from theory are flat. This is accounted for in terms of the molecular weight distribution (polydispersity) which is encountered in many practical systems. This effect has been explained by Fleer *et al.* [37]. With polydisperse polymer fractions, the larger molecules adsorb preferentially over the smaller molecules. At low polymer concentrations, nearly all molecular weights are adsorbed, leaving only a small fraction of polymer with the lowest molecular weight in solution. As the polymer concentration is increased, the higher molecular weight fractions displace the lower ones on the surface, which are now released in solution, thus shifting the molecular weight distribution of the polymer in bulk solution to lower values. This process continues with further increase in polymer concentration, leading to a fractionation process whereby the higher molecular weight fractions are adsorbed at the expense of the lower molecular weight fractions which are released in the bulk. However, in very concentrated solutions, monomers adsorb preferentially with respect to polymers and short chains with respect to longer chains. This is due to the fact that in this region, the conformational entropy term dominates the free energy, disfavoring the adsorption of long chains.

The bound fraction, $p$, is high at low polymer concentrations ($\phi_* < \phi_*^{c}$) approaching unity and it is relatively independent of molecular weight when

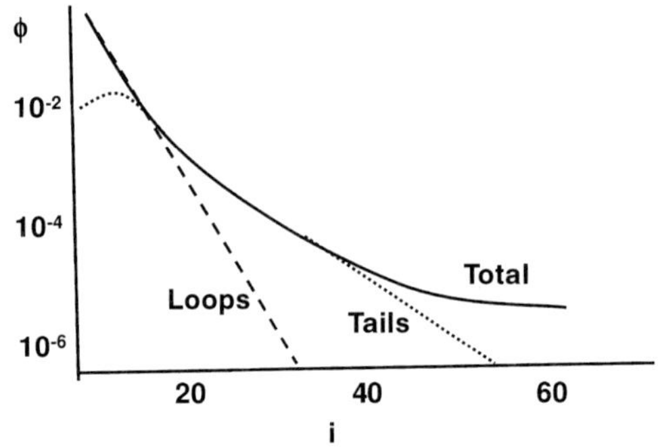


**Figure 4.19** Loop, tail and total segment concentration profiles.

$r > 20$; $p$ also increases with increase in the adsorption energy, $\chi^s$, but it decreases with increase in surface coverage and increase in the molecular weight of the polymer.

The structure of the adsorbed layer is described in terms of the segment density distribution, $\rho(z)$. As an illustration, Figure 4.19 shows some calculations by Scheutjens and Fleer [34–36] for loops and tails with $r = 1000$, $\phi_* = 10^{-6}$ and $\chi = 0.5$. In this example, 38% of the segments are in trains, 55.5% in loops and 6.5% in tails. This theory demonstrates the importance of tails, which dominate the total distribution in the outer region of the adsorbed layer. As we will discuss in the next section on experimental methods for the characterization of the adsorption and conformation of polymers at the solid/liquid interface, determination of the segment density distribution is not easy and usually assigns a value for the adsorbed layer thickness $\delta$. This increases with increase in the molecular weight of the polymer and increase in solvency of the medium for the chains.

## 4.8
## Experimental Methods for Measurement of Adsorption Parameters for Polymeric Surfactants

### 4.8.1
### Amount of Polymer Adsorbed, Γ – The Adsorption Isotherms

The amount of polymer adsorbed, Γ, can be directly determined in a similar way to that described for surfactants, except in this case one has to consider the relatively slow adsorption process, which may take several hours or even days to reach equilibrium. In addition, one needs very sensitive analytical methods for the determination of polymer concentration in the early stages of adsorption (which can be in the ppm range). As mentioned before, the amount of adsorption Γ can be calculated from a knowledge of the initial polymer concentration $C_1$ and that after reaching equilibrium $C_2$, the mass of the solid $m$ and the specific surface area $A_s$ as given by equation 33.

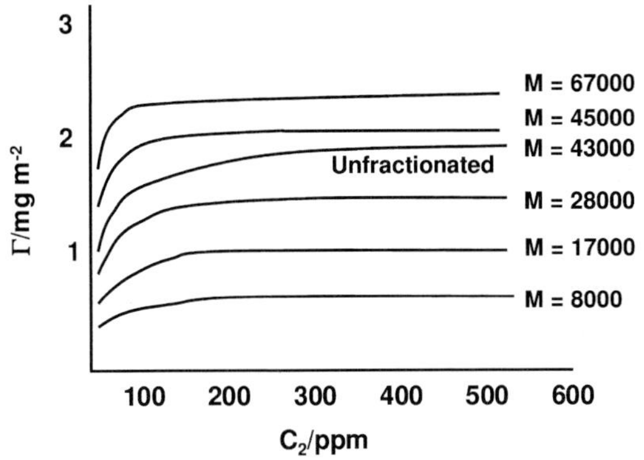


**Figure 4.20** Adsorption isotherms of poly(vinyl alcohol) on polystyrene latex at 25 °C.

As an illustration, Figure 4.20 shows the adsorption isotherms at 25 °C for poly(vinyl alcohol) (PVA) (containing 12% acetate) on polystyrene latex (a model hydrophobic surface similar to most agrochemical particles) [38]. The polymer was fractionated using preparative gel permeation chromatography [38] or by a sequential precipitation technique using acetone [39]. The fractions were characterized for their molecular weight using ultracentrifugation and later by intrinsic viscosity measurements. The intrinsic viscosity $[\eta]$ could be related to the weight-average molecular weight of the polymer (determined using ultracentrifugation) using the Mark–Houwink relationship:

$$[\eta] = KM^{\alpha} \tag{34}$$

The constants $K$ and $\alpha$ were established from knowledge of $[\eta]$ and $M$. The latter values could also be used to calculate the molecular dimensions (radius of gyration) and the polymer–solvent interaction parameter $\chi$ was also determined. The polystyrene latex used for the adsorption measurements was a model system prepared using surfactant-free polymerization and the particles were fairly monodisperse. Hence the specific surface area of the particles could be estimated from simple geometry using electron microscopy [38].

Figure 4.20 shows the high-affinity isotherms for the polymers and the increase in adsorption of the polymer with increase in molecular weight.

To study the effect of solvency on adsorption, measurements were carried out as a function of temperature [39] and addition of electrolyte (KCl or $Na_2SO_4$) [40]. Increasing temperature and/or addition of electrolyte reduce the solvency of the medium for the PVA chains (due to breakdown of the hydrogen bonds between the vinyl alcohol units and water). Figure 4.21 shows the adsorption isotherms for PVA with $M = 65\,000$ as a function of temperature. This shows a systematic increase in adsorption with increase in temperature, i.e. with reduction of solvency (increase in the value of the $\chi$), as expected from theory. The results obtained in the presence of KCl are shown in Figure 4.22. Addition of electrolyte

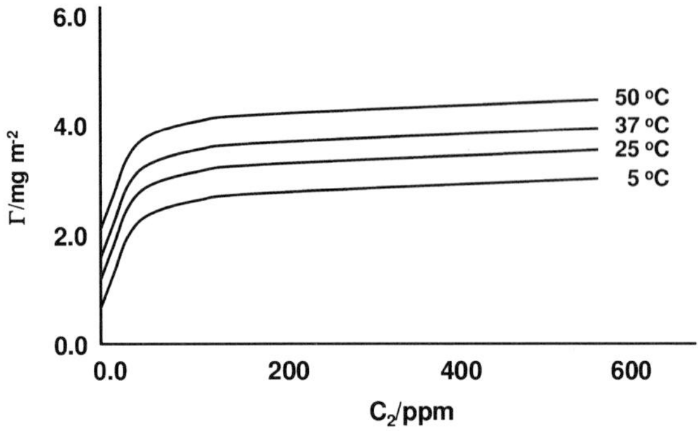


**Figure 4.21** Adsorption isotherms for PVA ($M = 65\,000$) on polystyrene latex at various temperatures.

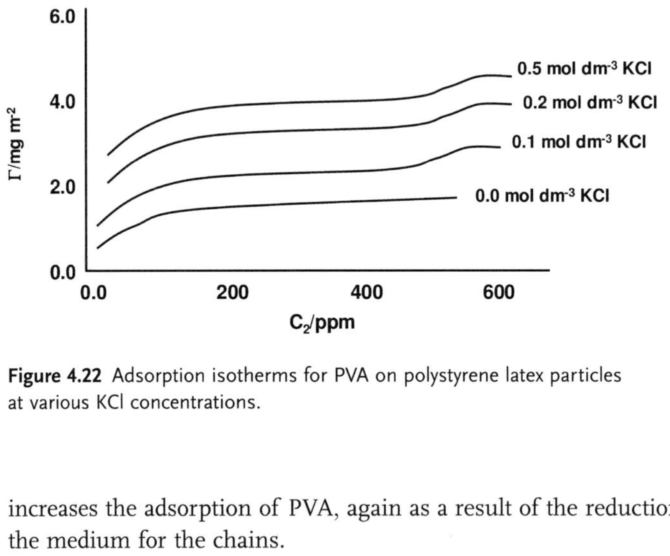


**Figure 4.22** Adsorption isotherms for PVA on polystyrene latex particles at various KCl concentrations.

increases the adsorption of PVA, again as a result of the reduction in solvency of the medium for the chains.

Results for polymeric surfactant adsorption on agrochemical particles are scarce. However, Tadros *et al.* [41, 42] showed similar trends for polymer adsorption on agrochemical particles. This is illustrated in Figures 4.23 and 4.24, which show the adsorption of PVA and a comb graft copolymer stabilizer [poly(methyl methacrylate) backbone with poly(ethylene oxide) side chains; Atlox 4913] on ethirimol (a fungicide) at room temperature. The high-affinity type of isotherm is clearly demonstrated and in both cases adsorption was irreversible, indicating strong and irreversible adsorption. However, the amount of adsorption per unit area (using the BET surface area of ethirimol of $0.22\,m^2\,g^{-1}$ obtained by Kr adsorption) was significantly higher than the values obtained on the model particles of polystyrene. This could be due to the errors involved in surface area determination of such coarse particles using BET gas adsorption.

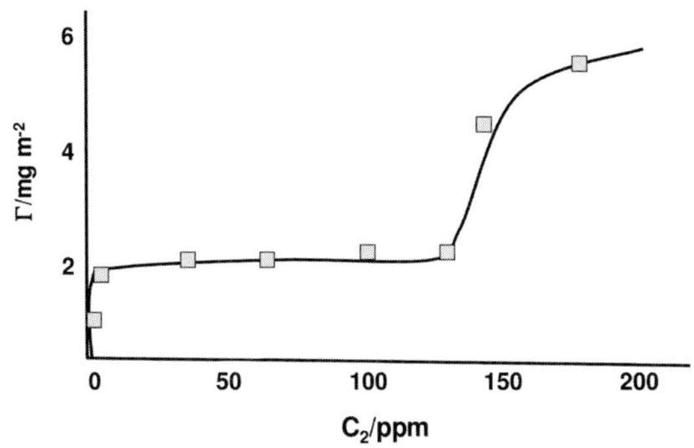


**Figure 4.23** Adsorption of PVA on ethirimol.

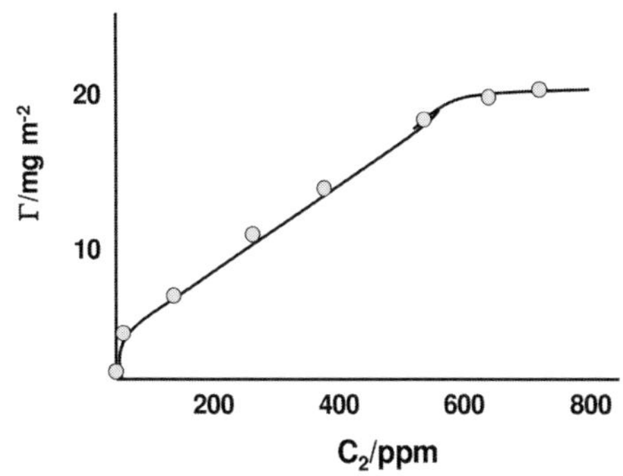


**Figure 4.24** Adsorption of the 'comb' graft copolymer on ethirimol.

### 4.8.2
### Polymer–bound Fraction $p$

The bound fraction $p$ represents the ratio of the number of segments in close contact with the surface (i.e. in trains) to the total number of segments in the polymer chain. The polymer-bound fraction, $p$, can be directly determined using spectroscopic methods such as infrared (IR), electron spin resonance (ESR) and nuclear magnetic resonance (NMR). The IR method depends on measuring the shift in some absorption peak for a polymer and/or surface group [43, 44]. The ESR and NMR methods depend on the reduction in the mobility of the segments that are in close contact with the surface (larger rotational correlation time for trains compared with loops). By using a pulsed NMR technique, one can estimate $p$ [45]. An indirect method for estimation of $p$ is to use microcalorimetry. Basically, one compares the enthalpy of adsorption per molecule with that per segment [46]. The latter may be obtained by using small molecules of similar structure to a polymer segment.

### 4.8.3
### Adsorbed Layer Thickness $\delta$ and Segment Density Distribution $\rho(z)$

Three direct methods can be applied for determination of adsorbed layer thickness: ellipsometry, attenuated total reflection (ATR) and neutron scattering. Both ellipsometry and ATR [47] depend on the difference between the refractive indices of the substrate, the adsorbed layer and the bulk solution and they require a flat reflecting surface. Ellipsometry [48] is based on the principle that light undergoes a change in polarizability when it is reflected at a flat surface (whether covered with a polymer layer or not).

The above limitations when using ellipsometry or ATR are overcome by the application of neutron scattering, which can be applied to both flat surfaces and particulate dispersions. The basic principle of neutron scattering is to measure the scattering due to the adsorbed layer, when the scattering length density of the particle is matched to that of the medium (the so-called 'contrast-matching' method). Contrast matching of particles and medium can be achieved by changing the isotopic composition of the system (using deuterated particles and a mixture of $D_2O$ and $H_2O$). It was also used for measurement of the adsorbed layer thickness of polymers, e.g. PVA or poly(ethylene oxide) on polystyrene latex [45]. Apart from obtaining $\delta$, one can also determine the segment density distribution $\rho(z)$.

Neutron scattering gives clearly a quantitative picture of the adsorbed polymer layer. However, its application in practice is limited since one needs to prepare deuterated particles or polymers for the contrast-matching procedure. The practical methods for determination of the adsorbed layer thickness are mostly based on hydrodynamic methods that are described below.

### 4.8.4
### Hydrodynamic Thickness Determination

Several methods may be applied to determine the hydrodynamic thickness of adsorbed polymer layers, of which viscosity, sedimentation coefficient (using an ultracentrifuge) and dynamic light scattering measurements are the most convenient. A less accurate method is to use zeta potential measurements, although this does not require the use of model monodisperse particles and hence it can be used for agrochemical suspensions. Below, the dynamic light scattering technique (referred to as photon correlation spectroscopy, PCS), which provides a rapid method for the determination of the hydrodynamic thickness, is described. This is followed by application of zeta potential measurements for the determination of hydrodynamic thickness, which can be applied to agrochemical particles.

PCS allows one to obtain the diffusion coefficient of the particles with and without the adsorbed layer ($D_\delta$ and $D$, respectively). This is obtained from measurement of the intensity fluctuation of scattered light as the particles undergo Brownian diffusion [49, 50]. When a light beam (e.g. monochromatic laser beam) passes through a dispersion, an oscillating dipole is induced in the particles, thus re-radiating the light. Due to the random arrangement of the particles (which are separated by a distance comparable to the wavelength of the light beam, i.e. the light is coherent with the interparticle distance), the intensity of the scattered light will, at any instant, appear as random diffraction or a 'speckle' pattern. As the particles undergo Brownian motion, the random configuration of the speckle pattern changes. The intensity at any one point in the pattern will, therefore, fluctuate such that the time taken for an intensity maximum to become a minimum (i.e. the coherence time) corresponds approximately to the time required for a particle to move one wavelength. Using a photomultiplier of active area about the size of a diffraction maximum, i.e. approximately one coherence area, this intensity fluctuation can be measured. A digital correlator is used to measure the photocount or intensity correlation function of the scattered light. The photocount correlation function can be used to obtain the diffusion coefficient $D$ of the particles. For monodisperse non-interacting particles (i.e. at sufficient dilution), the normalized correlation function $[g^{(1)}(\tau)]$ of the scattered electric field is given by the equation

$$[g^{(1)}(\tau)] = \exp(-\Gamma\tau) \tag{35}$$

where $\tau$ is the correlation delay time and $\Gamma$ is the decay rate or inverse coherence time. $\Gamma$ is related to $D$ by the equation

$$\Gamma = DK^2 \tag{36}$$

where $K$ is the magnitude of the scattering vector, given by

$$K = \left(\frac{4n}{\lambda_0}\right)\sin\left(\frac{\theta}{2}\right) \tag{37}$$

where $n$ is the refractive index of the solution, $\lambda$ is the wavelength of light in vacuum and $\theta$ is the scattering angle.

From $D$, the particle radius $R$ is calculated using the Stokes–Einstein equation:

$$D = \frac{kT}{6\pi\eta R} \tag{38}$$

where $k$ is the Boltzmann constant and $T$ is the absolute temperature. For a polymer-coated particle $R$ is denoted $R_\delta$, which is equal to $R + \delta_{\mathrm{h}}$. Thus, by measuring $D_\delta$ and $D$, one can obtain $\delta_{\mathrm{h}}$. It should be mentioned that the accuracy of the PCS method depends on the ratio $\delta_{\mathrm{h}}/R$, since $\delta_{\mathrm{h}}$ is determined by difference. Since the accuracy of the measurement is $\pm 1\%$, $\delta_{\mathrm{h}}$ should be at least 10% of the particle radius. This method can only be used with small particles and reasonably thick adsorbed layers.

Electrophoretic mobility ($u$) measurements can also be applied to measure $\delta_{\mathrm{h}}$ [51]. From $u$, the zeta potential $\zeta$, i.e. the potential at the slipping (shear) plane of the particles, can be calculated. Adsorption of a polymer causes a shift in the shear plane from its value in the absence of a polymer layer (which is close to the Stern plane) to a value that depends on the thickness of the adsorbed layer. Thus, by measuring $\zeta$ in the presence ($\zeta_\delta$) and absence ($\zeta$) of a polymer layer, one can estimate $\delta_h$. Assuming that the thickness of the Stern plane is $\Delta$, then $\zeta_\delta$ may be related to $\zeta$ (which may be assumed to be equal to the Stern potential $\psi_d$) by the equation

$$\tanh\left(\frac{e\psi_\delta}{4kT}\right) = \tanh\left(\frac{e\zeta}{4kT}\right)\exp[-\kappa(\delta_h - \Delta)] \quad (39)$$

where $\kappa$ is the Debye parameter, which is related to electrolyte concentration and valency.

It should be mentioned that the value of $\delta_h$ calculated using the above simple equation shows a dependence on electrolyte concentration, hence the method cannot be used in a straightforward manner. Cohen-Stuart *et al.* [51] showed that the measured electrophoretic thickness $\delta_e$ approaches $\delta_h$ only a t low electrolyte concentrations. Therefore, to obtain $\delta_h$ from electrophoretic mobility measurements, results should be obtained at various electrolyte concentrations and $\delta_e$ should be plotted versus the Debye length ($1/\kappa$) to obtain the limiting value at high ($1/\kappa$) (i.e. low electrolyte concentration) which now corresponds to $\delta_h$.

## References


1 E.A. Guggenheim, *Thermodynamics*, 5th edn, North-Holland, Amsterdam, p. 45 (1967).
2 J.W. Gibbs, *Collected Works*, Vol. 1, Longman, New York, p. 219 (1928).
3 S.S. Dukhin, G. Kretzscmar and R. Miller, *Dynamics of Adsorption at Liquid Interfaces*, Elsevier, Amsterdam (1995).
4 A.F.H. Ward and L. Tordai, *J. Phys. Chem., 14*, 453 (1946).
5 I. Panaitov and J.G. Petrov, *Ann. Univ. Sofia, Fac. Chem., 64*, 385 (1968–69).
6 R.S. Hansen, *J. Phys. Chem., 64*, 637 (1960).
7 R. Miller and K. Lunkenheimer, *Z. Phys. Chem., 259*, 863 (1978).
8 R. Zana, *Proc. NATO Adv. Study Inst., Ser. C, 18*, 133 (1974).
9 R. Miller, A. Hoffmann, R. Hartmann, K.H. Schano and A. Halbig, *Adv. Mater., 4*, 370 (1992).
10 J.T. Davies and E.K. Rideal, *Interfacial Phenomena*, Academic Press, New York (1969).
11 J. Kloubek, K. Friml and F. Krejci, *Collect. Czech. Chem. Commun., 41*, 1845 (1976).
12 D.B. Hough and H.M. Randall, in *Adsorption from Solution at the Solid/Liquid Interface*, ed. G.D. Parfitt and, C.H. Rochester, Academic Press, London, p. 247 (1983).
13 D.W. Fuerstenau and T.W. Healy, in *Adsorptive Bubble Separation Techniques*, ed. R. Lemlich, Academic Press, London, p. 91 (1972).
14 P. Somasundaran and E.D. Goddard, *Mod. Aspects Electrochem., 13*, 207 (1979).
15 T.W. Healy, *J. Macromol. Sci. Chem., 118*, 603 (1974).
16 P. Somasundaran and H.S. Hannah, in *Improved Oil Recovery by Surfactant and Polymer Flooding*, ed. D.O. Shah and

R.S. Schechter, Academic Press, London, p. 205 (1979).
17 F.G. Greenwood, G.D. Parfitt, N.H. Picton and D.G. Wharton, *Adv. Chem. Ser.*, *79*, 135 (1968).
18 R.E. Day, F.G. Greenwood and G.D. Parfitt, in *Proceedings of the 4th International Congress on Surface Active Substances*, Vol. 1, p. 1005 (1967).
19 F.Z. Saleeb and J.A. Kitchener, *J. Chem. Soc.*, 911 (1965).
20 P. Conner and R.H. Ottewill, *J. Colloid Interface Sci.*, *37*, 642 (1971).
21 D.W. Fuerstenau, in *The Chemistry of Biosurfaces*, ed. M.L. Hair, Marcel Dekker, New York, p. 91 (1971).
22 T. Wakamatsu and D.W. Fuerstenau, *Adv. Chem. Ser.*, *71*, 161 (1968).
23 A.M. Gaudin and D.W. Fuerstenau, *Trans. AIME*, *202*, 958 (1955).
24 J.S. Clunie and B.T. Ingram, in *Adsorption from Solution at the Solid/Liquid Interface*, ed. G.D. Parfitt and C.H. Rochester, Academic Press, London, p. 105 (1983).
25 I. Piirma, *Polymeric Surfactants*, Surfactant Science Series, No. 42, Marcel Dekker, New York (1992).
26 C.V. Stevens, A. Meriggi, M. Peristerpoulou, P.P. Christov, K. Booten, B. Levecke, A. Vandamme, N. Pittevils and Th.F. Tadros, *Biomacromolecules*, *2*, 1256 (2001).
27 E.L. Hirst, D.I. McGilvary and E.G. Percival, *J. Chem. Soc.*, 1297 (1950).
28 M. Suzuki, in *Science and Technology of Fructans*, ed. M. Suzuki and N.J. Chatterton, CRC Press, Boca Raton, FL, p. 21 (1993).
29 Th.F. Tadros, K. Booten, B. Levecke and A. Vandamme, to be published.
30 Th.F. Tadros, in *Polymer Colloids*, ed. R. Buscall, T. Corner and J. Stageman, Elsevier Applied Science, London, p. 105 (1985).
31 A. Silberberg, *J. Chem. Phys.*, *48*, 2835 (1968).
32 C.A. Hoeve, *J. Polym. Sci.*, *30*, 361 (1970), *34*, 1 (1971).
33 R.J. Roe, *J. Chem. Phys.*, *60*, 4192 (1974).
34 J.M.H.M. Scheutjens and G.J. Fleer, *J. Phys. Chem.*, *83*, 1919 (1979).
35 J.M.H.M. Scheutjens and G.J. Fleer, *J. Phys. Chem.*, *84*, 178 (1980).
36 J.M.H.M. Scheutjens and G.J. Fleer, *Adv. Colloid Interface Sci.*, *16*, 341 (1982).
37 G.J. Fleer, M.A. Cohen-Stuart, J.M.H.M. Scheutjens, T. Cosgrove and B. Vincent, *Polymers at Interfaces*, Chapman and Hall, London (1993).
38 M.J. Garvey, Th.F. Tadros and B. Vincent, *J. Colloid Interface Sci.*, *49*, 57 (1974).
39 Th. van den Boomgaard, T.A. King, Th.F. Tadros, H. Tang and B. Vincent, *J. Colloid Interface Sci.*, *61*, 68 (1978).
40 Th.F. Tadros and B. Vincent, *J. Colloid Interface Sci.*, *72*, 505 (1978).
41 Th.F. Tadros, *Adv. Chem. Ser.*, *9*, 173 (1975).
42 D. Heath, R.D. Knott, D.A. Knowles and Th.F. Tadros, *ACS Symp. Ser.*, *254*, 2 (1984).
43 B.J. Fontana and J.R. Thomas, *J. Phys. Chem.*, *65*, 480 (1961).
44 I.D. Robb and R. Smith, *Eur. Polym. J.*, *10*, 1005 (1974).
45 K.G. Barnett, T. Cosgrove, B. Vincent, A. Burgess, T.L. Crowley, J. Kims, J.D. Turner and Th.F. Tadros, *Polymer*, 22, 283 (1981).
46 M.A. Cohen-Stuart, G.J. Fleer and J. Bijesterbosch, *Colloid Interface Sci.*, *90*, 321 (1982).
47 F. Abeles, in *Ellipsometry in the Measurement of Surfaces and Thin Films*, ed. E. Passaglia, R.R. Stromberg and J. Kruger, Natl. Bur. Stand. Misc. Publ. No. 256 (1964).
48 T. Cosgrove, T.L. Crowley and T. Ryan, *Macromolecules*, *20*, 2879 (1987).
49 M.J. Garvey, Th.F. Tadros and B. Vincent, *J. Colloid Interface Sci.*, *55*, 440 (1976).
50 P.N. Pusey, in *Industrial Polymers: Characterization by Molecular Weights*, ed. J.H.S. Green and R. Dietz, Transcripta Books, London (1973).
51 M.A. Cohen-Stuart and J.W. Mulder, *Colloids Surf.*, *15*, 49 (1985).

# 5
# Interaction Forces Between Particles or Droplets in Agrochemical Dispersions (Suspension Concentrates or Emulsions, EWs) and Their Role in Colloid Stability

## 5.1
## Introduction

The interaction forces between particles of a suspension or droplets in an emulsion determine the colloid stability of the agrochemical formulation. For example, in a suspension concentrate one must ensure that the particles remain as individual units and any aggregation must be sufficiently weak that the system can be easily redispersed on shaking and/or dilution in the spray tank. Strong aggregation must be avoided since the resulting large units can cause blockage of the spray nozzles and uneven distribution of the agrochemical particles on the target, causing a reduction in biological efficacy. With EWs, aggregation of droplets must also be avoided for the same reasons as for suspensions. In addition, aggregation of emulsion may result in their coalescence with ultimate oil separation. Maintenance of colloid stability is also essential with suspoemulsions (mixtures of suspensions and emulsions). Aggregation of particles and droplets (referred to as heteroflocculation) must be prevented, otherwise the formulation loses its physical stability on storage.

The stability/instability of any agrochemical dispersion is determined by the balance of three main forces: (1) van der Waals attraction that is universal for all disperse systems and results mainly from the London dispersion forces between the particles or droplets; (2) double layer repulsion that arises when using ionic surfactants or polyelectrolytes; and (3) steric repulsion that arises when using adsorbed nonionic surfactants or polymers. Below, a description of these three interaction forces is first given, followed by the combination of these forces and a discussion of the theories of colloid stability. The latter can account for the stability/instability of the various dispersions.

## 5.2
## van der Waals Attraction

As is well known, atoms or molecules always attract each other at short distances of separation. The attractive forces are of three different types: dipole–dipole

Colloids and Interface Science Series, Vol. 5
*Colloids in Agrochemicals.* Tharwat F. Tadros

ISBN: 978-3-527-31465-2

interaction (Keesom), dipole-induced dipole interaction (Debye) and London dispersion force. The London dispersion force is the most important, since it occurs for polar and nonpolar molecules. It arises from fluctuations in the electron density distribution.

At small distances of separation $r$ in vacuum, the attractive energy between two atoms or molecules is given by

$$G_{aa} = -\frac{\beta_{11}}{r^6} \tag{1}$$

where $\beta_{11}$ is the London dispersion constant.

For colloidal particles which are made of atom or molecular assemblies, the attractive energies may be added and this results in the following expression for two spheres (at small h) [1]:

$$G_A = -\frac{AR}{12h} \tag{2}$$

where $A$ is the effective Hamaker constant:

$$A = (A_{11}^{\frac{1}{2}} - A_{22}^{\frac{1}{2}})^2 \tag{3}$$

where $A_{11}$ is the Hamaker constant between particles in vacuum and $A_{22}$ the Hamaker constant for equivalent volumes of the medium:

$$A = \pi q^2 \beta_{ii} \tag{4}$$

where $q$ is number of atoms or molecules per unit volume.

$G_A$ decreases with increase in $h$, as shown schematically in Figure 5.1.

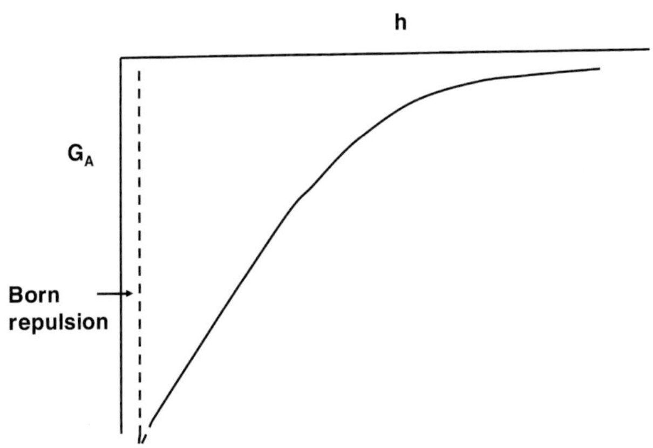


**Figure 5.1** Variation of the van der Waals attraction energy with separation distance.

## 5.3 Electrical Double Layer Repulsion

As mentioned in the Introduction, electrical double layers are produced when using ionic surfactants. On adsorption of these molecules on particles or droplets, a surface charge is produced from the head group of the ionic surfactant. This surface charge $\sigma_0$ is compensated by unequal distribution of counterions (opposite in charge to the surface) and co-ions (same sign as the surface) which extend to some distance from the surface. This forms the basis of the diffuse double layer proposed by Gouy and Chapman [2]. It is represented schematically in Figure 5.2 for a negatively charged surface produced by adsorption of an anionic surfactant.

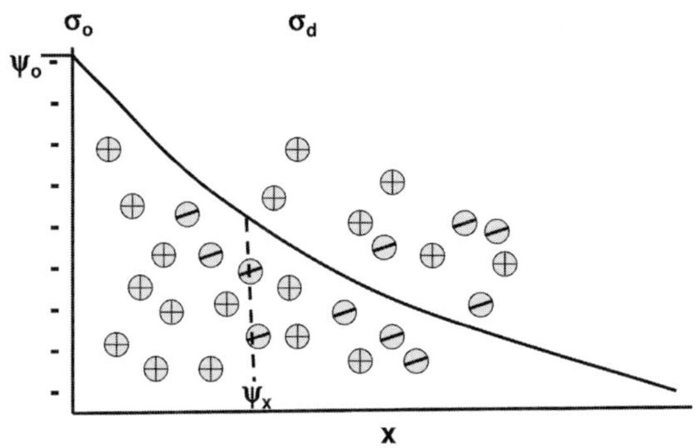


**Figure 5.2** Structure of the electrical double layer according to Gouy and Chapman.

The potential decays exponentially with distance $x$. At low potentials,

$$\psi = \psi_0 \exp(-\kappa x) \tag{5}$$

Note that when $x = 1/\kappa$, $\psi_x = \psi_0/e$; $1/\kappa$ is referred to as the 'thickness of the double layer'.

The double layer extension depends on the electrolyte concentration and valency of the counterions:

$$\frac{1}{\kappa} = \left(\frac{\varepsilon_r \varepsilon_0 kT}{2n_0 Z_i^2 e^2}\right)^{\frac{1}{2}} \tag{6}$$

where $\varepsilon_r$ is the permittivity (dielectric constant) (78.6 for water at 25 °C), $\varepsilon_0$ is the permittivity of free space, $k$ is the Boltzmann constant, $T$ is the absolute temperature, $n_0$ is the number of ions per unit volume of each type present in bulk solution, $Z_i$ is the valency of the ions and $e$ is the electronic charge.

Table 5.1 shows the values of $1/\kappa$ for a 1 : 1 electrolyte (e.g. KCl).

**Table 5.1** $1/\kappa$ for a 1 : 1 electrolyte (e.g. KCl).

| $C$/mol dm$^{-3}$ | $10^{-5}$ | $10^{-4}$ | $10^{-3}$ | $10^{-2}$ | $10^{-1}$ |
|---|---|---|---|---|---|
| $(1/\kappa)$/nm | 100 | 33 | 10 | 3.3 | 1 |

When charged colloidal particles in a dispersion approach each other such that the double layers begin to overlap (particle separation becomes less than twice the double layer extension), repulsion occurs. The individual double layers can no longer develop unrestrictedly, since the limited space does not allow complete potential decay [3]. This is illustrated in Figure 5.3 for two flat plates.

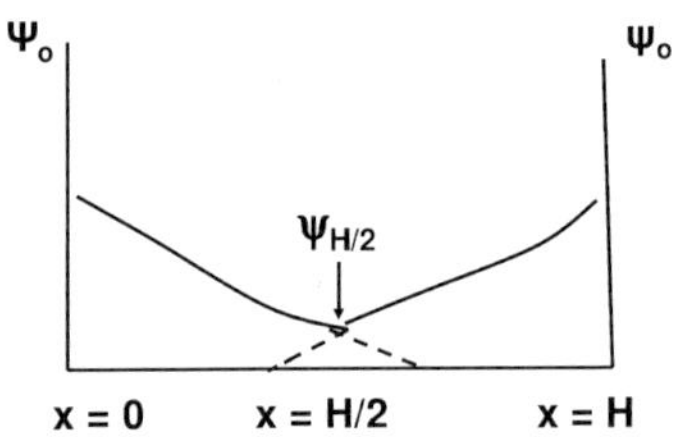


**Figure 5.3** Schematic representation of double layer interaction for two flat plates.

The potential $\psi_{H/2}$ half way between the plates is no longer zero (as would be the case for isolated particles at $x \rightarrow \infty$). For two spherical particles of radius $R$ and surface potential $\psi_0$ and condition $\kappa R < 3$, the expression for the electrical double layer repulsive interaction is given by

$$G_{\text{el}} = \frac{4\pi\varepsilon_r\varepsilon_0 R^2 {\psi_0}^2 \exp(-\kappa h)}{2R + h} \tag{7}$$

where $h$ is the closest distance of separation between the surfaces. This expression shows the exponential decay of $G_{\text{el}}$ with $h$. The higher the value of $\kappa$ (i.e. the higher the electrolyte concentration), the steeper is the decay, as shown schematically in Figure 5.4. This means that at any given distance $h$, the double layer repulsion decreases with increase in electrolyte concentration.

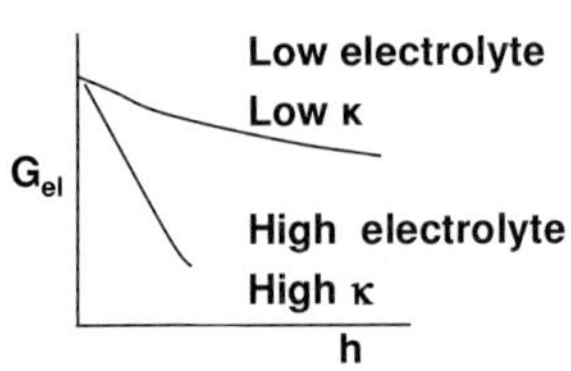


**Figure 5.4** Variation of $G_{\text{el}}$ with $h$ at different electrolyte concentrations.

## 5.4 Total Energy of Interaction: Deryaguin–Landau–Verwey–Overbeek (DLVO) Theory

Combination of $G_{el}$ and $G_A$ results in the well-known theory of stability of colloids (DLVO theory) [4, 5]:

$$G_T = G_{el} + G_A \tag{8}$$

A plot of $G_T$ versus $h$ is shown in Figure 5.5, which represents the case at low electrolyte concentrations, i.e. strong electrostatic repulsion between the particles. $G_{el}$ decays exponentially with $h$, i.e. $G_{el} \rightarrow 0$ as $h$ becomes large. $G_A$ is proportional to $1/h$, i.e. $G_A$ does not decay to zero at large $h$.

At long distances of separation, $G_A > G_{el}$, resulting in a shallow minimum (secondary minimum). At very short distances, $G_A \gg G_{el}$, resulting in a deep primary minimum. At intermediate distances, $G_{el} > G_A$, resulting in an energy maximum, $G_{max}$, whose height depends on $\psi_0$ (or $\psi_d$) and the electrolyte concentration and valency.

At low electrolyte concentrations ($<10^{-2}$ mol dm$^{-3}$ for a 1 : 1 electrolyte), $G_{max}$ is high ($>25kT$) and this prevents particle aggregation into the primary minimum. The higher the electrolyte concentration (and the higher the valency of the ions), the lower is the energy maximum.

Under some conditions (depending on electrolyte concentration and particle size), flocculation into the secondary minimum may occur. This flocculation is weak and reversible. By increasing the electrolyte concentration, $G_{max}$ decreases until at a given concentration it vanishes and particle coagulation occurs. This is illustrated in Figure 5.6, which shows the variation of $G_T$ with $h$ at various electrolyte concentrations.

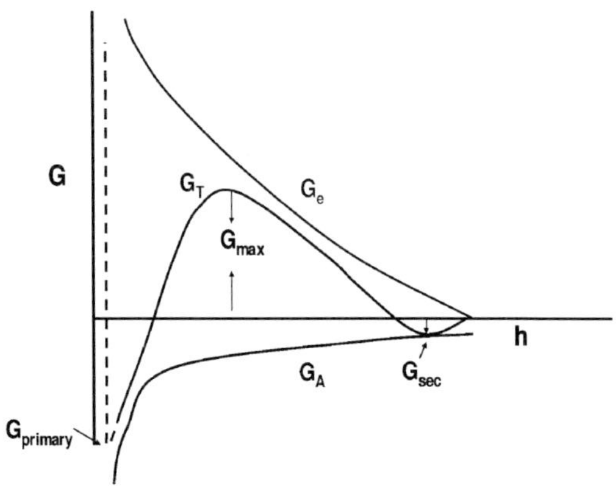


**Figure 5.5** Schematic representation of the variation of $G_T$ with $h$ according to the DLVO theory.

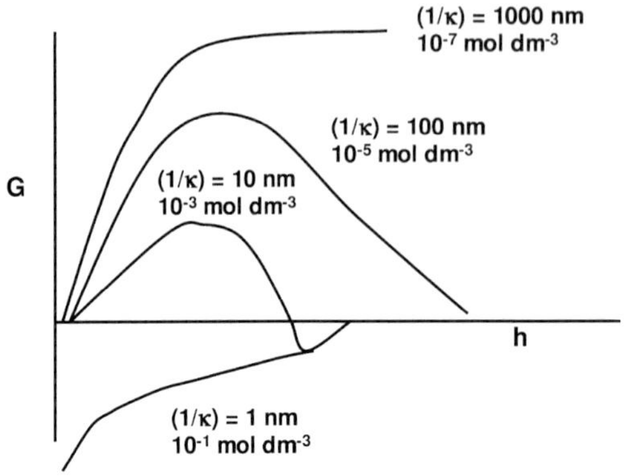


**Figure 5.6** Variation of $G_T$ at various electrolyte concentrations.

Coagulation occurs at a critical electrolyte concentration, the critical coagulation concentration (ccc), which depends on the electrolyte valency. At low surface potentials, ccc $\propto 1/Z^2$. This is referred to as the Schultze–Hardy rule.

One can define a rate constant for flocculation: $k_0$ = rapid rate of flocculation (in the absence of an energy barrier) and $k$ = slow rate of flocculation (in the presence of an energy barrier):

$$\frac{k_0}{k} = W \quad \text{(the stability ratio)} \tag{9}$$

Note that $W$ increases as $G_{max}$ increases.

The stability of colloidal dispersions can be quantitatively assessed from plots of log $W$ versus log $C$, as illustrated in Figure 5.7.

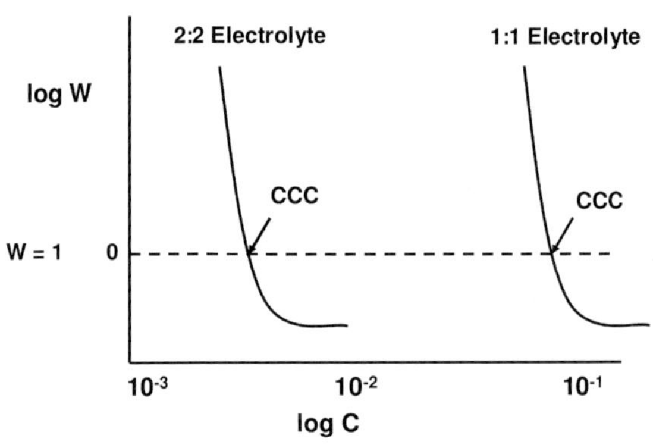


**Figure 5.7** Log $W$ versus log $C$ curves.

## 5.5 Criteria for Stabilization of Dispersions with Double Layer Interaction

The two main criteria for stabilization are (1) high surface or Stern potential (zeta potential), high surface charge and (2) low electrolyte concentration and low valency of counter- and co-ions. One should ensure that an energy maximum in excess of $25kT$ exists in the energy–distance curve. When $G_{max} \gg kT$, the particles in the dispersion cannot overcome the energy barrier, thus preventing coagulation. In some cases, particularly with large and asymmetric particles, flocculation into the secondary minimum may occur. This flocculation is usually weak and reversible and may be advantageous for preventing the formation of hard sediments.

### 5.5.1 The Zeta Potential

As mentioned above, one of the main criteria for electrostatic stability is the high surface or zeta potential, which can be measured experimentally [6], as will be discussed below. The zeta potential can be calculated from the measurable particle mobility $u$ ($m^2\,V^{-1}\,s^{-1}$):

$$u = \frac{v}{E/l} \tag{10}$$

where $v$ is the particle velocity, $E$ is the applied potential and $l$ is the distance between the two electrodes; $E/l$ is the field strength.

From $u$, the zeta potential $\zeta$ can be calculated using Smoluchowski equation (equation 11) that is applicable when $\kappa R \gg 1$ (which is the case with most agrochemical suspensions with $R > 0.5\,\mu m$):

$$u = \frac{\varepsilon_r \varepsilon_0 \xi}{\eta} \tag{11}$$

where $\varepsilon_r$ is the relative permittivity of the medium (78.6 for water at 25 °C), $\varepsilon_0$ is the permittivity of free space ($8.85 \times 10^{-12}\,F\,m^{-1}$), $\eta$ is the viscosity of the medium ($8.9 \times 10^{-4}$ Pa s for water at 25 °C), $\zeta$ is the zeta potential in volts ($=1.282 \times 10^6$ for water at 25 °C) and $u$ is expressed in $m^2\,V^{-1}\,s^{-1}$.

The most convenient and rapid method for measurement of the zeta potential is based on laser velocimetry. Laser light scattering can be used to measure the diffusion coefficients of particles by measuring the Doppler broadening of the frequency of the scattered light due to the velocity of the scattering centers. If an electric field is placed at right-angles to the incident light and in the plane defined by the incident and observation beams, the line broadening is unaffected but the center frequency of the scattered light is shifted to an extent determined by the electrophoretic mobility. The shift is very small compared with the incident frequency (~100 Hz for and incident frequency of ~$6 \times 10^{14}$ Hz), but with a laser source it can be detected by heterodyning (i.e. mixing) the scattered light with the incident beam and detecting the output of the difference frequency.

A homodyne method may be applied, in which case a modulator to generate an apparent Doppler shift at the modulated frequency is used. To increase the sensitivity of the laser Doppler method, the electric fields are much higher than those used in conventional electrophoresis. The Joule heating is minimized by pulsing of the electric field in opposite directions. The Brownian motion of the particles also contributes to the Doppler shift and an approximate correction can be made by subtracting the peak width obtained in the absence of an electric field from the electrophoretic spectrum. An He–Ne laser is used as the light source and the output of the laser is split into two coherent beams which are cross-focused in the cell to illuminate the sample. The light scattered by the particle, together with the reference beam, is detected by a photomultiplier. The output is amplified and analyzed to transform the signals to a frequency distribution spectrum. At the intersection of the beam, interferences of known spacing are formed.

The magnitude of the Doppler shift, $\Delta\nu$, is used to calculate the electrophoretic mobility $u$ using the following expression:

$$\Delta\nu = \left(\frac{2n}{\lambda_0}\right)\sin\left(\frac{\theta}{2}\right)uE \tag{12}$$

where $n$ is the refractive index of the medium, $\lambda_0$ is the incident wavelength in vacuum, $\theta$ is the scattering angle and $E$ is the field strength.

Several commercial instruments are available for measuring the electrophoretic light scattering. The Coulter Delsa 440SX (Beckman Coulter, Fullerton, CA, USA) is a multi-angle laser Doppler system employing heterodyning and autocorrelation signal processing. Measurements are made at four scattering angle (8, 17, 25 and 34°) and the temperature of the cell is controlled by a Peltier device. The instrument reports the electrophoretic mobility, zeta potential, conductivity and particle size distribution.

Malvern Instruments (Malvern, Worcs, UK) has two instruments: the ZetaSizer 3000 and ZetaSizer 5000: The ZetaSizer 3000 is a laser Doppler system using a crossed beam optical configuration and homodyne detection with photon correlation signal processing. The zeta potential is measured using laser Doppler velocimetry and the particle size is measured using photon correlation spectroscopy (PCS). The ZetaSizer 5000 uses PCS to measure both (1) the movement of the particles in an electric field for zeta potential determination and (2) random diffusion of particles at different measuring angles for size measurement on the same sample. The manufacturer claims that zeta potentials for particles in the range 50 nm–30 μm can be measured. In both instruments, a Peltier device is used for temperature control.

## 5.6
## Steric Stabilization of Dispersions

The use of natural and synthetic polymers (referred to as polymeric surfactants) for stabilization of suspension concentrates and emulsions (EWs) plays an important role in agrochemical formulations. Polymers are particularly important for the preparation of concentrated dispersions, i.e. at high volume fraction $\phi$ of the disperse phase [$\phi$ = (volume of all particles)/(total volume of dispersion)].

Polymers are also essential for the stabilization of nonaqueous dispersions, since in this case electrostatic stabilization is not possible (due to the low dielectric constant of the medium).

To understand the role of polymers in dispersion stability, it is essential to consider the adsorption and conformation of the macromolecule at the solid/liquid interface, which was discussed in Chapter 4.

## 5.7
## Interaction Between Particles or Droplets Containing Adsorbed Polymeric Surfactant Layers

When two particles or droplets each with a radius $R$ and containing an adsorbed polymer layer with a hydrodynamic thickness $\delta_h$ approach each other to a surface–surface separation distance $h$ that is smaller than $2\delta_h$, the polymer layers interact with each other, resulting in two main situations [8, 9]: the polymer chains may overlap with each other, or the polymer layer may undergo some compression. In both cases, there will be an increase in the local segment density of the polymer chains in the interaction region. This is illustrated schematically in Figure 5.8.

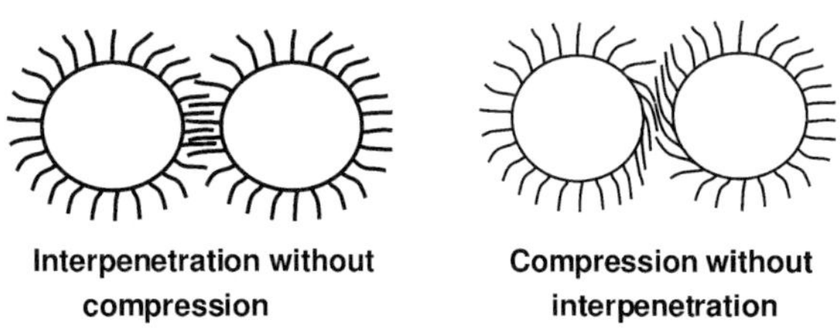


**Figure 5.8** Schematic representation of the interaction of two polymer layers.

The real situation is perhaps in between the above two cases, i.e. the polymer chains may undergo some interpenetration and some compression. Provided that the dangling chains (the A chains in A–B or A–B–A block or $BA_n$ graft copolymers) are in a good solvent, this local increase in segment density in the interaction zone will result in strong repulsion as a result of two main effects:

1. Increase in the osmotic pressure in the overlap region as a result of the unfavorable mixing of the polymer chains, when these are in good solvent

conditions [6, 7]. This is referred to as osmotic repulsion or mixing interaction and it is described by a free energy of interaction $G_{mix}$.

2. Reduction in the configurational entropy of the chains in the interaction zone; this entropy reduction results from the decrease in the volume available for the chains when these are either overlapped or compressed. This is referred to as volume restriction interaction or entropic or elastic interaction and it is described by a free energy of interaction $G_{el}$.

Combination of $G_{mix}$ and $G_{el}$ is usually referred to as the steric interaction free energy, $G_s$:

$$G_s = G_{mix} + G_{el} \tag{13}$$

The sign of $G_{mix}$ depends on the solvency of the medium for the chains. If in a good solvent, i.e. the Flory–Huggins interaction parameter $\chi$ is less than 0.5, then $G_{mix}$ is positive and the mixing interaction leads to repulsion (see below). In contrast, if $\chi > 0.5$ (i.e. the chains are in a poor solvent condition), $G_{mix}$ is negative and the mixing interaction becomes attractive.

$G_{el}$ is always positive and therefore in some cases one can produce stable dispersions in a relatively poor solvent (enhanced steric stabilization) [7].

### 5.7.1 Mixing Interaction, $G_{mix}$

This results from the unfavorable mixing of the polymer chains when these are in good solvent conditions. This is shown schematically in Figure 5.9. Consider two spherical particles with the same radius and each containing an adsorbed polymer layer with thickness $\delta$. Before overlap, one can define in each polymer layer a chemical potential for the solvent $\mu_i^{\alpha}$ and a volume fraction for the polymer in the layer $\phi_2$. In the overlap region (volume element d$V$), the chemical potential of the solvent is reduced to $\mu_i^{\beta}$; this results from the increase in polymer segment concentration in this overlap region.

In the overlap region, the chemical potential of the polymer chains is now higher than in the rest of the layer (with no overlap). This amounts to an increase in the

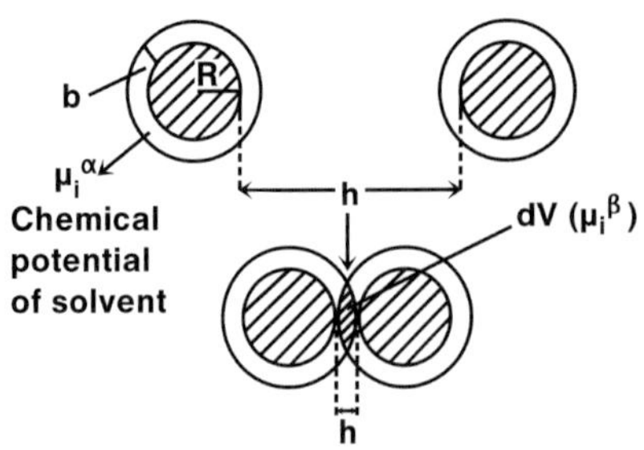


Figure 5.9 Schematic representation of polymer layer overlap.

osmotic pressure in the overlap region, As a result, solvent will diffuse from the bulk to the overlap region, thus separating the particles, and hence a strong repulsive energy arises from this effect.

The above repulsive energy can be calculated by considering the free energy of mixing of two polymer solutions, as for example treated by Flory and Krigbaum [10]. The free energy of mixing is given by two terms: an entropy term that depends on the volume fraction of polymer and solvent and an energy term that is determined by the Flory–Huggins interaction parameter $\chi$:

$$\delta(G_{mix}) = kT(n_1 \ln \phi_1 + n_2 \ln \phi_2 + \chi n_1 \phi_2) \tag{14}$$

where $n_1$ and $n_2$ are the number of moles of solvent and polymer with volume fractions $\phi_1$ and $\phi_2$, $k$ is the Boltzmann constant and $T$ is the absolute temperature.

The total change in free energy of mixing for the whole interaction zone, $V$, is obtained by summing over all the elements in $V$:

$$G_{mix} = \frac{2kTV_2{}^2}{V_1} \nu_2 \left(\frac{1}{2} - \chi\right) R_{mix}(h) \tag{15}$$

where $V_1$ and $V_2$ are the molar volumes of solvent and polymer, respectively, $\nu_2$ is the number of chains per unit area, $R_{mix}(h)$ is a geometric function which depends on the form of the segment density distribution of the chain normal to the surface, $\rho(z)$, $k$ is the Boltzmann constant and $T$ is the absolute temperature.

Using the above theory, one can derive an expression for the free energy of mixing of two polymer layers (assuming a uniform segment density distribution in each layer) surrounding two spherical particles as a function of the separation distance h between the particles [11].

The expression for $G_{mix}$ is

$$G_{mix} = \left(\frac{2V_2{}^2}{V_1}\right) \nu_2 \left(\frac{1}{2} - \chi\right) \left(\delta - \frac{h}{2}\right)^2 \left(3R + 2\delta + \frac{h}{2}\right) \tag{16}$$

The sign of $G_{mix}$ depends on the value of the Flory–Huggins interaction parameter $\chi$: if $\chi < 0.5$, $G_{mix}$ is positive and the interaction is repulsive; if $\chi > 0.5$, $G_{mix}$ is negative and the interaction is attractive. The condition $\chi = 0.5$ and $G_{mix} = 0$ is termed the $\theta$-condition. The latter corresponds to the case where the polymer mixing behaves as ideal, i.e. mixing of the chains does not lead to an increase or decrease in the free energy of the system.

### 5.7.2 Elastic Interaction, $G_{el}$

This arises from the decrease in configurational entropy of the chains on the approach of a second particle. As a result of this approach, the volume available

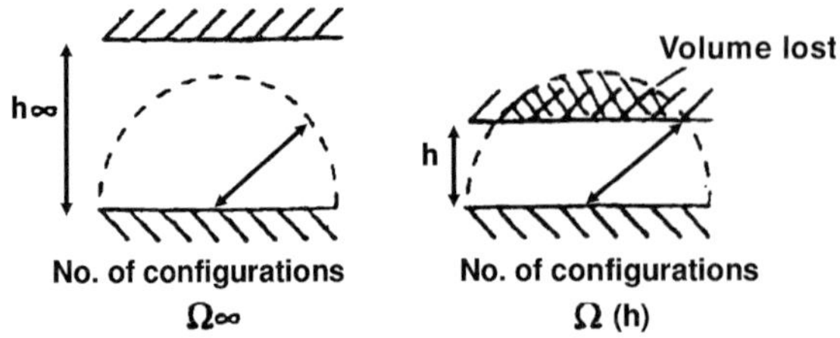


**Figure 5.10** Schematic representation of configurational entropy loss on approach of a second particle.

for the chains becomes restricted, resulting in decrease in the number of configurations. This is illustrated in Figure 5.10 by considering a simple molecule, represented by a rod that rotates freely in a hemisphere across a surface [12].

When the two surfaces are separated by an infinite distance, the number of configurations of the rod is $\Omega(\infty)$ which is proportional to the volume of the hemisphere. When a second particle approaches to a distance $h$ such that it cuts the hemisphere (losing some volume), the volume available to the chains is reduced and the number of configurations becomes $\Omega(h)$, which is less than $\Omega(\infty)$

For two flat plates, $G_{el}$ is given by the following expression:

$$\frac{G_{el}}{kT} = 2\nu_2 \ln\left[\frac{\Omega(h)}{\Omega(\infty)}\right] = 2\nu_2 R_{el}(h) \tag{17}$$

where $R_{el}(h)$ is a geometric function which depends on the segment density distribution. It should be stressed that $G_{el}$ is always positive and could play a major role in steric stabilization. It becomes very strong when the separation distance between the particles becomes comparable to the adsorbed layer thickness $\delta$.

Combination of $G_{mix}$ and $G_{el}$ with $G_A$ gives the total energy of interaction $G_T$ (assuming there is no contribution from any residual electrostatic interaction) [13], i.e.

$$G_T = G_{mix} + G_{el} + G_A \tag{18}$$

A schematic representation of the variation of $G_{mix}$, $G_{el}$, $G_A$ and $G_T$ with surface–surface separation distance $h$ is shown in Figure 5.11.

$G_{mix}$ increases very sharply with decrease in $h$ when $h < 2\delta$. $G_{el}$ increases very sharply with decrease in $h$ when $h < \delta$. $G_T$ versus $h$ shows a minimum, $G_{min}$, at separation distances comparable to $2\delta$; when $h < 2\delta$, $G_T$ shows a rapid increase with further decrease in $h$ [6, 7].

Unlike the $G_T$–$h$ curve predicted by the DLVO theory (which shows two minima and one energy maximum), the $G_T$–$h$ curves for systems that are sterically stabilized show only one minimum, $G_{min}$, followed by a sharp increase in $G_T$ with decrease in $h$ (when $h < 2\delta$). The depth of the minimum depends on the Hamaker constant $A$, the particle radius $R$ and adsorbed layer thickness $\delta$:

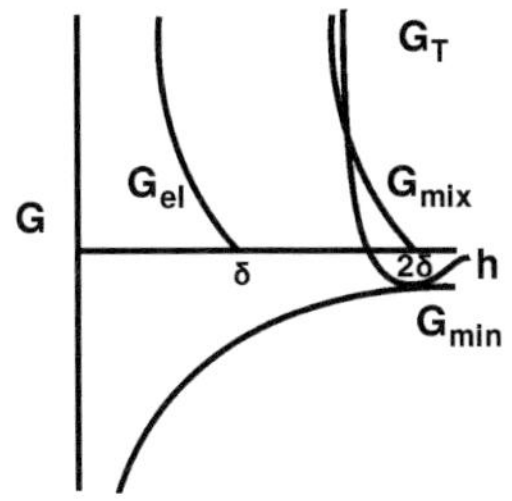


**Figure 5.11** Variation of $G_{mix}$, $G_{el}$, $G_A$ and $G_T$ with surface–surface distance between the particles.

$G_{min}$ increases with increase in $A$ and $R$. At a given $A$ and $R$, $G_{min}$ increases with decrease in $\delta$ (i.e. with decrease of the molecular weight, $M_w$, of the stabilizer). This is illustrated in Figure 5.12, which shows the energy-distance curves at various $\delta/R$ ratios. As the latter increases, $G_{min}$ decreases, and at sufficiently high value of $\delta/R$, $G_{min}$ becomes smaller than $kT$ and the dispersion approaches thermodynamic stability. This explains the very high stability of nano-suspensions and nano-emulsions.

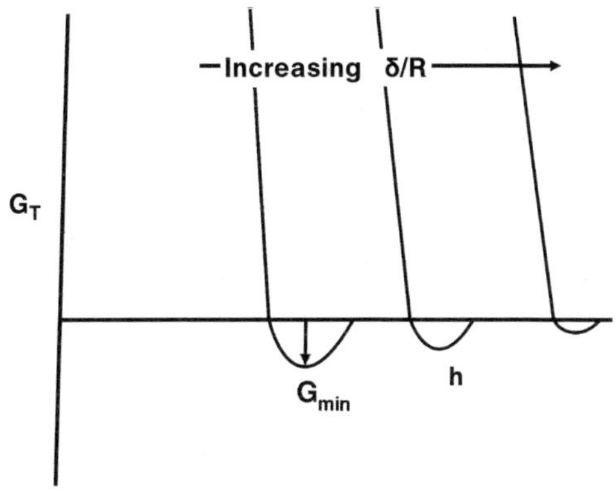


**Figure 5.12** Variation of $G_{min}$ with $\delta/R$.

## 5.7.3 Criteria for Effective Steric Stabilization

The criteria are as follows:

1. The particles should be completely covered by the polymer (the amount of polymer should correspond to the plateau value). Any bare patches may cause flocculation either by van der Waals attraction (between the bare patches) or by bridging flocculation (whereby a polymer molecule will become simultaneously adsorbed on two or more particles).

2. The polymer should be strongly 'anchored' to the particle surfaces, to prevent any displacement during particle approach. This is particularly important for concentrated suspensions and emulsions. For this purpose A–B and A–B–A block and $BA_n$ graft copolymers are the most suitable where the chain B is chosen to be highly insoluble in the medium and has a strong affinity to the surface. Examples of B groups for hydrophobic particles in aqueous media are polystyrene and poly(methyl methacrylate).
3. The stabilizing chain A should be highly soluble in the medium and strongly solvated by its molecules. Examples of A chains in aqueous media are poly(ethylene oxide), poly(vinyl alcohol) and polysaccharides.
4. $\delta$ should be sufficiently large (>5 nm) to prevent weak flocculation.

## 5.8 Flocculation of Sterically Stabilized Dispersions

Two main types of flocculation may be distinguished:

1. *Weak flocculation:* this occurs when the thickness of the adsorbed layer is small (usually <5 nm), particularly when the particle radius and Hamaker constant are large.
2. *Incipient flocculation:* this occurs when the solvency of the medium is reduced to become worse than a $\theta$-solvent (i.e. $\chi > 0.5$). This is illustrated in Figure 5.13, where $\chi$ was increased from <0.5 (good solvent) to >0.5 (poor solvent).

When $\chi > 0.5$, $G_{mix}$ becomes negative (attractive), which, when combined with the van der Waals attraction at this separation distance, gives a deep minimum, causing flocculation. In most cases, there is a correlation between the critical flocculation point and the $\theta$ condition of the medium. Good correlation is found in

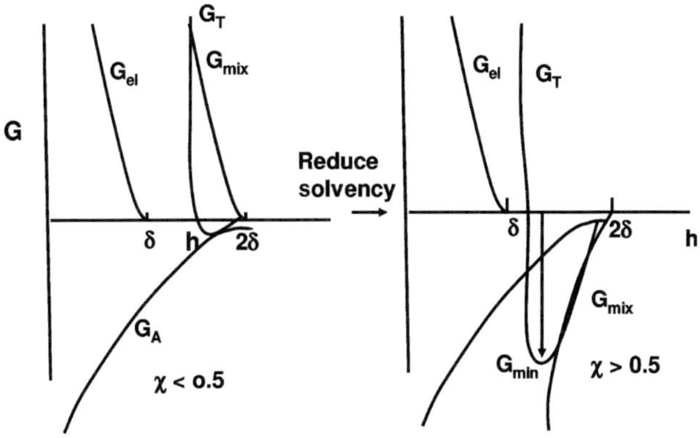


**Figure 5.13** Influence of reduction in solvency on the energy–distance curves for sterically stabilized dispersions.

many cases between the critical flocculation temperature (CFT) and $\theta$-temperature of the polymer in solution (with block and graft copolymers one should consider the $\theta$-temperature of the stabilizing chains A) [8, 9]. Good correlation is also found between the critical volume fraction (CFV) of a non-solvent for the polymer chains and their $\theta$-point under these conditions. However, in some cases such correlations may break down, particularly the case for polymers which adsorb by multi-point attachment. This situation has been described by Napper [9], who referred to it as 'enhanced' steric stabilization.

Thus, by measuring the $\theta$-point (CFT or CFV) for the polymer chains (A) in the medium under investigation (which could be obtained from viscosity measurements), one can establish the stability conditions for a dispersion, before its preparation. This procedure helps also in designing effective steric stabilizers such as block and graft copolymers.

## References


1 H.C. Hamaker, *Physica*, *4*, 1058 (1937).
2 J. Lyklema, Structure of the solid/liquid interface and the electrical double layer, in *Solid/Liquid Dispersions*, ed. Th.F. Tadros, Academic Press, London (1987).
3 B.H. Bijesterbosch, Stability of Solid/liquid dispersions, in *Solid/Liquid Dispersions*, ed. Th.F. Tadros, Academic Press, London (1987).
4 B.V. Deryaguin and L. Landau, *Acta Physicochim. URSS*, *14*, 633 (1941).
5 E.J.W. Verwey and J.Th.G. Overbeek, *Theory of Stability of Lyophobic Colloids*, Elsevier, Amsterdam (1948).
6 R.J. Hunter, *Zeta Potential in Colloid Science; Principles and Applications*, Academic Press, London (1981).
7 M.V. von Smoluchowski, *Handbuch der Electricität und des Magnetismus* Vol. II, Barth, Leipzig (1914).
8 Th.F. Tadros, Polymer adsorption and dispersion stability, in *The Effect of Polymers on Dispersion Properties*, ed. Th.F. Tadros, Academic Press, London (1981).
9 D.H. Napper, *Polymeric Stabilization of Colloidal Dispersions*, Academic Press, London (1981).
10 P.J. Flory and W.R. Krigbaum, *J. Chem. Phys.*, *18*, 1086 (1950).
11 E.W. Fischer, *Kolloid Z.*, *160*, 120 (1958).
12 E.L. Mackor and J.H. van der Waals, *J. Colloid Sci.*, *7*, 535 (1951).
13 F.Th. Hesselink, A. Vrij and J.Th.G. Overbeek, *J. Phys. Chem.*, *75*, 2094 (1971).

# 6
# Emulsion Concentrates (EWs)

## 6.1
## Introduction

Recently, many agrochemicals have been formulated as oil-in-water (O/W) emulsion concentrates (EWs). These systems offer many advantages over the more traditionally used emulsifiable concentrates (ECs). By using an O/W system, one can reduce the amount of oil in the formulation since in most cases a small proportion of oil is added to the agrochemical oil (if this has a high viscosity) before emulsification. In some cases, if the agrochemical oil has a low to medium viscosity, one can emulsify the active ingredient directly into water. With many agrochemicals with low melting point, which is not suitable for the preparation of a suspension concentrate, one can dissolve the active ingredient in a suitable oil and the oil solution is then emulsified into water. EWs which are aqueous based present less hazard to the operator, reducing any skin irritation. In addition, in most cases EWs are less phytotoxic to plants compared with ECs. The O/W emulsion is convenient for incorporation of water-soluble adjuvants (mostly surfactants). EWs can also be less expensive than ECs since a lower surfactant concentration is used to produce the emulsion and also one replaces a large proportion of oil by water. The only drawback of EWs compared with ECs is the need to use high-speed stirrers and/or homogenizers to obtain the required droplet size distribution. In addition, EWs require control and maintenance of their physical stability. As will be discussed later, EWs are only kinetically stable and one has to control the breakdown processes that occur on storage such as creaming or sedimentation, flocculation, Ostwald ripening, coalescence and phase inversion.

This chapter starts with the principles of the formation of emulsions and the role of the surfactants. This is followed by a section on the procedures that can be applied to select the emulsifiers. The subsequent section deals with the breakdown processes that may occur on storage and methods of their prevention. The last section is concerned with the assessment and prediction of the long-term physical stability of EWs.

Colloids and Interface Science Series, Vol. 5
*Colloids in Agrochemicals.* Tharwat F. Tadros

ISBN: 978-3-527-31465-2

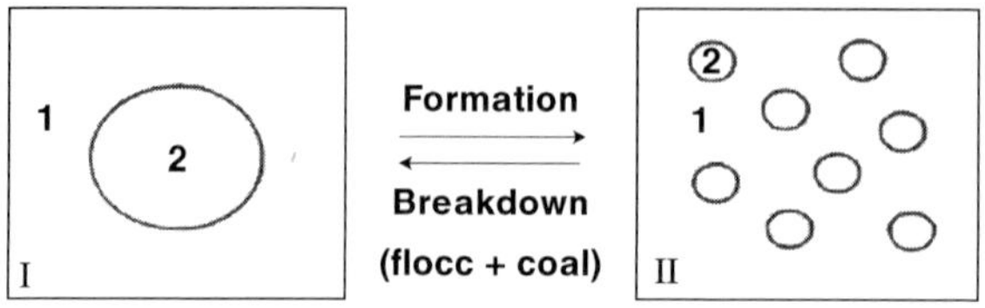


**Figure 6.1** Schematic representation of emulsion formation and breakdown.

## 6.2 Formation of Emulsions

Consider a system in which an oil is represented by a large drop 2 of area $A_1$ immersed in a liquid 2, which is now subdivided into a large number of smaller droplets (1) with total area $A_2$ ($A_2 \gg A_1$) as shown in Figure 6.1. The interfacial tension $\gamma_{12}$ is the same for the large and smaller droplets since the latter are generally in the range from 0.1 to a few μm.

The change in free energy in going from state I to state II is made up of two contributions: a surface energy term (which is positive) that is equal to $\Delta A\gamma_{12}$ (where $\Delta A = A_2 - A_1$), and an entropy of dispersions term, which is also positive (since producing a large number of droplets is accompanied by an increase in configurational entropy) and which is equal to $T\Delta S^{\text{conf}}$.

From the second law of thermodynamics:

$$\Delta G^{\text{form}} = \Delta A\gamma_{12} - T\Delta S^{\text{conf}} \tag{1}$$

In most cases $\Delta A\gamma_{12} \gg T\Delta S^{\text{conf}}$, which means that $\Delta G^{\text{form}}$ is positive, i.e. the formation of emulsions is non-spontaneous and the system is thermodynamically

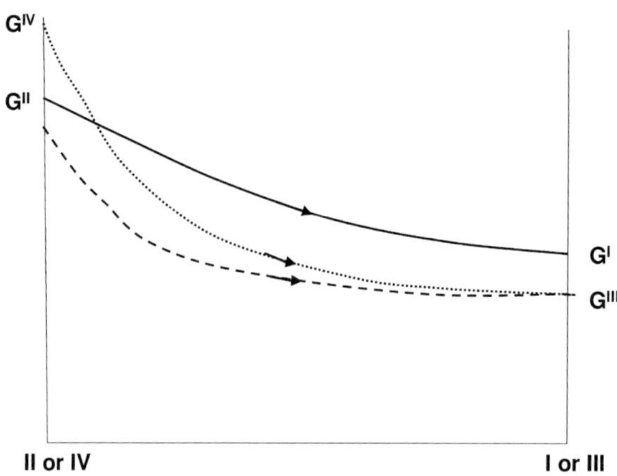


**Figure 6.2** Free energy path in emulsion breakdown: — flocculation + coalescence; - - - flocculation + coalescence + sedimentation; ........ flocculation + coalescence + sedimentation + Ostwald ripening.

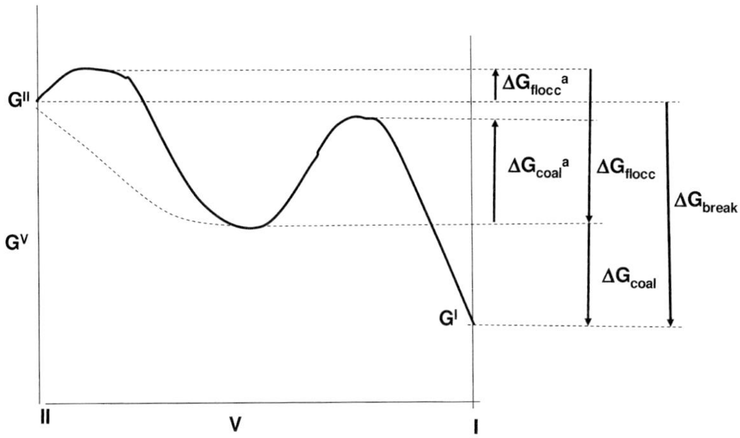


**Figure 6.3** Schematic representation of free energy path for breakdown (flocculation and coalescence) for systems containing an energy barrier.

unstable. In the absence of any stabilization mechanism, the emulsion will break by flocculation, coalescence, Ostwald ripening or a combination of all these processes. This is illustrated in Figure 6.2, which shows several paths for emulsion breakdown processes.

In the presence of a stabilizer (surfactant and/or polymer), an energy barrier is created between the droplets and therefore the reversal from state II to state I becomes non-continuous as a result of the presence of these energy barriers, as illustrated in Figure 6.3. In the presence of the above energy barriers, the system becomes kinetically stable [1]. As discussed in Chapter 5, the energy barrier can be created by electrostatic and/or steric repulsion, which will overcome the van der Waals attraction.

## 6.3 Mechanism of Emulsification

To prepare an emulsion, oil, water, surfactant and energy are needed [2, 3]. This can be considered by taking into account the energy required to expand the interface, $\Delta A\gamma$ (where $\Delta A$ is the increase in interfacial area when the bulk oil with area $A_1$ produces a large number of droplets with area $A_2$; $A_2 \gg A_1$; $\gamma$ is the interfacial tension). Since $\gamma$ is positive, the energy to expand the interface is large and positive. This energy term cannot be compensated by the small entropy of dispersion $T\Delta S$ (which is also positive) and, as discussed before, the total free energy of formation of an emulsion, $\Delta G$, is positive.

Thus, emulsion formation is non-spontaneous and energy is required to produce the droplets. The formation of large droplets (a few microns), as is the case for macroemulsions, is fairly easy and hence high-speed stirrers such as the Ultra

Turrax or Silverson mixer are sufficient to produce the emulsion. In contrast, the formation of small drops (submicrons, as is the case with nano-emulsions) is difficult and this requires a large amount of surfactant and/or energy. The high energy required for formation of nanoemulsions can be understood from consideration of the Laplace pressure $p$ (the difference in pressure between inside and outside the droplet [2, 3]:

$$\Delta p = \gamma \left( \frac{1}{R_1} + \frac{1}{R_2} \right) \tag{2}$$

where $R_1$ and $R_2$ are the principal radii of curvature of the drop. For a spherical drop, $R_1 = R_2 = R$ and

$$\Delta p = \frac{\gamma}{2R} \tag{3}$$

To break up a drop into smaller ones, it must be strongly deformed and this deformation increases $p$ [1, 2]. Surfactants play major roles in the formation of emulsions: By lowering the interfacial tension, $p$ is reduced and hence the stress needed to break up a drop is reduced [2, 3]. Surfactants prevent coalescence of newly formed drops.

To describe emulsion formation, one has to consider two main factors: hydrodynamics and interfacial science. To assess emulsion formation, one usually measures the droplet size distribution using, for example, laser diffraction techniques. A useful average diameter $d$ is

$$d_{nm} = \left( \frac{S_m}{S_n} \right)^{1/(n-m)} \tag{4}$$

In most cases $d_{32}$ (the volume/surface average or Sauter mean) is used. The width of the size distribution can be given as the coefficient of variation $c_m$, which is the standard deviation of the distribution weighted with $d^m$ divided by the corresponding average $d$. Generally $C_2$ will be used, which corresponds to $d_{32}$.

An alternative way to describe the emulsion quality is to use the specific surface area $A$ (surface area of all emulsion droplets per unit volume of emulsion):

$$A = \pi s^2 = \frac{6\phi}{d_{32}} \tag{5}$$

where $\phi$ is the volume fraction of the emulsion.

## 6.4 Methods of Emulsification

Several procedures [2, 3] may be applied for emulsion preparation, including simple pipe flow (low agitation energy L), static mixers and general stirrers (low to medium energy, L–M), high-speed mixers such as the Ultra Turrax (M), colloid mills and high-pressure homogenizers (high energy, H) and ultrasound generators (M–H). The method of preparation can be continuous (C) or batchwise (B): pipe flow and static mixers. C; stirrers and Ultra Turrax, B, C; colloid mill and high-pressure homogenizers, C; ultrasound, B, C.

In all methods, there is liquid flow [4, 5]: unbounded and strongly confined flow. In unbounded flow, any droplet is surrounded by a large amount of flowing liquid (the confining walls of the apparatus are far away from most of the droplets). The forces can be frictional (mostly viscous) or inertial. Viscous forces cause shear stresses to act on the interface between the droplets and the continuous phase (primarily in the direction of the interface). The shear stresses can be generated by laminar flow (LV) [6] or turbulent flow (TV) [7].

Within each regime, an essential variable is the intensity of the forces acting:

$$\text{Viscous stress during laminar flow} = \eta G \tag{6}$$

where $G$ is the velocity gradient.

The intensity in turbulent flow [6, 7] is expressed by the power density $\varepsilon$ (the amount of energy dissipated per unit volume per unit time):

$$\varepsilon = \eta G^2 \tag{7}$$

The viscosity of the oil plays an important role in the break-up of droplets: the higher the viscosity, the longer it will take to deform a drop. The deformation time $\tau_{\text{def}}$ is given by the ratio of oil viscosity to the external stress acting on the drop:

$$\tau_{\text{def}} = \frac{\eta_{\text{D}}}{\sigma_{\text{ext}}} \tag{8}$$

The viscosity of the continuous phase $\eta_{\text{C}}$ plays an important role in some regimes. For the turbulent viscous regime, a larger $\eta_{\text{C}}$ leads to smaller droplets. For the laminar viscous regime, the effect is even stronger.

## 6.5 Role of Surfactants in Emulsion Formation

Surfactants lower the interfacial tension $\gamma$ and this causes a reduction in droplet size. The latter decreases with decrease in $\gamma$. For the turbulent regime, the droplet diameter is proportional to $\gamma^{3/5}$. The effect of reducing $\gamma$ on the droplet size is illustrated in Figure 6.4, which shows a plot of the droplet surface area $A$ and mean drop size $d_{32}$ as a function of surfactant concentration $m$ for various systems [2, 3].

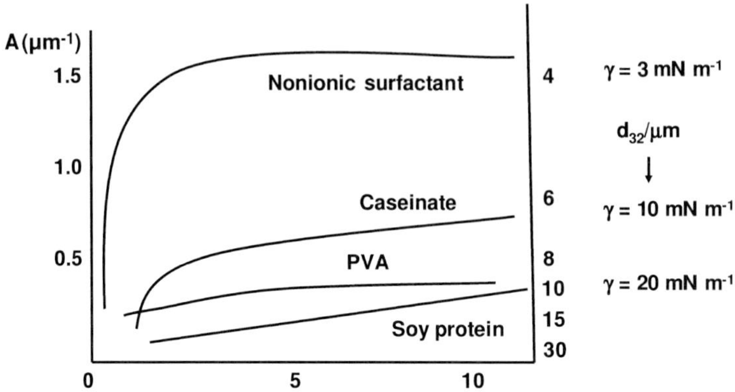


**Figure 6.4** Variation of $A$ and $d_{32}$ with $m$ for various surfactant systems.

The amount of surfactant required to produce the smallest drop size will depend on its activity $a$ (concentration) in the bulk, which determines the reduction in $\gamma$, as given by the Gibbs adsorption equation (see Chapter 3):

$$-\mathrm{d}\gamma = RT\Gamma \mathrm{d} \ln a \tag{9}$$

where $R$ is the gas constant, $T$ is the absolute temperature and $\Gamma$ is the surface excess (number of moles adsorbed per unit area of the interface). $\Gamma$ increases with increase in surfactant concentration and eventually it reaches a plateau value (saturation adsorption). This is illustrated in Figure 6.5 for various emulsifiers.

The value of $\gamma$ obtained depends on the nature of the oil and surfactant used. Small molecules such as nonionic surfactants lower $\gamma$ more than polymeric surfactants such as PVA. Another important role of the surfactant is its effect on the

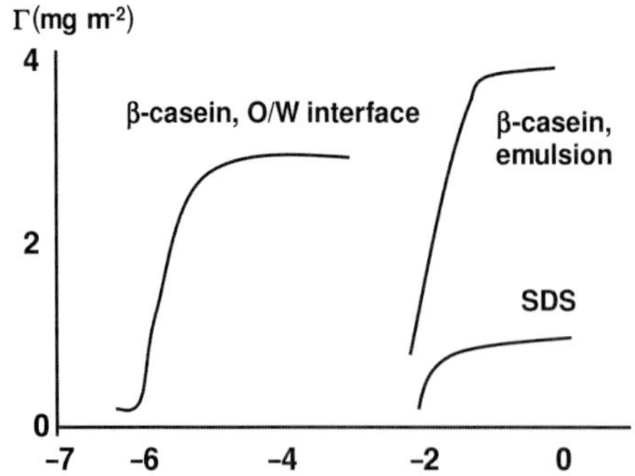


**Figure 6.5** Variation of $\Gamma$ (mg m$^{-2}$) with log $C_{eq}$ (wt%). The oils are $\beta$-casein (O/W interface), toluene, $\beta$-casein (emulsion), soybean, SDS and benzene [2, 3].

interfacial dilational modulus $\varepsilon$ [8]:

$$\varepsilon = \frac{d\gamma}{d \ln A} \tag{10}$$

During emulsification, an increase in the interfacial area $A$ takes place and this causes a reduction in $\Gamma$. The equilibrium is restored by adsorption of surfactant from the bulk, but this takes time (shorter times occur at higher surfactant activity). Thus $\varepsilon$ is small at small $a$ and also at large $a$. Because of the lack or slowness of equilibrium with polymeric surfactants, $\varepsilon$ will not be the same for expansion and compression of the interface.

In practice, surfactant mixtures are used and these have pronounced effects on $\gamma$ and $\varepsilon$. Some specific surfactant mixtures give lower $\gamma$ values than either of the two individual components [2, 3]. The presence of more than one surfactant molecule at the interface tends to increase $\varepsilon$ at high surfactant concentrations. The various components vary in surface activity. Those with the lowest $\gamma$ tend to predominate at the interface, but if present at low concentrations, it may take a long time before reaching the lowest value. Polymer–surfactant mixtures may show some synergetic surface activity.

### 6.5.1 Role of Surfactants in Droplet Deformation

In addition to their effect on reducing $\gamma$, surfactants play major roles in deformation and break-up of droplets [2, 3]. This is summarized as follows. Surfactants allow the existence of interfacial tension gradients, which is crucial for the formation of stable droplets. In the absence of surfactants (clean interface), the interface cannot withstand a tangential stress; the liquid motion will be continuous. If a liquid flows along the interface with surfactants, the latter will be swept downstream, causing an interfacial tension gradient. The interface will then drag some of the bordering liquid with it (the Marangoni effect).

Interfacial tension gradients [9–11] are very important in stabilizing the thin liquid film between the droplets, which is very important during the beginning of emulsification (films of the continuous phase may be drawn through the disperse phase and collisions are very extensive). The magnitude of the $\gamma$ gradients and of the Marangoni effect depends on the surface dilational modulus $\varepsilon$.

Another important role of the emulsifier is to prevent coalescence during emulsification. This is certainly not due to the strong repulsion between the droplets, since the pressure at which two drops are pressed together is much greater than the repulsive stresses. The counteracting stress must be due to the formation of $\gamma$ gradients.

Closely related to the above mechanism is the Gibbs–Marangoni effect. The depletion of surfactant in the thin film between approaching drops results in a $\gamma$ gradient without liquid flow being involved. This results in an inward flow of liquid that tends to drive the drops apart [2, 3].

The Gibbs–Marangoni effect also explains the Bancroft rule, which states that the phase in which the surfactant is most soluble forms the continuous phase. If the surfactant is in the droplets, a $\gamma$ gradient cannot develop and the drops would be prone to coalescence. Thus, surfactants with HLB > 7 tend to form O/W emulsions and those with HLB < 7 tend to form W/O emulsions. The Gibbs–Marangoni effect also explains the difference between surfactants and polymers for emulsification. Polymers give larger drops than surfactants. Polymers give a smaller value of $\varepsilon$ at low concentrations when compared with surfactants.

Various other factors should also be considered for emulsification, such as the disperse phase volume fraction $\phi$. An increase in $\phi$ leads to an increase in droplet collision and hence coalescence during emulsification. With increase in $\phi$, the viscosity of the emulsion increases and could change the flow from being turbulent to laminar.

The presence of many particles results in a local increase in velocity gradients. This means that $G$ increases. In turbulent flow, an increase in $\phi$ will induce turbulence depression, which will result in larger droplets. Turbulence depression by added polymers tends to remove the small eddies, resulting in the formation of larger droplets.

If the mass ratio of surfactant to continuous phase is kept constant, an increase in $\phi$ results in a decrease in surfactant concentration and hence an increase in $\gamma_{eq}$. This results in larger droplets. If the mass ratio of surfactant to disperse phase is kept constant, the above changes are reversed.

## 6.6 Selection of Emulsifiers

### 6.6.1 The Hydrophilic–Lipophilic Balance (HLB) Concept

The selection of different surfactants in the preparation of either O/W or W/O emulsions is often still made on an empirical basis. A semi-empirical scale for selecting surfactants is the hydrophilic–lipophilic balance (HLB number) developed by Griffin [12, 13]. This scale is based on the relative percentage of hydrophilic to lipophilic (hydrophobic) groups in the surfactant molecule(s). For an O/W emulsion droplet, the hydrophobic chain resides in the oil phase whereas the hydrophilic head group resides in the aqueous phase. For a W/O emulsion droplet, the hydrophilic group(s) reside in the water droplet, whereas the lipophilic groups reside in the hydrocarbon phase. A summary of HLB ranges and their application is given in Table 6.1.

Table 6.1 gives a guide to the selection of surfactants for a particular application. The HLB number depends on the nature of the oil [12, 13]. As an illustration, Table 6.2 gives the required HLB numbers to emulsify various oils.

The relative importance of the hydrophilic and lipophilic groups was first recognized when using mixtures of surfactants containing varying proportions

**Table 6.1** Summary of HLB ranges and their applications.

| HLB range | Application |
|---|---|
| 3–6 | W/O emulsifier |
| 7–9 | Wetting agent |
| 8–18 | O/W emulsifier |
| 13–15 | Detergent |
| 15–18 | Solubilizer |

**Table 6.2** HLB numbers required to emulsify various oils.

| Oil | W/O emulsion | O/W emulsion |
|---|---|---|
| Paraffin oil | 4 | 10 |
| Beeswax | 5 | 9 |
| Lanolin, anhydrous | 8 | 12 |
| Cyclohexane | – | 15 |
| Toluene | – | 15 |

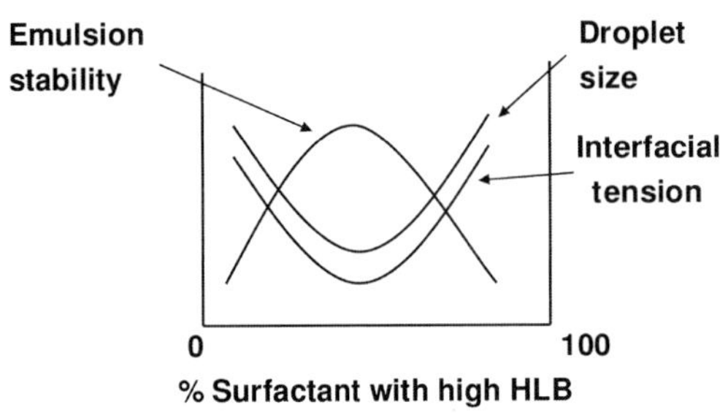


**Figure 6.6** Variation of emulsion stability, droplet size and interfacial tension with percentage of surfactant with high HLB number.

of a low and high HLB number [12, 13]. The efficiency of any combination (as judged by phase separation) was found to pass through a maximum when the blend contained a particular proportion of the surfactant with the higher HLB number. This is illustrated in Figure 6.6, which shows the variation of emulsion stability, droplet size and interfacial tension as a function of the percentage of surfactant with high HLB number.

The average HLB number may be calculated from additivity:

$$\mathrm{HLB} = x_1 \mathrm{HLB}_1 + x_2 \mathrm{HLB}_2 \qquad (11)$$

where $x_1$ and $x_2$ are the weight fractions of the two surfactants with $HLB_1$ and $HLB_2$.

Griffin [12, 13] developed simple equations for calculation of the HLB number of relatively simple nonionic surfactants. For a polyhydroxy fatty acid ester:

$$\mathrm{HLB} = 20\left(1 - \frac{S}{A}\right) \tag{12}$$

where $S$ is the saponification number of the ester and $A$ is the acid number. For a glyceryl monostearate, $S = 161$ and $A = 198$, hence the HLB is 3.8 (suitable for a W/O emulsion). For a simple alcohol ethoxylate, the HLB number can be calculated from the weight percent of ethylene oxide ($E$) and polyhydric alcohol ($P$):

$$\mathrm{HLB} = \frac{E + P}{5} \tag{13}$$

If the surfactant contains PEO as the only hydrophilic group contribution from one OH group neglected, then

$$\mathrm{HLB} = \frac{E}{5} \tag{14}$$

For a nonionic surfactant, $C_{12}H_{25}{-}O{-}(CH_2CH_2O)_6$, the HLB is 12 (suitable for an O/W emulsion).

The above simple equations cannot be used for surfactants containing propylene oxide or butylene oxide. They also cannot be applied for ionic surfactants. Davies [14] devised a method for calculating the HLB number for surfactants from their chemical formulae, using empirically determined group numbers. A group number is assigned to various component groups. A summary of the group numbers for some surfactants is given in Table 6.3.

The HLB is given by the following empirical equation,

$$\mathrm{HLB} = 7 + \sum(\text{hydrophilic group Nos}) - \sum(\text{lipohilic group Nos}) \tag{15}$$

Davies [14] has shown that the agreement between HLB numbers calculated from this equation and those determined experimentally is fairly satisfactory.

Various other procedures have been developed to obtain a rough estimate of the HLB number. Griffin [12] found a good correlation between the cloud point of a 5% solution of various ethoxylated surfactants and their HLB number.

Davies [14] attempted to relate the HLB values to the selective coalescence rates of emulsions. Such correlations were not realized since it was found that the emulsion stability and even its type depend to a large extent on the method of dispersing the oil into the water and vice versa. At best the HLB number can only be used as a guide for selecting optimum compositions of emulsifying agents.

**Table 6.3** HLB group numbers for surfactants.

| Surfactant type | HLB group number |
|---|---|
| **Hydrophilic** | |
| $-SO_4Na^+$ | 38.7 |
| $-COO^-$ | 21.2 |
| $-COONa$ | 19.1 |
| N(tertiary amine) | 9.4 |
| Ester (sorbitan ring) | 6.8 |
| $-O-$ | 1.3 |
| CH– (sorbitan ring) | 0.5 |
| **Lipophilic** | |
| $(-CH-)$, $(-CH_2-)$, $CH_3$ | 0.475 |
| **Derived** | |
| $-CH_2-CH_2-O$ | 0.33 |
| $-CH_2-CH_2-CH_2-O-$ | −0.15 |

One may take any pair of emulsifying agents, which fall at opposite ends of the HLB scale, e.g. Tween 80 (sorbitan monooleate with 20 mol EO, HLB = 15) and Span 80 (sorbitan monooleate, HLB = 5), using them in various proportions to cover a wide range of HLB numbers. The emulsions should be prepared in the same way, with a few percent of the emulsifying blend. The stability of the emulsions is then assessed at each HLB number from the rate of coalescence or qualitatively by measuring the rate of oil separation. In this way, one may be able to find the optimum HLB number for a given oil. Having found the most effective HLB value, various other surfactant pairs are compared at this HLB value, to find the most effective pair.

### 6.6.2 The Phase Inversion Temperature (PIT) Concept

This concept, which was developed by Shinoda and co-workers [15, 16], is closely rated to the HLB balance concept described above. Shinoda and co-workers found that many O/W emulsions stabilized with nonionic surfactants undergo a process of inversion at a critical temperature (PIT). The PIT can be determined by following the emulsion conductivity (a small amount of electrolyte is added to increase the sensitivity) as a function of temperature. The conductivity of the O/W emulsion increases with increase in temperature until the PIT is reached, above which there will be a rapid reduction in conductivity (W/O emulsion is formed).

Shinoda and co-workers [15, 16] found that the PIT is influenced by the HLB number of the surfactant. The size of the emulsion droplets was found to depend on the temperature and HLB number of the emulsifiers. The droplets are less stable towards coalescence close to the PIT. However, by rapid cooling of the emulsion, a stable system may be produced. Relatively stable O/W emulsions were obtained when the PIT of the system was 20–65 °C higher than the storage temperature. Emulsions prepared at a temperature just below the PIT followed by rapid cooling generally have smaller droplet sizes. This can be understood if one considers the change in interfacial tension with temperature. The interfacial tension decreases with increase in temperature, reaching a minimum close to the PIT, after which it increases. Thus, the droplets prepared close to the PIT are smaller than those prepared at lower temperatures. These droplets are relatively unstable towards coalescence near the PIT, but by rapid cooling of the emulsion one can retain the smaller size. The above procedure can be applied to prepare mini- or nano-emulsions.

The optimum stability of the emulsion was found to be relatively insensitive to changes in the HLB value or the PIT of the emulsifier, but instability was very sensitive to the PIT of the system. It is essential, therefore, to measure the PIT of the emulsion as a whole (with all other ingredients). At a given HLB value, the stability of the emulsions against coalescence increases markedly as the molar mass of both the hydrophilic and lipophilic components increases.

The enhanced stability using high molecular weight surfactants (polymeric surfactants) can be understood from consideration of the steric repulsion, which produces more stable films. Films produced using macromolecular surfactants resist thinning and disruption, thus reducing the possibility of coalescence.

The emulsions showed maximum stability when the distribution of the PEO chains was broad. The cloud point is lower but the PIT is higher than in the corresponding case for narrow size distributions. The PIT and HLB number are directly related parameters.

Addition of electrolytes reduces the PIT and hence an emulsifier with a higher PIT value is required when preparing emulsions in the presence of electrolytes. Electrolytes cause dehydration of the PEO chains and in effect this reduces the cloud point of the nonionic surfactant. One needs to compensate for this effect by using a surfactant with higher HLB. The optimum PIT of the emulsifier is fixed if the storage temperature is fixed. In view of the above correlation between PIT and HLB and the possible dependence of the kinetics of droplet coalescence on the HLB number. It has been suggested that PIT measurements can be used as a rapid method for assessment of emulsion stability [14]. However, one should be careful in using such methods for assessment of the long-term stability since the correlations were based on a very limited number of surfactants and oils. Measurement of the PIT can at best be used as a guide for the preparation of stable emulsions. Assessment of the stability should be evaluated by following the droplet size distribution as a function of time using a Coulter Counter or light diffraction techniques. Following the rheology of the emulsion as a function of time and temperature may also be used for assessment of the stability against coalescence. Care should be taken in analyzing the rheological results. Coalescence results in an increase in the droplet size and this is usually followed by a reduction in the viscosity of the emulsion. This trend is only observed if the coalescence is not accompanied by flocculation of the emulsion droplets (which results in an increase in viscosity). Ostwald ripening can also complicate the analysis of the rheological data.

### 6.6.3 The Cohesive Energy Ratio (CER) Concept for Emulsifier Selection

Beerbower and Hill [17] considered the dispersing tendency on the oil and water interfaces of the surfactant or emulsifier in terms of the ratio of the cohesive energies of the mixtures of oil with the lipophilic portion of the surfactant and the water with the hydrophilic portion. They used the Winsor $R_0$ concept, which is the ratio of the intermolecular attraction of oil molecules (O) and lipophilic portion of surfactant (L), $C_{LO}$, to that of water (W) and hydrophilic portion (H), $C_{HW}$:

$$R_0 = \frac{C_{LO}}{C_{HW}} \tag{16}$$

Several interaction parameters may be identified at the oil and water sides of the interface. One can identify at least nine interaction parameters, as illustrated in Figure 6.7. In the absence of emulsifier, there will be only three interaction parameters: $C_{OO}$, $C_{WW}$ and $C_{OW}$; if $C_{OW} << C_{WW}$, the emulsion breaks.

The above interaction parameters may be related to the Hildebrand solubility parameter $\delta$ [18] (at the oil side of the interface) and the Hansen [19] nonpolar, hydrogen bonding and polar contributions to $\delta$ at the water side of the interface. The solubility parameter of any component is related to its heat of vaporization, $\Delta H$, by the expression

$$\delta^2 = \frac{\Delta \mathrm{H} - RT}{V_M} \tag{17}$$

where $V_M$ is the molar volume.

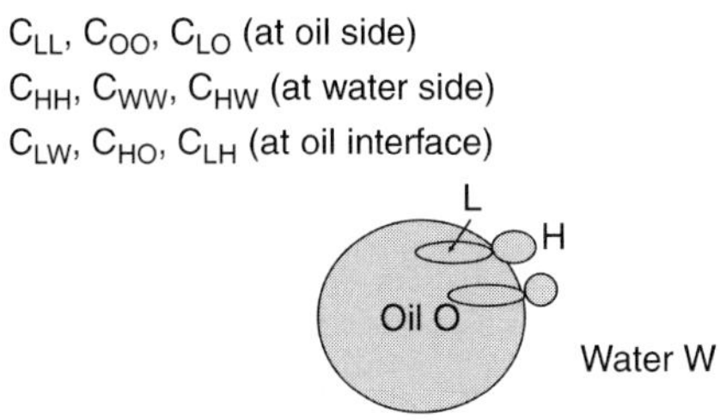


**Figure 6.7** Schematic representation of interaction parameters.

Hansen [19] considered $\delta$ (at the water side of the interface) to consist of three main contributions, a dispersion contribution, $\delta_d$, a polar contribution, $\delta_p$, and a hydrogen bonding contribution, $\delta_h$. These contributions have different weighting factors:

$$\delta^2 = \delta_d{}^2 + \delta_p{}^2 + \delta_h{}^2 \tag{18}$$

Beerbower and Hill [17] used the following expression for the HLB number,

$$\mathrm{HLB} = 20 \frac{M_H}{M_L + M_H} = 20 \frac{V_H \rho_H}{V_L \rho_L + V_H \rho_H} \tag{19}$$

where $M_H$ and $M_L$ are the molecular weights of the hydrophilic and lipophilic portions of the surfactants, respectively, $V_L$ and $V_H$ are their corresponding molar volumes and $\rho_H$ and $\rho_L$ are the densities.

The cohesive energy ratio was originally defined by Winsor (equation 16). When $C_{LO} > C_{HW}$, $R > 1$ and a W/O emulsion forms. If $C_{LO} < C_{HW}$, $R < 1$ and an O/W emulsion forms. If $C_{LO} = C_{HW}$, $R = 1$ and a planar system results; this denotes the inversion point.

$R_0$ can be related to $V_L$, $\delta_L$ and $V_H$, $\delta_H$ by the expression

$$R_0 = \frac{V_L \delta_L{}^2}{V_H \delta_H{}^2} \tag{20}$$

Using equation (16)

$$R_0 = \frac{V_L(\delta_d{}^2 + 0.25\delta_p{}^2 + 0.25\delta_h{}^2)_L}{V_H(\delta_d 2 + 0.25\delta_p{}^2 + 0.25\delta_h{}^2)_H} \tag{21}$$

Combining equations 20 and 21, one obtains the following general expression for the cohesive energy ratio:

$$R_0 = \left(\frac{20}{\mathrm{HLB}} - 1\right) \frac{\rho_H(\delta_d{}^2 + 0.25\delta_p{}^2 + 0.25\delta_h{}^2)_L}{\rho_L(\delta_d{}^2 + 0.25\delta_p{}^2 + 0.25\delta_h{}^2)_H} \tag{22}$$

For an O/W system, HLB = 12–15 and $R_0 = 0.58$–$0.29$ ($R_0 < 1$); for a W/O system, HLB = 5–6 and $R_0 = 2.3$–$1.9$ ($R_0 > 1$); for a planar system, HLB = 8–10 and $R_0 = 1.25$–$0.85$ ($R_0 \approx 1$).

The $R_0$ equation combines both the HLB and cohesive energy densities. It gives a more quantitative estimate of emulsifier selection. $R_0$ considers HLB, molar volume and chemical match. The success of the above approach depends on the availability of data on the solubility parameters of the various surfactant portions. Some values are tabulated in the book by Barton [20].

### 6.6.4
### The Critical Packing Parameter (CPP) for Emulsifier Selection

The critical packing parameter (CPP) is a geometric expression relating the hydrocarbon chain volume ($v$) and length ($l$) and the interfacial area occupied by the head group (a) [21]:

$$\mathrm{CPP} = \frac{v}{l_c a_0} \tag{23}$$

where $a_0$ is the optimal surface area per head group and $l_c$ is the critical chain length. Regardless of the shape of any aggregated structure (spherical or cylindrical micelle or a bilayer), no point within the structure can be further from the hydrocarbon–water surface than $l_c$. The critical chain length, $l_c$, is roughly equal to but less than the fully extended length of the alkyl chain.

The CPP for any micelle shape can be calculated from simple packing constraints. Consider a spherical micelle:

$$\text{volume of the micelle} = \left(\frac{4}{3}\right)\pi r^3 = nv \tag{24}$$

$$\text{area of the micelle} = 4\pi r^2 = na \tag{25}$$

where $v$ is the volume of the hydrocarbon chain, $n$ is the aggregation number and $a$ is the cross-sectional area of the hydrocarbon chain.

From equations 24 and 25:

$$a_0 = \frac{v}{l_c} \tag{26}$$

$$a = \frac{3v}{r} \tag{27}$$

Since $r$ has to be less than $l_c$, packing constraints imply that $a > 3a_0$ or $a_0/a < 1/3$. Hence the CPP for a spherical micelle is $<1/3$.

Surfactants that form spherical micelles with the above packing constraints are more suitable for O/W emulsions.

For a cylindrical micelle:

$$\text{volume} = \pi r^2 l = nv \tag{28}$$

$$\text{area} = 2\pi r l = na \tag{29}$$

$$a = \frac{2v}{r} \tag{30}$$

Since $r$ has to be less than the extended length of the hydrocarbon chain $l_c$, then packing constraints imply that $a > 2a_0$ or $a_0/a < 1/2$. Hence the CPP for a cylindrical micelle is <1/2. When the CPP exceeds 1/2, but is less than 1, spherical bilayers (vesicles) can be produced. When the CPP is ~1, the bilayers may remain planar. When the CPP > 1, inverted micelles are produced. Surfactants that produce these structures are suitable for the formation of W/O emulsions.

### 6.6.5
### Emulsion Stability

Several breakdown processes may occur on storage, depending on: (1) particle size distribution and density difference between the droplets and the medium; (2) magnitude of the attractive versus repulsive forces, which determines flocculation; (3) solubility of the disperse droplets and the particle size distribution, which determines Ostwald ripening; (4) stability of the liquid film between the droplets, which determines coalescence; and (5) phase inversion.

The various breakdown processes are illustrated in Figure 6.8. This is followed by description of each of the breakdown processes and methods that can be applied to prevent such instability.

It should be mentioned that understanding each of these breakdown processes at a molecular lever is far from being achieved at present. This is because the breakdown processes can occur simultaneously. In addition, the preparation of model emulsions that are monodisperse is difficult. Also, the adsorption and conformation of surfactants and polymers at the O/W interface involve various inter-

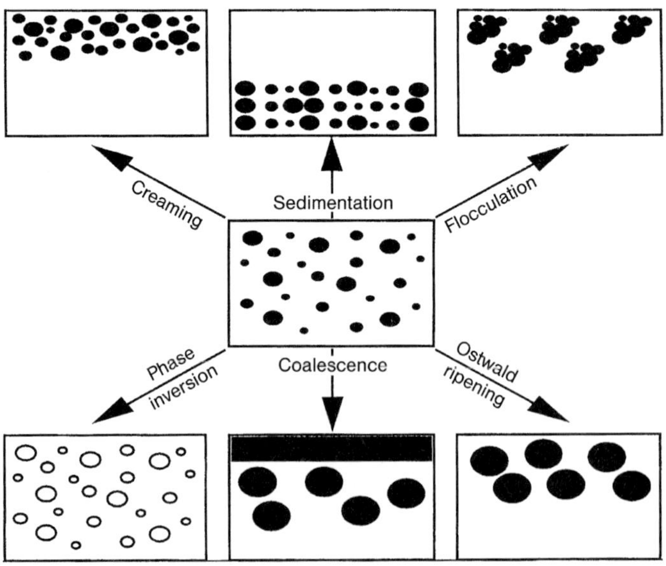


**Figure 6.8** Schematic representation of the various breakdown processes in emulsions.

actions with the oil and continuous phases and it is difficult to understand such interactions at a molecular level.

## 6.7 Creaming or Sedimentation of Emulsions

This is the result of gravity, when the density of the droplets and the medium are not equal. Figures 6.9 and 6.10 give a schematic representation of creaming or sedimentation for three cases [1].

Case (a) represents the situation for small droplets (<0.1 μm, i.e. nano-emulsions), whereby the Brownian diffusion $kT$ (where $k$ is the Boltzmann constant and $T$ is the absolute temperature) exceeds the force of gravity (mass × acceleration due to gravity $g$):

$$kT \ll \frac{4}{3}\pi R^3 \Delta\rho g L \tag{31}$$

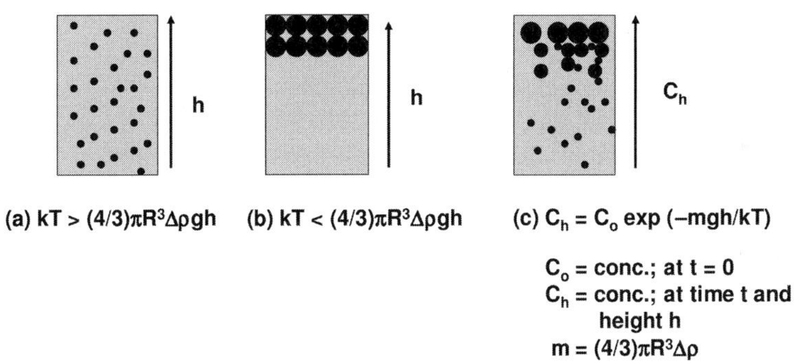


**Figure 6.9** Schematic representation of emulsion creaming.

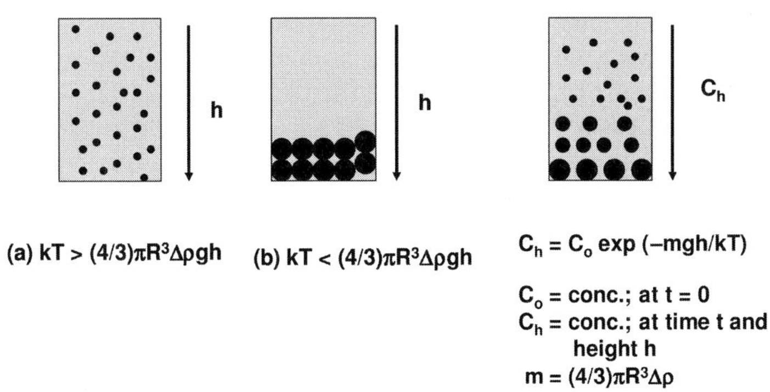


**Figure 6.10** Schematic representation of sedimentation of emulsions.

where $R$ is the droplet radius, $\Delta\rho$ is the density difference between the droplets and the medium and $L$ is the height of the container).

Case (b) represents emulsions consisting of 'monodisperse' droplets with radius $>1\,\mu m$. In this case, the emulsion separates into two distinct layers with the droplets forming a cream or sediment, leaving a clear supernatant liquid. This situation is seldom observed in practice.

Case (c) is for polydisperse (practical) emulsions, in which case the droplets will cream or sediment at various rates. In this case, a concentration gradient builds up, with the larger droplets staying at the top of the cream layer or the bottom:

$$C(h) = C_0 \exp\left(-\frac{mgh}{kT}\right) \tag{32}$$

$$m = \frac{4}{3}\pi R^3 \Delta\rho g \tag{33}$$

where $C(h)$ is the concentration (or volume fraction $\phi$) of droplets at height $h$ and $C_0$ is the concentration at zero time, which is the same at all heights.

### 6.7.1
### Creaming or Sedimentation Rates

#### 6.7.1.1 Very Dilute Emulsions ($\phi < 0.01$)

In this case, the rate could be calculated using Stokes' law, which balances the hydrodynamic force with gravity force:

$$\text{hydrodynamic force} = 6\pi\eta R v_0 \tag{34}$$

$$\text{gravity force} = \frac{4}{3}\pi R^3 \Delta\rho g \tag{35}$$

$$v_0 = \frac{2}{9}\frac{\Delta\rho g R^2}{\eta_0} \tag{36}$$

where $v_0$ is the Stokes velocity and $\eta_0$ is the viscosity of the medium.

For an O/W emulsion with $\Delta\rho = 0.2$ in water ($\eta_0 \approx 10^{-3}$ Pa s), the rate of creaming or sedimentation is $\sim 4.4 \times 10^{-5}$ m s$^{-1}$ for $10\,\mu m$ droplets and $\sim 4.4 \times 10^{-7}$ m s$^{-1}$ for $1\,\mu m$ droplets. This means that in a 0.1 m container, creaming or sedimentation of the $10\,\mu m$ droplets is complete in $\sim 0.6$ h and for the $1\,\mu m$ droplets this takes $\sim 60$ h.

#### 6.7.1.2 Moderately Concentrated Emulsions ($0.2 < \phi < 0.1$)

In this case, one has to take into account the hydrodynamic interaction between the droplets, which reduces the Stokes velocity to a value $v$ given by the following expression:

$$v = v_0(1 - k\phi) \tag{37}$$

where $k$ is a constant that accounts for hydrodynamic interaction; $k$ is of the order of 6.5, which means that the rate of creaming or sedimentation is reduced by about 65%.

#### 6.7.1.3 Concentrated Emulsions ($\phi > 0.2$)

The rate of creaming or sedimentation becomes a complex function of $\phi$, as is illustrated in Figure 6.11, which also shows the change in relative viscosity $\eta_r$ with $\phi$. As can be seen, $v$ decreases with increase in $\phi$ and ultimately it approaches zero when $\phi$ exceeds a critical value, $\phi_p$, which is the so-called 'maximum packing fraction'. The value of $\phi_p$ for monodisperse 'hard spheres' ranges from 0.64 (for random packing) to 0.74 (for hexagonal packing). The value of $\phi_p$ exceeds 0.74 for polydisperse systems. Also for emulsions which are deformable, $\phi_p$ can be much larger than 0.74.

Figure 6.11 also shows that when $\phi$ approaches $\phi_p$, $\eta_r$ approaches $\infty$. In practice, most emulsions are prepared at $\phi$ values well below $\phi_p$, usually in the range 0.2–0.5, and under these conditions creaming or sedimentation is the rule rather than the exception.

Several procedures may be applied to reduce or eliminate creaming or sedimentation, and these are discussed below.

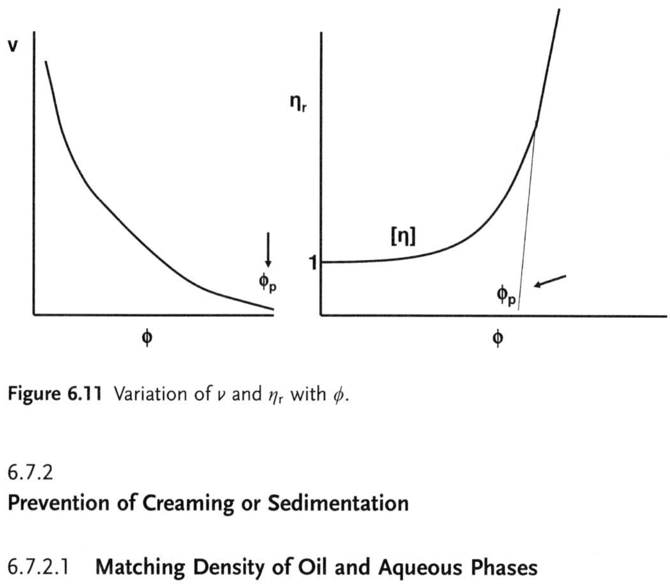


**Figure 6.11** Variation of $v$ and $\eta_r$ with $\phi$.

### 6.7.2
### Prevention of Creaming or Sedimentation

#### 6.7.2.1 Matching Density of Oil and Aqueous Phases

Clearly, if $\Delta\rho = 0$, $v = 0$; however, this method is seldom practical. Density matching, if possible, only occurs at one temperature.

#### 6.7.2.2 Reduction of Droplet Size

Since the gravity force is proportional to $R^3$, then if $R$ is reduced by a factor of 10, the gravity force is reduced 1000-fold. Below a certain droplet size (which also depends on the density difference between oil and water), the Brownian diffusion may exceed gravity and creaming or sedimentation is prevented. This is the principle of the formulation of nano-emulsions (with size range 50–200 nm), which may show very little or no creaming or sedimentation. The same applies for micro-emulsions (size range 5–50 nm).

#### 6.7.2.3 Use of 'Thickeners'

These are high molecular weight polymers, natural or synthetic, such as xanthan gum, hydroxyethylcellulose, alginates and carrageenans. To understand the role of these 'thickeners', let us consider the gravitational stresses exerted during creaming or sedimentation:

$$\text{stress} = \text{mass of drop} \times \text{acceleration due to gravity} = \frac{4}{3}\pi R^3 \Delta\rho g \tag{38}$$

To overcome such stress one needs a restoring force:

$$\text{restoring force} = \text{area of drop} \times \text{stress of drop} = 4\pi R^2 \sigma_\text{p} \tag{39}$$

Thus, the stress exerted by the droplet, $\sigma_\text{p}$, is given by

$$\sigma_\text{p} = \frac{\Delta\rho R g}{3} \tag{40}$$

Simple calculation shows that $\sigma_\text{p}$ is in the range $10^{-3}$–$10^{-1}$ Pa, which implies that for the prediction of creaming or sedimentation one needs to measure the viscosity at such low stresses. This can be achieved by using constant stress or creep measurements.

The above-described thickeners' satisfy the criteria for obtaining very high viscosities at low stresses or shear rates. This can be illustrated from plots of shear stress $\tau$ and viscosity $\eta$ versus shear rate (or shear stress), as shown in Figure 6.12. These systems are described as pseudoplastic' or shear thinning.

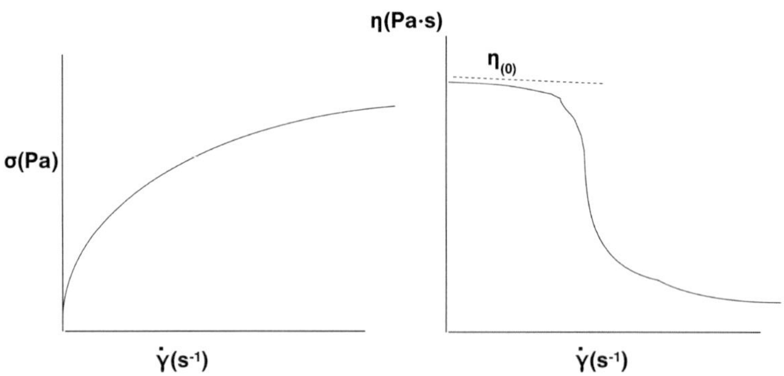


**Figure 6.12** Variation of (stress) $\sigma$ and viscosity $\eta$ with shear rate $\gamma$.

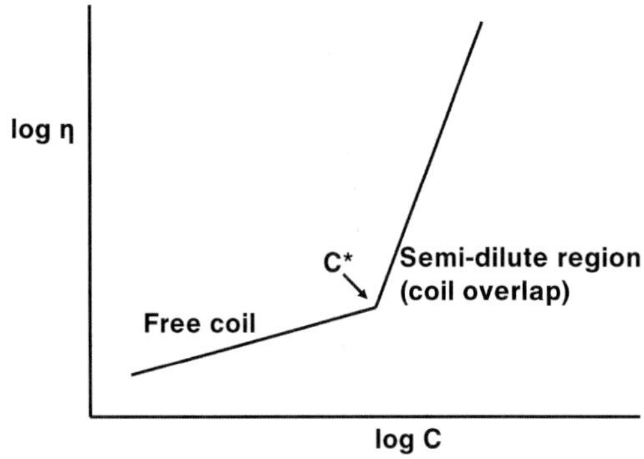


**Figure 6.13** Log $\eta$ versus log $C$ for polymer solutions.

The low shear (residual or zero shear rate) viscosity $\eta(0)$ can reach several thousand Pa s and such high values prevent creaming or sedimentation.

The above behavior is obtained above a critical polymer concentration ($C^*$) which can be located from plots of log $\eta$ versus log $C$ as is illustrated in Figure 6.13. Below $C^*$ the log $\eta$–log $C$ curve has a slope in the region of 1, whereas above $C^*$ the slope of the line exceeds 3.

#### 6.7.2.4 **Controlled Flocculation**

As described in Chapter 5, the total energy–distance of separation curve for electrostatically stabilized particles shows a shallow minimum (secondary minimum) at a relatively large separation between the droplets. By addition of small amounts of electrolyte, such a minimum can be made sufficiently deep for weak flocculation to occur. The same applies for sterically stabilized emulsions, which show only one minimum, the depth of which can be controlled by reducing the thickness of the adsorbed layer. This can be achieved by reducing the molecular weight of the stabilizer and/or addition of a non-solvent for the chains (e.g. electrolyte).

The above phenomenon of weak flocculation may be applied to reduce creaming or sedimentation, although in practice this is not easy since one has also to control the droplet size.

#### 6.7.2.5 **Depletion Flocculation**

This is obtained by addition of 'free' (non-adsorbing) polymer to the continuous phase [22, 23]. At a critical concentration or volume fraction of free polymer, $\phi_p^+$, weak flocculation occurs, since the free polymer coils become 'squeezed out' from between the droplets. This is illustrated in Figure 6.14, which shows the situation when the polymer volume fraction exceeds the critical concentration.

The osmotic pressure outside the droplets is higher than that between the droplets and this results in attraction whose magnitude depends on the concentration of the free polymer and its molecular weight, and also the droplet size and

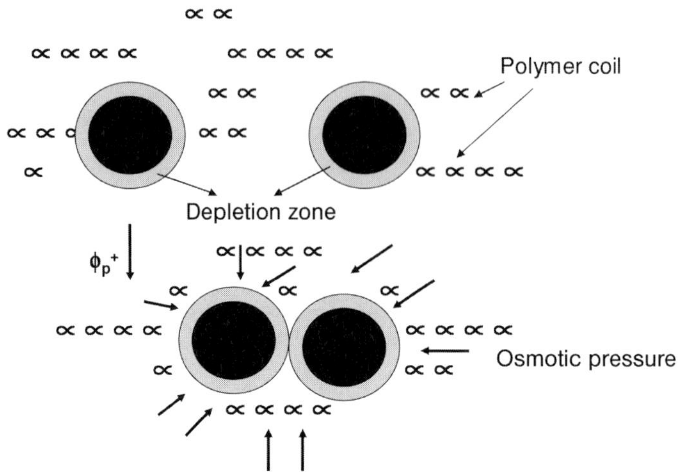


**Figure 6.14** Schematic representation of depletion flocculation.

$\phi$. The value of $\phi_p{}^+$ decreases with increase in the molecular weight of the free polymer. It also decreases as the volume fraction of the emulsion increases.

The above weak flocculation can be applied to reduce creaming or sedimentation, although it suffers from the drawback of temperature dependence: as the temperature increases, the hydrodynamic radius of the free polymer decreases (due to dehydration) and hence more polymer will be required to achieve the same effect at lower temperatures. Also, if the free polymer concentration is increased above a certain limit, phase separation may occur and the flocculated emulsion droplets may cream or sediment faster than in the absence of the free polymer.

## 6.8 Flocculation of Emulsions

Flocculation is the result of van der Waals attraction that is universal for all disperse systems. The van der Waals attraction $G_A$ was described in detail in Chapter 5. It was shown that $G_A$ is inversely proportional to the droplet–droplet separation distance $h$ and it depends on the effective Hamaker constant $A$ of the emulsion system. One way to overcome the van der Waals attraction is by electrostatic stabilization using ionic surfactants, which results in the formation of electrical double layers that introduce a repulsive energy that overcomes the attractive energy. Emulsions stabilized by electrostatic repulsion become flocculated at intermediate electrolyte concentrations. The second and most effective method of overcoming flocculation is by 'steric stabilization' using nonionic surfactants or polymers. Stability may be maintained in electrolyte solutions (as high as 1 mol dm$^{-3}$ depending on the nature of the electrolyte) and up to high temperatures (in excess of 50 °C), provided that the stabilizing chains (e.g. PEO) are still in better than $\theta$ conditions ($\chi < 0.5$).

### 6.8.1
### General Rules for Reducing (Eliminating) Flocculation

A summary of the criteria that are required to reduce (eliminate) flocculation is given below.

#### 6.8.1.1 Charge-stabilized Emulsions, e.g. Using Ionic Surfactants
The most important criterion is to make $G_{max}$ as high as possible; this is achieved by three main conditions: high surface or zeta potential, low electrolyte concentration and low valency of ions.

#### 6.8.1.2 Sterically Stabilized Emulsions

Four main criteria are necessary in this case:

1. Complete coverage of the droplets by the stabilizing chains.
2. Firm attachment (strong anchoring) of the chains to the droplets. This requires the chains to be insoluble in the medium and soluble in the oil. However, this is incompatible with stabilization, which requires a chain that is soluble in the medium and strongly solvated by its molecules. These conflicting requirements are solved by the use of A–B or A–B–A block or $BA_n$ graft copolymers [B is the anchor' chain and A is the stabilizing chain(s)]:

```
B~~~~~~~~~~~~~A    A~~~~~B~~~~A    ~~~~~B~~~~~~
                                   | | | | | | |
                                   A A A A A A A
```

   Examples of the B chains for O/W emulsions are polystyrene, poly(methyl methacrylate), poly(propylene oxide) and alkyl poly(propylene oxide). For the A chain(s), poly(ethylene oxide) (PEO) and poly(vinyl alcohol) are good examples.
3. Thick adsorbed layers: the adsorbed layer thickness should be in the region of 5–10 nm. This means that the molecular weight of the stabilizing chains could be in the region of 1000–5000.
4. The stabilizing chain should be maintained in good solvent conditions ($\chi < 0.5$) under all conditions of temperature changes on storage.

6.9
Ostwald Ripening

The driving force for Ostwald ripening is the difference in solubility between the small and large droplets (the smaller droplets have higher Laplace pressure and

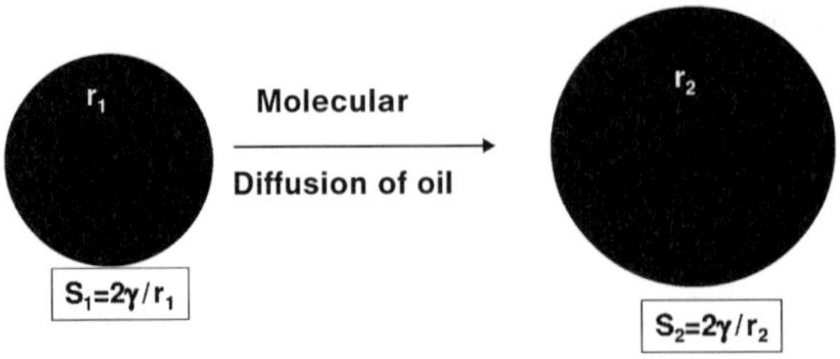

Figure 6.15 Schematic representation of Ostwald ripening.

higher solubility than the larger droplets). This is illustrated Figure 6.15, where r_1 decreases and r_2 increases as a result of diffusion of molecules from the smaller to the larger droplets.

The difference in chemical potential between different sized droplets was given by Lord Kelvin [24]:

$$S(r) = S(\infty)\exp\left(\frac{2\gamma V_m}{rRT}\right) \tag{41}$$

where $S(r)$ is the solubility surrounding a particle of radius r, $S(\infty)$ is the bulk solubility, V_m is the molar volume of the dispersed phase, R is the gas constant and T is the absolute temperature. The quantity $2\gamma V_m/RT$ is termed the characteristic length. It has an order of ~1 nm or less, indicating that the difference in solubility of a 1 μm droplet is of the order of 0.1% or less. Theoretically, Ostwald ripening should lead to condensation of all droplets into a single drop [24]. This does not occur in practice since the rate of growth decreases with increase inelectrostatically stabilized particles droplet size.

For two droplets with radii r_1 and r_2 ($r_1 < r_2$):

$$\frac{RT}{V_m}\ln\left[\frac{S(r_1)}{S(r_2)}\right] = 2\gamma\left(\frac{1}{r_1} - \frac{1}{r_2}\right) \tag{42}$$

This equation shows that the larger the difference between r_1 and r_2, the higher is the rate of Ostwald ripening.

Ostwald ripening can be quantitatively assessed from plots of the cube of the radius versus time t [25–27]:

$$r^3 = \frac{8}{9}\left[\frac{S(\infty)\gamma V_m D}{\rho RT}\right]t \tag{43}$$

where D is the diffusion coefficient of the disperse phase in the continuous phase.

Several methods may be applied to reduce Ostwald ripening:

1. Addition of a second disperse phase component which is insoluble in the continuous medium (e.g. squalane) [25]. In this case, partitioning between different droplet sizes occurs, with the component having low solubility expected to be concentrated in the smaller droplets. During Ostwald ripening in a two-component system, equilibrium is established when the difference in chemical potential between different sized droplets (which results from curvature effects) is balanced by the difference in chemical potential resulting from partitioning of the two components. This effect reduces further growth of droplets.

2. Modification of the interfacial film at the O/W interface. According to equation 43, a reduction in γ results in reduction in the Ostwald ripening rate. By using surfactants that are strongly adsorbed at the O/W interface (e.g. polymeric surfactants) and which do not desorb during ripening (by choosing a molecule that is insoluble in the continuous phase), the rate could be significantly reduced [28]. An increase in the surface dilational modulus ε ($= d\gamma/d \ln A$) and a decrease in γ would be observed for the shrinking drop, and this tends to reduce further growth.

A–B–A block copolymers such as PHS–PEO–PHS (which is soluble in the oil droplets but insoluble in water) can be used to achieve the above effect. This polymeric emulsifier enhances the Gibbs elasticity and causes a decrease in γ to very low values.

6.10 Coalescence of Emulsions

When two emulsion droplets come in close contact in a floc or creamed layer or during Brownian diffusion, thinning and disruption of the liquid film may occur, resulting in eventual rupture. On close approach of the droplets, film thickness fluctuations may occur. Alternatively, the liquid surfaces undergo some fluctuations forming surface waves, as illustrated in Figure 6.16.

The surface waves may grow in amplitude and the apices may join as a result of the strong van der Waals attraction (at the apex, the film thickness is the smallest). The same applies if the film thins to a small value (critical thickness for coalescence).

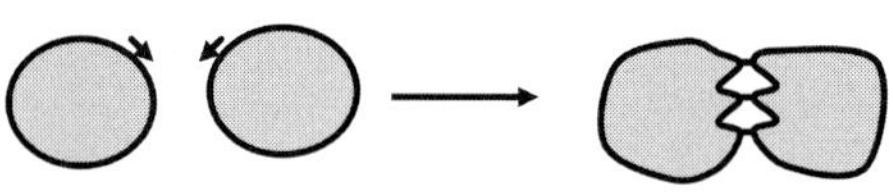

Figure 6.16 Schematic representation of surface fluctuations.

A very useful concept was introduced by Deryaguin and Scherbaker [29], who suggested that a 'disjoining pressure' $\pi(h)$ is produced in the film which balances the excess normal pressure:

$$\pi(h) = P(h) - P_0 \tag{44}$$

where $P(h)$ is the pressure of a film with thickness h and P_0 is the pressure of a sufficiently thick film such that the net interaction free energy is zero.

$\pi(h)$ may be equated to the net force (or energy) per unit area acting across the film:

$$\pi(h) = -\frac{dG_T}{dh} \tag{45}$$

where G_T is the total interaction energy in the film.

$\pi(h)$ is made up of three contributions, due to electrostatic repulsion (π_E), steric repulsion (π_S) and van der Waals attraction (π_A):

$$\pi(h) = \pi_E + \pi_S + \pi_A \tag{46}$$

To produce a stable film, $\pi_E + \pi_S > \pi_A$, and this is the driving force for the prevention of coalescence, which can be achieved by two mechanisms and their combination: (1) increased repulsion, both electrostatic and steric, and (2) damping of the fluctuation by enhancing the Gibbs elasticity. In general, smaller droplets are less susceptible to surface fluctuations and hence coalescence is reduced. This explains the high stability of nano-emulsions.

Several methods may be applied to achieve the above effects:

1. *Use of mixed surfactant films:* in many cases, using mixed surfactants, say anionic and nonionic or long-chain alcohols, can reduce coalescence as a result of several effects: high Gibbs elasticity, high surface viscosity and hindered diffusion of surfactant molecules from the film.

2. *Formation of lamellar liquid crystalline phases at the O/W interface:* this mechanism was suggested by Friberg *et al.* [30], who suggested that surfactant or mixed surfactant film can produce several bilayers that 'wrap' the droplets. As a result of these multilayer structures, the potential drop is shifted to longer distances, thus reducing the van der Waals attraction. A schematic representation of the role of liquid crystals is shown in Figure 6.17, which illustrates the difference between having a monomolecular layer and a multilayer as is the case with liquid crystals. For coalescence to occur, these multilayers have to be removed 'two-by-two', and this forms an energy barrier preventing coalescence.

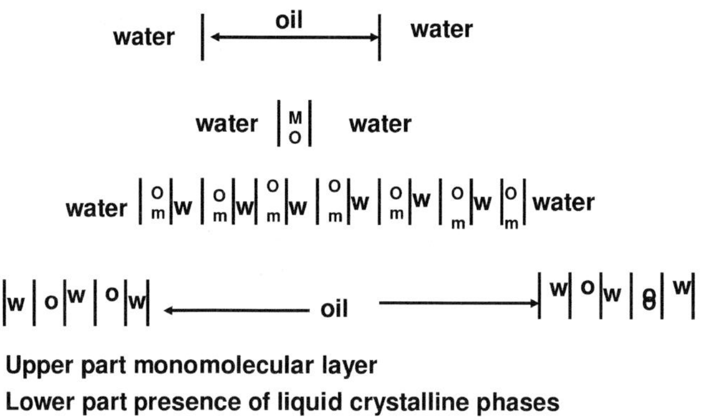

Figure 6.17 Schematic Representation of the role of liquid crystalline phases.

6.10.1
Rate of Coalescence

Since film drainage and rupture are a kinetic process, coalescence is also a kinetic process. If one measures the number of particles n (flocculated or not) at time t:

$$n = n_t + n_v m \tag{47}$$

where n_t is the number of primary particles remaining and n_v is the number of aggregates consisting of m separate particles.

For studying emulsion coalescence, one should consider the rate constant of flocculation and coalescence. If coalescence is the dominant factor, then the rate K follows first-order kinetics:

$$n - \frac{n_0}{Kt}[1 + \exp(-Kt)] \tag{48}$$

which shows that a plot of log n versus t should give a straight line from which K can be calculated.

6.11
Phase Inversion

Phase inversion of emulsions can be one of two types: transitional inversion induced by changing factors that affect the HLB of the system, e.g. temperature and/or electrolyte concentration, and catastrophic inversion, which is induced by increasing the volume fraction of the disperse phase.

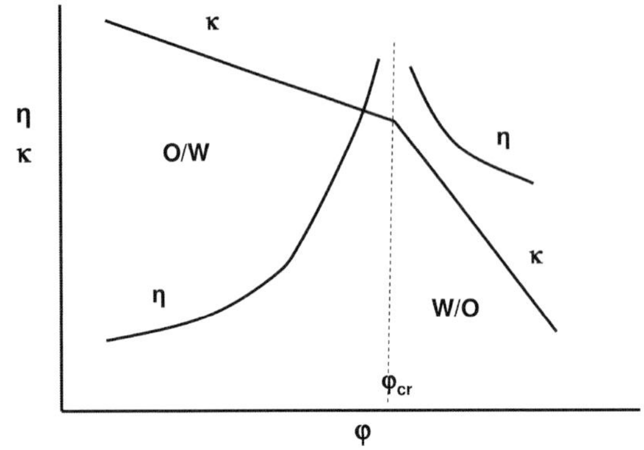

Figure 6.18 Variation of viscosity (η) and conductivity (κ) with oil volume fraction.

Catastrophic inversion is illustrated in Figure 6.18, which shows the variation of viscosity and conductivity with the oil volume fraction ϕ. As can be seen, inversion occurs at a critical ϕ, which may be identified with the maximum packing fraction. At ϕ_{cr}, η suddenly decreases; the inverted W/O emulsion has a much lower volume fraction; κ also decreases sharply at the inversion point since the continuous phase is now oil, which has very low conductivity.

Earlier theories of phase inversion were based on packing parameters. When ϕ exceeds the maximum packing ($\sim$0.64 for random packing and $\sim$0.74 for hexagonal packing of monodisperse spheres; for polydisperse systems, the maximum packing exceeds 0.74), inversion occurs. However, these theories are not adequate, since many emulsions invert at ϕ values well below the maximum packing as a result of the change in surfactant characteristics with variation of conditions. For example, when using a nonionic surfactant based on PEO, the latter chain changes its solvation with increase in temperature and/or addition of electrolyte. Many emulsions show phase inversion at a critical temperature (the phase inversion temperature) that depends on the HLB number of the surfactant and also the presence of electrolytes. With increase in temperature and/or addition of electrolyte, the PEO chains become dehydrated and finally they become more soluble in the oil phase. Under these conditions, the O/W emulsion will invert to a W/O emulsion. This dehydration effect amounts to a decrease in the HLB number and, when the latter reaches a value that is more suitable for a W/O emulsion, inversion will occur. At present, there is no quantitative theory that accounts for phase inversion of emulsions.

6.12
Experimental Methods for Assessment of Emulsion Stability

As mentioned before, understanding the emulsion breakdown processes at a molecular level is far from being achieved at present. It is therefore necessary to develop methods of assessment of each process and attempt to predict the long-term physical stability of emulsions.

6.12.1 Assessment of Creaming or Sedimentation

Several methods may be applied to assess the creaming or sedimentation of emulsions: (1) measurement of the rate by direct observation of emulsion separation using graduated cylinders that are placed at constant temperature; this method allows one to obtain both the rate and the equilibrium cream or sediment volume; (2) turbidity measurements as a function of height at various times, using for example the Turboscan (that measures turbidity from the backscattering of near-IR light); and (3) measurement of ultrasonic velocity and absorption at various heights in the cream or sedimentation tubes.

Centrifugation may be applied to accelerate the rate of creaming or sedimentation, and this method is sometimes used for the prediction of emulsion stability. The assumption is made that by increasing the g force the rate of sedimentation or creaming is significantly increased, and this could be applied to predict the process from measurement at short time periods.

In a centrifuge, the gravity force is given by

$$g = \omega^2 x \tag{49}$$

where x is the mean distance of the centrifuge tube from the axis of rotation and ω is the angular velocity ($\omega = 2\pi\nu$, where ν is the number of revolutions per second). Note that if the centrifuge tube is not small compared with x, then the applied centrifugal field cannot be considered to be uniform over the length of the tube.

Modern analytical ultracentrifuges allows one to follow the separation of emulsions in a quantitative manner. With typical O/W emulsions, three layers are generally observed: a clear aqueous phase, an opaque phase consisting of distorted polyhedral oil droplets. and a clear separated oil phase, resulting from coalescence of the polyhedra. The degree of emulsion stability may be taken as the volume of the opaque phase remaining after time t. Alternatively, one may use the volume of oil separated at infinite time as an index of stability.

A simple expression may be used to treat the data in a quantitative manner:

$$\frac{t}{V} = \frac{1}{bV_\infty} + \frac{t}{V_\infty} \tag{50}$$

where V is the volume of oil separated at time t, V_∞ is the extrapolated volume at infinite time and b is a constant. A plot of t/V versus t should give a straight line from which b and V_∞ can be calculated. These two parameters may be taken as indices of emulsion stability.

A more rigorous procedure to study emulsion stability using the ultracentrifuge is to observe the system at various speeds of rotation. At relatively low centrifuge speeds, one may observe the expected opaque cream layer. At sufficiently high centrifuge speeds, one may observe a coalesced oil layer and a cream layer which are separated by an extra layer of deformed oil droplets. This deformed layer looks like a 'foam', i.e. it consists of oil droplets separated by thin aqueous films.

For certain emulsions, one may find that by increasing the centrifuge speed, the 'foam'/cream layer boundary does not move. Under conditions where there is an equilibrium between the 'foam'/cream layer, one may conclude that there is no barrier to be overcome in forming the foam layer from the cream layer. This implies that in the foam layer, the aqueous film separating two oil droplets thins to a 'black' film under the action of van der Waals forces. The boundary between the foam layer and the coalesced layer is associated with a force (or pressure) barrier. One may observe the minimum centrifuge speed that is necessary to produce a visible amount of coalesced oil after say 30 min of centrifugation. This centrifuge speed may be used to calculate the 'critical pressure' that needs to be applied to induce coalescence.

6.12.2
Assessment of Emulsion Flocculation

For dilute emulsions (which may be obtained by carefully diluting the concentrate in the supernatant liquid), the rate of flocculation can be determined by measuring turbidity, τ, as a function of time:

$$\tau = A n_0 V_1^2 (1 + n_0 k t) \tag{51}$$

where A is an optical constant, n_0 is the number of droplets at time $t = 0$, V_1 is the volume of the droplets and k is the rate constant of flocculation. Thus, a plot of τ versus t gives a straight line, in the initial time of flocculation, and k can be calculated from the slope of the line. Flocculation of emulsions can also be assessed by direct droplet counting using optical microscopy (with image analysis), using a Coulter Counter and light diffraction techniques (e.g. using the Mastersizer, Malvern Instruments, Malvern, UK).

The flocculation of emulsion concentrates can be followed using rheological methods. In the absence of any Ostwald ripening and/or coalescence, flocculation of the emulsion concentrates is accompanied by an increase in the viscosity, yield value or elastic modulus. These rheological parameters can easily be measured using rotational viscometers. Clearly, if Ostwald ripening and/or coalescence occurs at the same time as emulsion flocculation, the viscosity, yield value or elastic modulus will show a complex dependence of these parameters on time and this makes the analysis of the rheological results very difficult.

6.12.3
Assessment of Ostwald Ripening

As mentioned above, the best procedure to follow Ostwald ripening is to plot r^3 versus time, following equation 43. This gives a straight line from which the rate of Ostwald ripening can be calculated. In this way, one can assess the effect of the various additives that may reduce Ostwald ripening, e.g. addition of highly insoluble oil and/or an oil-soluble polymeric surfactant.

6.12.4
Assessment of Coalescence

The rate of coalescence is measured by following the droplet number n or average droplet size d (diameter) as a function of time. Plots of the logarithm of droplet number or average diameter versus time give straight lines (at least in the initial stages of coalescence) from which the rate of coalescence K can be estimated using equation 48. In this way, one can compare the different stabilizers, e.g. mixed surfactant films, liquid crystalline phase and macromolecular surfactants.

6.12.5
Assessment of Phase Inversion

The most common procedure to assess phase inversion is to measure the conductivity or the viscosity of the emulsion as a function of ϕ, increase in temperature and/or addition of electrolyte. For example, for an O/W emulsion, phase inversion to a W/O emulsion is accompanied by a rapid decrease in conductivity and viscosity.

References

1 Th.F. Tadros and B. Vincent, in *Encyclopedia of Emulsion Technology*, ed. P. Becher, Marcel Dekker, New York (1983).

2 P. Walstra and P.E.A. Smolders, in *Modern Aspects of Emulsions*, ed. B.P. Binks, Royal Society of Chemistry, Cambridge (1998).

3 P. Walstra and P.E.A. Smolders, in *Modern Aspects of Emulsions*, ed. B.P. Binks, Royal Society of Chemistry, Cambridge (1998).

4 H.A. Stone, *Annu. Rev. Fluid Mech.*, *226*, 95 (1994).

5 J.A. Wierenga, F. van Dieren, J.J.M. Janssen and W.G.M. Agterof, *Trans. Inst. Chem. Eng.*, *74A*, 554 (1996).

6 V.G. Levich, *Physicochemical Hydrodynamics*, Prentice-Hall, Englewood Cliffs, NJ (1962).

7 J.T. Davies, *Turbulent Phenomena*, Academic Press, London (1972).

8 E.H. Lucasses-Reynders, in *Encyclopedia of Emulsion Technology*, ed. P. Becher, Marcel Dekker, New York (1996).

9 E.H. Lucasses-Reynders, *Colloids Surf. A*, *91*, 79 (1994).

10 J. Lucassen, in *Anionic Surfactants*, ed. E.H. Lucassen-Reynders, Marcel Dekker, New York (1981).
11 M. van den Tempel, in *Proceedings of International Congress on Surface Activity*, Vol. 2, p. 573 (1960).
12 W.C. Griffin, *J. Cosmet. Chem.*, *1*, 311 (1949); *5*, 249 (1954).
13 P. Becher, in *Nonionic Surfactants*, ed. M.J. Schick, Surfactant Science Series, Vol. 1, Marcel Dekker, New York (1987).
14 J.T. Davies, in *Proceedings of International Congress on Surface Activity*, Vol. 1, p. 426 (1959).
15 K. Shinoda, *J. Colloid Interface Sci.*, *25*, 396 (1967).
16 K. Shinoda and H. Saito, *J. Colloid Interface Sci.*, *30*, 258 (1969).
17 A. Beerbower and M.W. Hill, *Am. Cosmet. Perfum.*, *87*, 85 (1972).
18 J.H. Hildebrand, *Solubility of Non-Electrolytes*, 2nd edn, Reinhold, New York (1936).
19 C.M. Hansen, *J. Paint Technol.*, *39*, 505 (1967).
20 A.F.M. Barton, *Handbook of Solubility Parameters and Other Cohesive Parameters*, CRC Press, Boca Raton, FL (1983).
21 J.N. Israelachvili, J.N. Mitchell and B.W. Ninham, *J. Chem. Soc., Faraday Trans. 2*, *72*, 1525 (1976).
22 S. Asakura and F. Oosawa, *J. Polym. Sci.*, *33*, 245 (1958).
23 G.J. Fleer, J.H.M.H. Scheutjens and B. Vincent, *ACS Symp. Ser.*, *240*, 245 (1984).
24 W. Thompson (Lord Kelvin), *Philos. Mag.*, *42*, 448 (1871).
25 A.S. Kabalanov and E.D. Shchukin, *Adv. Colloid Interface Sci.*, *38*, 69 (1992); A.S. Kabalanov, *Langmuir*, *10*, 680 (1994).
26 I.M. Lifshitz and V.V. Slesov, *Sov. Phys. JETP*, *35*, 331 (1959).
27 C. Wagner, *Z. Electrochem.*, *35*, 581 (1961).
28 P. Walstra, in *Encyclopedia of Emulsion Technology*, Vol. 4, ed. P. Becher, Marcel Dekker, New York (1996).
29 B.V. Deryaguin and R.L. Scherbaker, *Kolloid Zh.*, *23*, 33 (1961).
30 S. Friberg, P.O. Jansson and E. Cederberg, *J. Colloid Interface Sci.*, *55*, 614 (1976).

7
Suspension Concentrates (SCs)

7.1
Introduction

The formulation of agrochemicals as dispersions of solids in aqueous solution (to be referred to as suspension concentrates or SCs) has attracted considerable attention in recent years. Such formulations are natural replacements for wettable powders (WPs). The latter are produced by mixing the active ingredient with a filler (usually a clay material) and a surfactant (dispersing and wetting agent). These powders are dispersed into the spray tank to produce a coarse suspension which is applied to the crop. Although wettable powders are simple to formulate, they are not the most convenient for the farmer. Apart from being dusty (and occupying a large volume due to their low bulk density), they tend to settle fast in the spray tank and they do not provide optimum biological efficiency as a result of the large particle size of the system. In addition, one cannot incorporate the necessary adjuvants (mostly surfactants) in the formulation. These problems are overcome by formulating the agrochemical as an aqueous SC.

Several advantages may be quoted for SCs. First, one can control the particle size by controlling the milling conditions and proper choice of the dispersing agent. Second, it is possible to incorporate high concentrations of surfactants in the formulation, which is sometimes essential for enhancing wetting, spreading and penetration (see Chapter 13). Stickers may also be added to enhance adhesion and in some cases to provide slow release.

In recent years there has been considerable research into the factors that govern the stability of suspension concentrates [1–4]. The theories of colloid stability could be applied to predict the physical states of these systems on storage. In addition, analysis of the problem of sedimentation of SCs at a fundamental level has been undertaken [3]. Since the density of the particles is usually larger than that of the medium (water), SCs tend to separate as a result of sedimentation. The sedimented particles tend to form a compact layer at the bottom of the container (sometimes referred to as clay or cake), which is very difficult to redisperse. It is essential, therefore, to reduce sedimentation and formation of clays by incorporation of rheology modifiers, referred to as suspending or antisettling agents.

Colloids and Interface Science Series, Vol. 5
Colloids in Agrochemicals. Tharwat F. Tadros

ISBN: 978-3-527-31465-2

This chapter addresses the above phenomena at a fundamental level. It starts with a section on the preparation of suspension concentrates and the role of surfactants (dispersing agents). The problem of powder wetting is analyzed by considering the interfacial phenomena involved. This is followed by a section on the control of the physical stability of suspensions. A distinction is made between colloid stability, which requires the presence of net repulsion between the particles, and physical stability, which implies the absence of separation on storage and spontaneous dispersion on dilution in the spray tank on application. The problem of Ostwald ripening (crystal growth) is also briefly described and particular attention is paid to the role of surfactants. The subsequent section deals with the problem of sedimentation and prevention of claying. The various methods that may be applied to reduce sedimentation and prevention of the formation of hard clays are summarized. The last section is concerned with the methods that may be applied for the assessment of the physical stability of SCs. For the assessment of flocculation and crystal growth, particle size analysis techniques are commonly applied. The bulk properties of the suspension, such as sedimentation and separation, and redispersion on dilution may be assessed using rheological techniques. The latter is summarized with particular emphasis on their application in the prediction of the long-term physical stability of suspension concentrates.

7.2 Preparation of Suspension Concentrates and the Role of Surfactants/Dispersing Agents

Suspension concentrates are usually formulated using a wet milling process, which requires the addition of a surfactant/dispersing agent. The latter should satisfy the following criteria: (1) a good wetting agent for the agrochemical powder (both external and internal surfaces of the powder aggregates or agglomerates must be spontaneously wetted); (2) a good dispersing agent to break such aggregates or agglomerates into smaller units and subsequently help in the milling process (one usually aims at a dispersion with a volume mean diameter of 1–2 μm); and (3) it should provide good stability in the colloid sense (this is essential for maintaining the particles as individual units once formed). Powerful dispersing agents are particularly important for the preparation of highly concentrated suspensions (sometimes required for seed dressing). Any flocculation will cause a rapid increase in the viscosity of the suspension and this makes the wet milling of the agrochemical a difficult process.

7.3 Wetting of the Agrochemical Powder

Dry powders of organic compounds usually consist of particles of various degrees of complexity, depending on the isolation stages and the drying process. Gener-

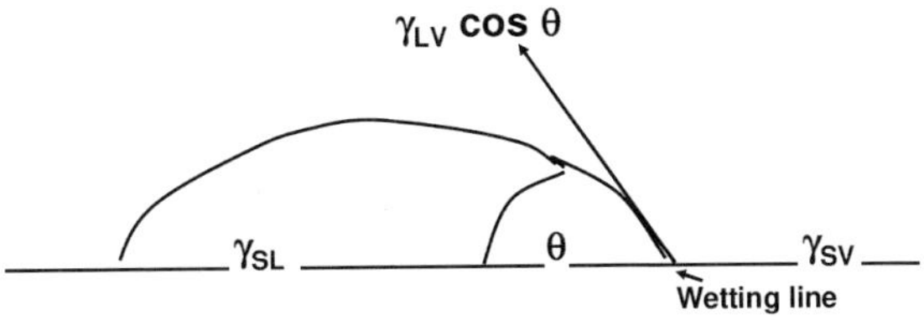

Figure 7.1 Schematic representation of the contact angle and wetting line.

ally, the particles in a dry powder form aggregates (in which the particles are joined together with their crystal faces) or agglomerates (in which the particles touch at edges or corners) forming a looser, more open structure. It is essential in the dispersion process to wet both the external and the internal surfaces and displace the air trapped between the particles. This is usually achieved by the use of surface-active agents of the ionic or nonionic type. In some cases, macromolecules or polyelectrolytes may be efficient in this wetting process. This may be the case since these polymers contain a very wide distribution of molecular weights and the low molecular weight fractions may act as efficient wetting agents. For efficient wetting, the molecules should lower the surface tension of water (see below) and they should diffuse fast in solution and become quickly adsorbed at the solid/solution interface.

Wetting of a solid is usually described in terms of the equilibrium contact angle θ and the appropriate interfacial tensions. This is illustrated in Figure 7.1.

Using the classical Young's equation:

$$\gamma_{SV} - \gamma_{SL} = \gamma_{LV} \cos \theta \tag{1}$$

or

$$\cos \theta = \frac{\gamma_{SV} - \gamma_{SL}}{\gamma_{LV}} \tag{2}$$

where γ represents the interfacial tension and the symbols S, L and V refer to the solid, liquid and vapor, respectively. It is clear from equation 2 that if $\theta < 90°$, a reduction in γ_{LV} improves wetting. Hence the use of surfactants which reduce both γ_{LV} and γ_{SL} to aid wetting is clear. However, the process of wetting of particulate solids is more complex and it involves at least three distinct types of wetting [5], namely adhesional wetting, spreading wetting and immersional wetting. All these processes are determined by the liquid surface tension and the contact angle. The difference between γ_{SV} and γ_{SL} or $\gamma_{LV} \cos \theta$ is referred to as the adhesion or wetting tension.

Let us consider an agrochemical powder with surface area A. Before the powder is dispersed in the liquid it has a surface tension γ_{SV} and after immersion γ_{SL}. The work of dispersion W_d is simply given by the difference in adhesion or wetting tension of the SL and SV:

$$W_d = A(\gamma_{SL} - \gamma_{SV}) = -A\gamma_{LV}\cos\theta \tag{3}$$

It is clear that if $\theta < 90^\circ$ $\cos\theta$ is positive and W_d is negative, i.e. wetting of the powder is spontaneous. Since surfactants are added in sufficient amounts ($\gamma_{dynamic}$ is lowered sufficiently), spontaneous dispersion is the rule rather than the exception.

Wetting of the internal surface requires penetration of the liquid into channels between and inside the agglomerates. The process is similar to forcing a liquid through fine capillaries. To force a liquid through a capillary with radius r, a pressure p is required that is given by

$$p = -\frac{2\gamma_{LV}\cos\theta}{r} = \left[\frac{-2(\gamma_{SV} - \gamma_{SL})}{r\gamma_{LV}}\right] \tag{4}$$

γ_{SL} has to be made as small as possible; rapid surfactant adsorption on the solid surface, low θ. When $\theta = 0$, $p \propto \gamma_{LV}$. Hence for penetration into pores one requires a high γ_{LV}. Thus, wetting of the external surface requires a low contact angle θ and low surface tension γ_{LV}. Wetting of the internal surface (i.e. penetration through pores) requires low θ but high γ_{LV}. These two conditions are incompatible and a compromise has to be made: $\gamma_{SV} - \gamma_{SL}$ must be kept at a maximum; γ_{LV} should be kept as low as possible but not too low.

The above conclusions illustrate the problem of choosing the best dispersing agent for a particular powder. This requires measurement of the above parameters and also testing of the efficiency of the dispersion process.

The next stage to be considered is the wetting of the internal surface, which implies penetration of the liquid into channels between and inside the agglomerates. This is more difficult to define precisely. However, one may make use of the equation derived for capillary phenomena as discussed by Rideal [6] and Washburn [7] who considered the penetration of liquids in capillaries.

For horizontal capillaries (gravity neglected), the depth of penetration l in time t is given by the Rideal–Washburn equation [6, 7]:

$$l = \left(\frac{rt\gamma_{LV}\cos\theta}{2\eta}\right)^{\frac{1}{2}} \tag{5}$$

To enhance the rate of penetration, γ_{LV} has to be made as high as possible, θ as low as possible and η as low as possible.

For dispersion of powders into liquids, one should use surfactants that lower θ while not reducing γ_{LV} too much. The viscosity of the liquid should also be kept at a minimum. Thickening agents (such as polymers) should not be added during the dispersion process. It is also necessary to avoid foam formation during the dispersion process.

For a packed bed of particles, r may be replaced by k, which contains the effective radius of the bed and a tortuosity factor, which takes into account the complex path formed by the channels between the particles, i.e.

$$l^2 = \frac{kt\gamma_{LV} \cos \theta}{2\eta} \tag{6}$$

Thus a plot of l^2 versus t gives a straight line and from the slope of the line one can obtain θ.

The Rideal–Washburn equation can be applied to obtain the contact angle of liquids (and surfactant solutions) in powder beds; k should first be obtained using a liquid that produces zero contact angle and this is discussed below.

7.3.1 Measurement of Contact Angles of Liquids and Surfactant Solutions on Powders

A packed bed of powder is prepared say in a tube fitted with sintered glass at the end (to retain the powder particles). It is essential to pack the powder uniformly in the tube (a plugger may be used in this case). The tube containing the bed is immersed in a liquid that gives spontaneous wetting (e.g. a lower alkane), i.e. the liquid gives a zero contact angle and $\cos \theta = 1$. By measuring the rate of penetration of the liquid (this can be carried out gravimetrically using, for example, a microbalance or a Kruss instrument), one can obtain k. The tube is then removed from the lower alkane liquid and left to stand for evaporation of the liquid. It is then immersed in the liquid in question and the rate of penetration is measured again as a function of time. Using equation 6, one can calculate $\cos \theta$ and hence θ.

Hence, in summary, the dispersion of a powder in a liquid depends on three main factors, the energy of wetting of the external surface, the pressure involved in the liquid penetrating inside and between the agglomerates and the rate of penetration of the liquid into the powder. All these factors are related to two main parameters, γ_{LV} and θ. In general, the process is likely to be more spontaneous the lower is θ and the higher is γ_{LV}. Since these two factors tend to operate in opposite senses, the choice of a suitable surfactant (dispersing agent) can be a difficult task.

7.3.2 Dispersion and Milling

For the dispersion of aggregates and agglomerates into smaller units, one requires high-speed mixing, e.g. a Silverson mixer. In some cases the dispersion process is easy and the capillary pressure may be sufficient to break up the aggregates and agglomerates into primary units. The process is aided by the surfactant, which becomes adsorbed on the particle surface. However, one should be careful during the mixing process not to entrap air (foam), which causes an increase in the viscosity of the suspension and prevents easy dispersion and subsequent grinding. If foam formation becomes a problem, one should add antifoaming agents such as a polysiloxane.

After completion of the dispersion process, the suspension is transferred to a ball or bead mill for size reduction. Milling or comminution (the generic term for size reduction) is a complex process and there is little fundamental information on its mechanism. For the breakdown of single crystals into smaller units, mechanical energy is required. This energy in a bead mill, for example, is supplied by impaction of the glass beads with the particles. As a result, permanent deformation of the crystals and crack initiation result. This will eventually lead to the fracture of the crystals into smaller units. However, since the milling conditions are random, it is inevitable that some particles receive impacts that are far in excess of those required for fracture, whereas others receive impacts that are insufficient to fracture them. This makes the milling operation grossly inefficient and only a small fraction of the applied energy is actually used in comminution. The rest of the energy is dissipated as heat, vibration, sound, interparticulate friction, friction between the particles and beads and elastic deformation of unfractured particles. For these reasons, milling conditions are usually established by a trial and error procedure. Of particular importance is the effect of various surface-active agents and macromolecules on the grinding efficiency. The role played by these agents in the comminution process is far from being understood, although Rehbinder and co-workers [8–10] have given this problem particular consideration. As a result of adsorption of surfactants at the solid/liquid interface, the surface energy at the boundary is reduced and this facilitates the process of deformation or destruction. The adsorption of the surfactant at the solid/solution interface in cracks facilitates their propagation. This is usually referred to as the 'Rehbinder effect' [10]. The surface energy manifests itself in destructive processes on solids, since the generation and growth of cracks and separation of one part of a body from another are directly connected with the development of a new free surface. Therefore, as a result of adsorption of surface-active agents at structural defects in the surface of the crystals, fine grinding is facilitated. In the extreme case where there is a very great reduction in surface energy at the sold/liquid boundary, spontaneous dispersion may take place, resulting in the formation of colloidal particles ($<1\ \mu m$). Schukin and Rehbinder [10] have developed a theory for such spontaneous dispersion.

7.3.3 Effect of Surfactant Adsorption

Surfactants lower the surface tension of water, γ, and they adsorb at the solid/liquid interface. A plot of γ_{LV} versus log C (where C is the surfactant concentration) results in a gradual reduction in γ_{LV} followed by a linear decrease in γ_{LV} with log C [just below the critical micelle concentration (cmc)] and when the cmc is reached γ_{LV} remains virtually constant. This was discussed in detail in Chapter 4.

From the slope of the linear portion of the γ–log C curve (just below the cmc), one can obtain the surface excess (number of moles of surfactant per unit area at the L/A interface). Using the Gibbs adsorption isotherm:

$$\frac{d\gamma}{d \log C} = -2.303RT\Gamma \tag{7}$$

where Γ is the surface excess (mol m^{-2}), R is the gas constant and T is the absolute temperature. From Γ, one can obtain the area per molecule:

$$\text{area per molecule} = \frac{1}{\Gamma N_{av}}(m^2) = \frac{10^{18}}{\Gamma N_{av}}(nm^2) \tag{8}$$

Most surfactants produce a vertically oriented monolayer just below the cmc. The area per molecule is usually determined by the cross-sectional area of the head group. For ionic surfactants containing say an $-OSO_3^-$ or $-SO_3^-$ head group, the area per molecule is in the region of 0.4 nm^2. For nonionic surfactants containing several moles of ethylene oxide [11–13], the area per molecule can be much larger (1–2 nm^2). Surfactants will also adsorb at the solid/liquid interface. For hydrophobic surfaces, the main driving force for adsorption is by hydrophobic bonding. This results in lowering of the contact angle of water on the solid surface. For hydrophilic surfaces, adsorption occurs via the hydrophilic group, e.g. cationic surfactants on silica. Initially the surface becomes more hydrophobic and the contact angle θ increases with increase in surfactant concentration. However, at higher cationic surfactant concentration, a bilayer is formed by hydrophobic interaction between the alkyl groups and the surface becomes more and more hydrophilic and eventually the contact angle reaches zero at high surfactant concentrations.

Smolders [11] suggested the following relationship for change of θ with C:

$$\frac{d\gamma_{LV} \cos\theta}{d \ln C} = \frac{d\gamma_{SV}}{d \ln C} - \frac{d\gamma_{SL}}{d \ln C} \tag{9}$$

Using the Gibbs equation:

$$\sin\theta\left(\frac{d\gamma}{d \ln C}\right) = RT(\Gamma_{SV} - \Gamma_{SL} - \gamma_{LV}\cos\theta) \tag{10}$$

Since $\gamma_{LV} \sin\theta$ is always positive, then $d\theta/d \ln C$ will always have the same sign as the right-hand side of equation 10. Three cases may be distinguished: $d\theta/d \ln C < 0$, $\Gamma_{SV} < \Gamma_{SL} + \Gamma_{LV}\cos\theta$, addition of surfactant improves wetting; $d\theta/d \ln C = 0$, $\Gamma_{SV} = \Gamma_{SL} + \Gamma_{LV}\cos\theta$, surfactant has no effect on wetting; $d\theta/d \ln C > 0$, $\Gamma_{SV} > \Gamma_{SL} + \Gamma_{LV}\cos\theta$, surfactant causes dewetting.

7.4
Control of the Physical Stability of Suspension Concentrates

When considering the stability of suspension concentrates, one must distinguish between the colloid stability and the overall physical stability. Colloid stability implies the absence of aggregation between the particles, which requires the presence of an energy barrier that is produced by electrostatic, steric repulsion or combination of the two (electrosteric). Physical stability implies the absence of a sedimentation and/or separation, ease of dispersion on shaking and/or dilution in the spray tanks. As discussed before, to achieve the overall physical stability one may apply control and reversible flocculation methods and/or use a rheology modifier.

To distinguish between colloid stability/instability and physical stability, one must consider the state of the suspension on standing, as illustrated schematically in Figure 7.2.

These states are determined by (1) the magnitude and balance of the various interaction forces, electrostatic repulsion, steric repulsion and van der Waals

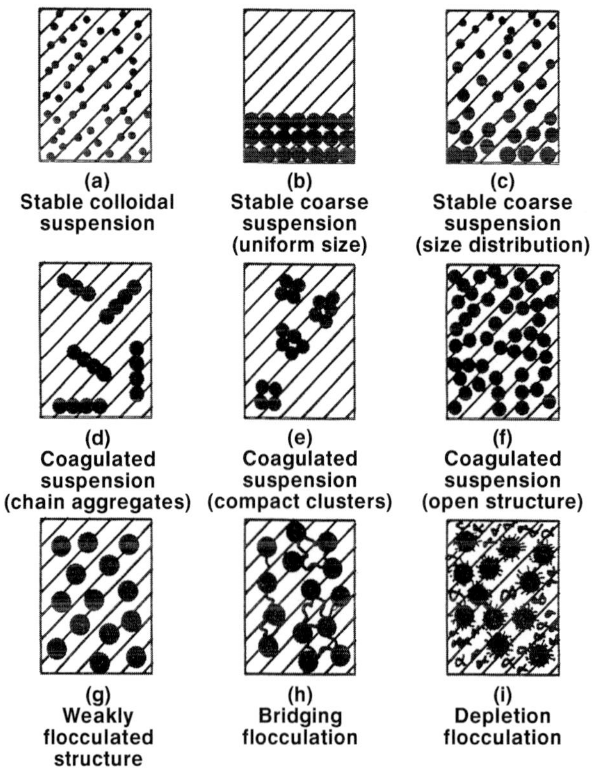

Figure 7.2 States of the suspension.

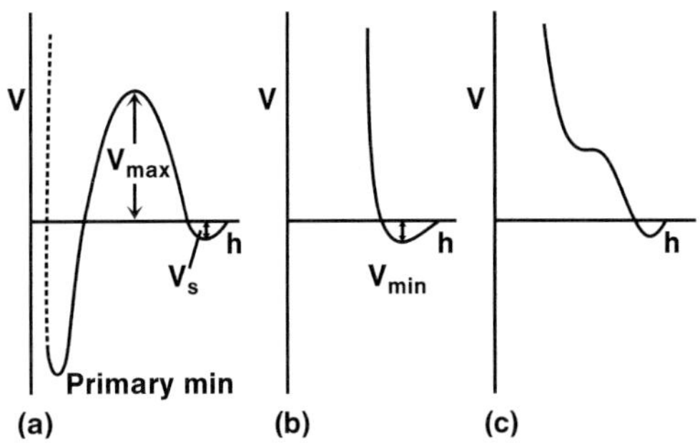

Figure 7.3 Energy–distance curves for the three stabilization mechanisms.

attraction; (2) particle size and shape distribution; (3) the density difference between the disperse phase and medium, which determines the sedimentation characteristics; (4) conditions and prehistory of the suspension, e.g. agitation, which determines the structure of the flocs formed (chain aggregates, compact clusters, etc.); and (5) the presence of additives, e.g. high molecular weight polymers that may cause bridging or depletion flocculation.

These states may be described in terms of three different energy–distance curves: (a) electrostatic, produced for example by the presence of ionogenic groups on the surface of the particles, or adsorption of ionic surfactants; (b) steric, produced for example by adsorption of nonionic surfactants or polymers; and (c) electrostatic + steric (electrosteric), produced for example by polyelectrolytes. These are illustrated in Figure 7.3.

A brief description of the various states shown in Figure 7.2 is as follows. States (a) to (c) correspond to a suspension that is stable in the colloid sense. The stability is obtained as a result of net repulsion due to the presence of extended double layers (i.e. at low electrolyte concentration), the result of steric repulsion produced adsorption of nonionic surfactants or polymers, or the result of combination of double layer and steric repulsion (electrosteric). State (a) represents the case of a suspension with small particle size (submicron), whereby the Brownian diffusion overcomes the gravity force producing uniform distribution of the particles in the suspension, i.e.

$$kT > \frac{4}{3}\pi R^3 \Delta\rho g h \tag{11}$$

where k is the Boltzmann constant, T is the absolute temperature, R is the particle radius, $\Delta\rho$ is the buoyancy (difference in density between the particles and the medium), g is the acceleration due to gravity and h is the height of the container.

A good example of the above case is a nano-suspension with particle size well below 1 μm that is stabilized by an ionic surfactant or nonionic surfactant or polymer. This suspension will show no separation on storage for long periods of time.

States (b) and (c) represent the case of suspensions, where the particle size range is outside the colloid range (>1 μm). In this case, the gravity force exceeds the Brownian diffusion.

With state (b), the particles are uniform and they will settle under gravity, forming a hard sediment (technically referred to 'clay' or 'cake'). The repulsive forces between the particles allow them to move past each other until they reach small distances of separation (which are determined by the location of the repulsive barrier). Due to the small distances between the particles in the sediment, it is very difficult to redisperse the suspension by simple shaking.

With case (c), consisting of a wide distribution of particle sizes, the sediment may contain larger proportions of the larger size particles, but still a hard 'clay' is produced. These 'clays' are dilatant (i.e. shear thickening) and they can be easily detected by inserting a glass rod in the suspension. Penetration of the glass rod into these hard sediments is very difficult.

States (d) to (f) represent the case for coagulated suspensions which have either a small repulsive energy barrier or its complete absence. State (d) represents the case of coagulation under conditions of no stirring, in which case chain aggregates are produced that will settle under gravity, forming a relatively open structure. State (e) represents the case of coagulation under stirring conditions whereby compact aggregates are produced that will settle faster than the chain aggregates and the sediment produced is more compact. State (f) represents the case of coagulation at high volume fraction of the particles, ϕ. In this case, the whole particles will form a 'one-floc' structure that is formed from chains and cross chains that extend from one wall to the other in the container. Such a coagulated structure may undergo some compression (consolidation) under gravity, leaving a clear supernatant liquid layer at the top of the container. This phenomenon is referred to as syneresis. State (g) represents the case of weak and reversible flocculation. This occurs when the secondary minimum in the energy–distance curve (Figure 7.3a) is deep enough to cause flocculation. This can occur at moderate electrolyte concentrations, in particular with larger particles. The same occurs with sterically and electrosterically stabilized suspensions (Figure 7.3b and c). This occurs when the adsorbed layer thickness is not very large, particularly with large particles. The minimum depth required for causing weak flocculation depends on the volume fraction of the suspension. The higher the volume fraction, the lower is the minimum depth required for weak flocculation.

The above flocculation is weak and reversible, i.e. on shaking the container redispersion of the suspension occurs. On standing, the dispersed particles aggregate to form a weak 'gel'. This process (referred to as sol–gel transformation) leads to reversible time dependence of viscosity (thixotropy). On shearing the suspension, the viscosity decreases and when the shear is removed, the viscosity is recovered. This phenomenon is applied in paints. On application of the paint (by a brush or roller), the gel is fluidized, allowing uniform coating of the paint. When shearing is stopped, the paint film recovers its viscosity and this avoids any dripping.

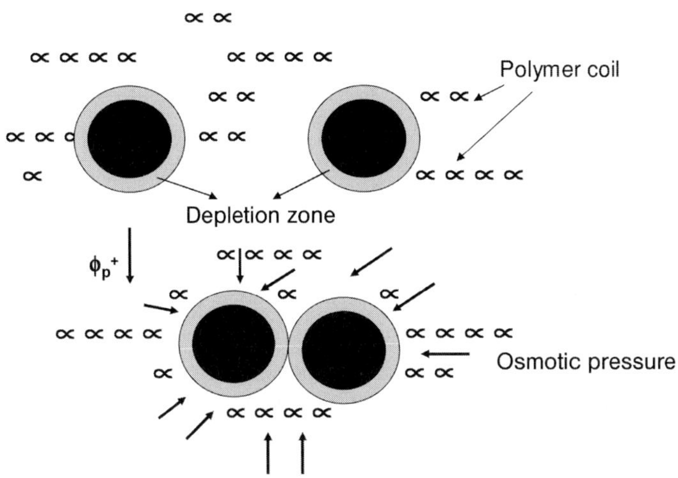

Figure 7.4 Schematic representation of depletion flocculation.

State (h) represents the case where the particles are not completely covered by the polymer chains. In this case, simultaneous adsorption of one polymer chain on more than one particle occurs, leading to bridging flocculation. If the polymer adsorption is weak (low adsorption energy per polymer segment), the flocculation could be weak and reversible. In contrast, if the adsorption of the polymer is strong, tough flocs are produced and the flocculation is irreversible. The last phenomenon is used for solid/liquid separation, e.g. in water and effluent treatment.

Case (i) represents a phenomenon, referred to as depletion flocculation, produced by addition of 'free' non-adsorbing polymer [12]. In this case, the polymer coils cannot approach the particles to a distance Δ (that is determined by the radius of gyration of free polymer, R_G), since the reduction of entropy on close approach of the polymer coils is not compensated by adsorption energy. The suspension particles will be surrounded by a depletion zone with thickness Δ. Above a critical volume fraction of the free polymer, ϕ_p^+, the polymer coils are 'squeezed out' from between the particles and the depletion zones begin to interact. The interstices between the particles are now free from polymer coils and hence an osmotic pressure is exerted outside the particle surface (the osmotic pressure outside is higher than in between the particles), resulting in weak flocculation (11). A schematic representation of depletion flocculation is shown in Figure 7.4.

The magnitude of the depletion attraction free energy, G_{dep}, is proportional to the osmotic pressure of the polymer solution, which in turn is determined by ϕ_p and molecular weight M. The range of depletion attraction is proportional to the thickness of the depletion zone, Δ, which is roughly equal to the radius of gyration, R_G, of the free polymer. A simple expression for G_{dep} is (12)

$$G_{dep} = \frac{2\pi R\Delta^2}{V_1}(\mu_1 - \mu_1{}^0)\left(1 + \frac{2\Delta}{R}\right) \tag{12}$$

where V_1 is the molar volume of the solvent, μ_1 is the chemical potential of the solvent in the presence of free polymer with volume fraction ϕ_p and $\mu_1{}^0$ is the chemical potential of the solvent in the absence of free polymer; $(\mu_1 - \mu_1{}^0)$ is proportional to the osmotic pressure of the polymer solution.

7.4.1 Stability Against Aggregation

The control of stability against irreversible flocculation (where the particles are held together into aggregates that cannot be redispersed by shaking or on dilution) is achieved by the use of powerful dispersing agents, e.g. surfactants of the ionic or nonionic type, nonionic polymers or polyelectrolytes. These dispersing agents must be strongly adsorbed on the particle surfaces and fully cover them. With ionic surfactants, irreversible flocculation is prevented by the repulsive force generated from the presence of an electrical double layer at the particle/solution interface (see Chapter 5). Depending on the conditions, this repulsive force can be made sufficiently large to overcome the ubiquitous van der Waals attraction between the particles, at intermediate distances of separation. With nonionic surfactants and macromolecules, repulsion between the particles is ensured by the steric interaction of the adsorbed layers on the particle surfaces (see Chapter 5). With polyelectrolytes, both electrostatic and steric repulsion exist. Below, a summary of the role of surfactants in the stabilization of particles against flocculation is presented.

Ionic surfactants such as sodium dodecylbenzene sulfonate (NaDBS) and cetyltrimethylammonium chloride (CTACl) adsorb on hydrophobic particles of agrochemicals, as a result of the hydrophobic interaction between the alkyl group of the surfactant and the particle surface. As a result, the particle surface will acquire a charge that is compensated by counterions (Na^+ in the case of NaDBS and Cl^- in the case of CTACl), forming an electrical double layer.

The adsorption of ionic surfactants at the solid/solution interface is of vital importance in determining the stability of suspension concentrates. As discussed in Chapter 4, the adsorption of ionic surfactants on solid surfaces can be directly measured by equilibrating a known amount of solid (with known surface area) with surfactant solutions of various concentrations. After reaching equilibrium, the solid particles are removed (for example, by centrifugation) and the concentration of surfactant in the supernatant liquid is determined analytically. From the difference between the initial and final surfactant concentrations (C_1 and C_2, respectively), the number of moles of surfactant adsorbed, Γ, per unit area of solid is determined and the results may be fitted to a Langmuir isotherm:

$$\Gamma = \frac{\Delta C}{mA} = \frac{abC_2}{1 + bC_2} \tag{13}$$

where $\Delta C = C_1 - C_2$, m is the mass of the solid with surface area A, a is the saturation adsorption and b is a constant that is related to the free energy of adsorption, ΔG [$b \propto \exp(\Delta G/RT)$]. From a, the area per surfactant ion on the surface can be calculated (area per surfactant ion = $1/aN_{av}$).

Results on the adsorption of ionic surfactants on pesticides are scarce. However, Tadros [4] obtained some results on the adsorption of NaDBS and CTABr on a fungicide, namely ethirimol. For NaDBS, the shape of the isotherm was of a Langmuir type, giving an area per DBS^- at saturation of ~0.14 nm^2. The adsorption of CTA^+ showed a two-step isotherm with areas per CTA^+ of 0.27 and 0.14 nm^2, respectively. These results suggest full saturation of the surface with surfactant ions which are vertically oriented.

The above discussion shows that ionic surfactants can be used to stabilize agrochemical suspensions by producing sufficient electrostatic repulsion. When two particles with adsorbed surfactant layers approach each other to a distance where the electrical double layers begin to overlap, strong repulsion occurs, preventing any particle aggregation (see Chapter 5). The energy–distance curve for such electrostatically stabilized dispersions is shown schematically in Figure 7.3a. This shows an energy maximum, which if high enough ($>25kT$) prevents particle aggregation into the primary minimum. However, ionic surfactants are the least attractive dispersing agents for the following reasons. Adsorption of ionic surfactants is seldom strong enough to prevent some desorption, with the result of production of 'bare' patches which may induce particle aggregation. The system is also sensitive to ionic impurities which are present in the water used for suspension preparation. In particular, the system will be sensitive to divalent ions (Ca^{2+} or Mg^{2+}) which produce flocculation at relatively low concentrations.

Nonionic surfactants of the ethoxylate type, e.g. $R(CH_2CH_2O)_nOH$ and $RC_6H_5(CH_2CH_2O)_2OH$, provide a better alternative provided that the molecule contains sufficient hydrophobic groups to ensure their adsorption and enough ethylene oxide units to provide an adequate energy barrier. As discussed in Chapter 5, the origin of steric repulsion arises from two main effects. The first effect arises from the unfavorable mixing of the poly(ethylene oxide) chains which are in good solvent conditions (water as the medium). This effect is referred to as the mixing or osmotic repulsion. The second effect arises from the decrease in configurational entropy of the chains when these are forced to overlap on approach of the particles. This is referred to as the elastic or volume restriction effect. The energy–distance curve for such systems (Figure 7.3b) clearly demonstrates the attraction of steric stabilization. Apart from a small attractive energy minimum (which can be reasonably shallow with sufficiently long poly(ethylene oxide) chains), strong repulsion occurs and there is no barrier to overcome. A better option is to use block and graft copolymers (polymeric surfactants) consisting of A and B units combined together in A–B, A–B–A or BA_n fashion. B represents units with high affinity for the particle surface and basically insoluble in the continuous medium, thus providing strong adsorption ('anchoring units'). A, on the other hand, represents units with high affinity to the medium (strong chain–solvent interaction) and little or no affinity to the particle surface. An example of such a powerful dispersant is a graft copolymer of poly(methyl methacrylate)–methacrylic acid (the anchoring portion) and methoxy poly(ethylene oxide) (the stabilizing chain) methacrylate [4]. Adsorption measurements of such a polymer on a pesticide, namely ethirimol (a fungicide), showed a high affinity isotherm with no desorption. Using such a macromolecular surfactant, a suspension of high volume fractions could be prepared.

The third class of dispersing agents which is commonly used in SC formulations is that of polyelectrolytes. Of these, sulfonated naphthalene–formaldehyde condensates and lignosulfonates are the most commonly used in agrochemical formulations. These systems show a combined electrostatic and steric repulsion and the energy–distance curve is illustrated schematically in Figure 7.3c. It shows a shallow minimum and maximum at intermediate distances (characteristic of electrostatic repulsion) and also strong repulsion at relatively short distances (characteristic of steric repulsion). The stabilization mechanism of polyelectrolytes is sometimes referred to as electrosteric. These polyelectrolytes offer some versatility in SC formulations. Since the interaction is fairly long range in nature (due to the double layer effect), one does not obtain the 'hard sphere' type of behavior which may lead to the formation of hard sediments. The steric repulsion ensures the colloid stability and prevention of any aggregation on storage.

7.4.2
Ostwald Ripening (Crystal Growth)

There are several ways in which crystals can grow in an aqueous suspension. One of the most familiar is the phenomenon of 'Ostwald ripening', which occurs as a result of the difference in solubility between the small and large crystals:

$$\frac{RT}{M}\ln\left(\frac{S_1}{S_2}\right) = \frac{2\sigma}{\rho}\left(\frac{1}{r_1} - \frac{1}{r_2}\right) \tag{14}$$

where S_1 and S_2 are the solubilities of crystals of radii r_1 and r_2, respectively, σ is the specific surface energy, ρ is the density and M is the molecular weight of the solute molecules, R is the gas constant and T is the absolute temperature. Since r_1 is smaller than r_2 then S_1 is larger than S_2. Another mechanism for crystal growth is related to polymorphic changes in solutions, and again the driving force is the difference in solubility between the two polymorphs. In other words, the less soluble form grows at the expense of the more soluble phase. This is sometimes also accompanied by changes in the crystal habit. Different faces of the crystal may have different surface energies and deposition may take place preferentially on one of the crystal faces, modifying its shape. Other important factors are the presence of crystal dislocations, kinks, surface impurities, etc. Most of these effects have been discussed in detail in monographs on crystal growth [12, 13, 15, 16].

The growth of crystals in suspension concentrates may create undesirable changes. As a result of the drastic change in particle size distribution, the settling of the particles may be accelerated, leading to caking and cementing together of some particles in the sediment. Moreover, an increase in particle size may lead to a reduction in biological efficiency. Hence prevention of crystal growth or at least reducing it to an acceptable level is essential in most suspension concentrates. Surfactants affect crystal growth in a number of ways. The surfactant may affect the rate of dissolution by affecting the rate of transport away from the boundary layer at the crystal/solution interface. On the other hand, if the surfactant forms micelles that can solubilize the solute, crystal growth may be enhanced as a result of the increasing concentration gradient. Thus, by proper choice of dispersing agent, one may reduce the crystal growth of suspension concentrates. This has been demonstrated by Tadros [15] for terbacil suspensions. When using Pluronic P75 [poly(ethylene oxide)–poly(propylene oxide) block copolymer], crystal growth was significant. On replacing the Pluronic surfactant with poly(vinyl alcohol), the rate of crystal growth was greatly reduced and the suspension concentrate was acceptable.

It should be mentioned that many surfactants and polymers may act as crystal growth inhibitors if they adsorb strongly on the crystal faces, thus preventing solute deposition. However, the choice of an inhibitor is still an art and there are not many rules that can be used for the selection of crystal growth inhibitors.

7.4.3 Stability Against Claying or Caking

Once a dispersion that is stable in the colloid sense has been prepared, the next task is to eliminate claying or caking. This is a consequence of settling of the colloidally stable suspension particles. The repulsive forces necessary to ensure this colloid stability allows the particles to move past each other, forming a dense sediment which is very difficult to redisperse. Such sediments are dilatant (shear thickening; see Section 7.6.2) and hence the SC becomes unusable. Before describing the methods used for controlling settling and prevention of formation of dilatant clays, an account is given of the settling of suspensions and the effect of increasing the volume fraction of the suspension on the settling rate.

7.4.4 Settling of Suspensions

The sedimentation velocity v_0 of a very dilute suspension of rigid non-interacting particles with radius a can be determined by equating the gravitational force with the opposing hydrodynamic force as given by Stokes' law, i.e.

$$\frac{4}{3}\pi a^3(\rho - \rho_0)g = 6\pi\eta_0 a v_0 \tag{15}$$

$$v_0 = \frac{2a^2(\rho - \rho_0)}{9\eta_0} \tag{16}$$

where ρ is the density of the particles and ρ_0 that of the medium, η_0 is the viscosity of the medium and g is the acceleration due to gravity. Equation 16 predicts a sedimentation rate for particles with radius 1 μm in a medium with a density difference of $0.2\,\mathrm{g\,cm^{-3}}$ and a viscosity of 1 mPa s (i.e. water at 20 °C) of $4.4 \times 10^{-7}\,\mathrm{m\,s^{-1}}$. Such particles will sediment to the bottom of a 0.1 m high container in about 60 h. For 10 μm particles, the sedimentation velocity is $4.4 \times 10^{-5}\,\mathrm{m\,s^{-1}}$ and such particles will sediment to the bottom of a 0.1 m high container in about 40 min.

The above treatment using Stokes' law applied only to very dilute suspensions (volume fraction $\phi < 0.01$). For more concentrated suspensions, the particles no longer sediment independently of each other and one has to take into account both the hydrodynamic interaction between the particles (which applies for moderately concentrated suspensions) and other higher order interactions at relatively high volume fractions. A theoretical relationship between the sedimentation velocity v of non-flocculated suspensions and particle volume fraction has been derived by Maude and Whitmore [15] and by Batchelor [17]. Such theories apply to relatively low volume fractions ($\phi < 0.1$) and they show that the sedimentation velocity v at a volume fraction ϕ is related to that at infinite dilution v_0 (the Stokes' velocity) by an equation of the form

$$v = v_0(1 - k\phi) \tag{17}$$

where k is a constant in the range 5–6. Batchelor [17] derived a rigorous theory for sedimentation in a relatively dilute dispersion of spheres. He considered that the reduction in Stokes' velocity with increase in particle number concentration arises from hydrodynamic interactions. The value of k in equation 17 was calculated and found to be equal to 6.55. This theory applies up to a volume fraction of 0.1. At higher volume fractions, the sedimentation velocity becomes a complex function of ϕ and only empirical equations are available to describe the variation of v with ϕ. For example, Reed and Anderson [18] developed a virial expansion technique to describe the settling rate of concentrated suspensions. They derived the following expression for the average velocity, v_{av}:

$$v_{av} = v_0 \frac{1 - 1.83\phi}{1 + 4.70\phi} \tag{18}$$

Good agreement between experimental settling rates and those calculated using equation 18 was obtained up to $\phi = 0.4$.

More recently, Buscall *et al.* [19] measured the rate of settling of polystyrene latex particles with $a = 1.55\,\mu\mathrm{m}$ in $10^{-3}\,\mathrm{mol\,dm^{-3}}$ KCl up to $\phi = 0.5$. The results are shown in Figure 7.5. It can be seen that v/v_0 decreases exponentially with increase in ϕ, approaching zero at $\phi > 0.5$, i.e. in the region of close packing. An empirical equation for the relative settling rate has been derived using the Dougherty–Krieger equation for the relative viscosity, η_r $(=\eta/\eta_0)$:

$$\eta_r = \left(1 - \frac{\phi}{\phi_p}\right)^{-[\eta]\phi_p} \tag{19}$$

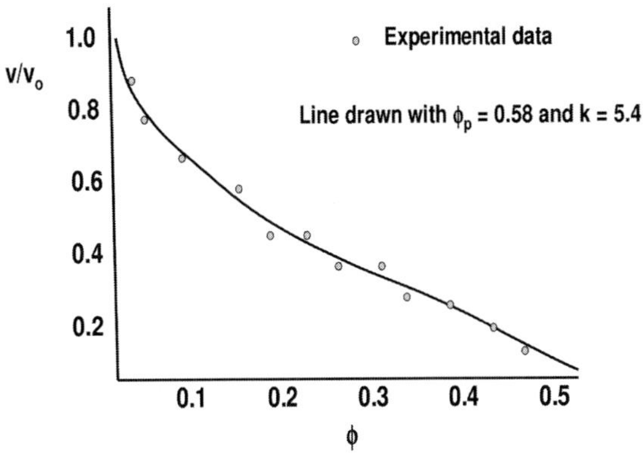

Figure 7.5 Variation of sedimentation rate with volume fraction for polystyrene dispersions.

where $[\eta]$ is the intrinsic viscosity (which is equal to 2.5 for hard spheres) and ϕ_p is the maximum packing fraction (which is close to 0.6). Assuming that $v/v_0 = \alpha(\eta_0/\eta)$, it is easy to derive the following empirical relationship for the relative sedimentation velocity, v_r $(=v/v_0)$:

$$v_r = \left(1 - \frac{\phi}{\phi_p}\right)^{\alpha[\eta]\phi_p} = \left(1 - \frac{\phi}{\phi_p}\right)^{k\phi_p} \qquad (20)$$

By allowing the latex to settle completely and then determining the volume concentration of the packed bed, a value of 0.58 was obtained for ϕ_p (close to random packing). Using this value and $k = 5.4$, the relative rate of sedimentation was calculated and this gave the full line shown in Figure 7.5, whereas the circles represent the experimental points. Hence agreement between the calculations using the empirical equation and the experimental results is reasonable.

It seems from the above discussion that there is a correlation between the reduction in sedimentation rate and the increase in relative viscosity of the suspension as the volume fraction of the suspension is increased. This is shown schematically in Figure 7.6, which shows that $v \to 0$ and $\eta_r \to \infty$ as $\phi \to \phi_p$. This implies that suspension concentrates with volume fractions approaching the maximum packing do not show any appreciable settling.

However, such dense suspensions have extremely high viscosities and are not a practical solution for reduction of settling. In most cases, one prepares a suspension concentrate at practical volume fractions (0.2–0.4) and then uses an antisettling agent to reduce settling. As we will discuss in the next section, most of the antisettling agents used in practice are high molecular weight polymers. These materials show an increase in the viscosity of the medium with increase in their concentration. However, at a critical polymer concentration (which depends on the nature of the polymer and its molecular weight) they show a very rapid increase in viscosity with further increase in their concentration. This critical concentration (sometimes denoted by C^*) represents the situation where the

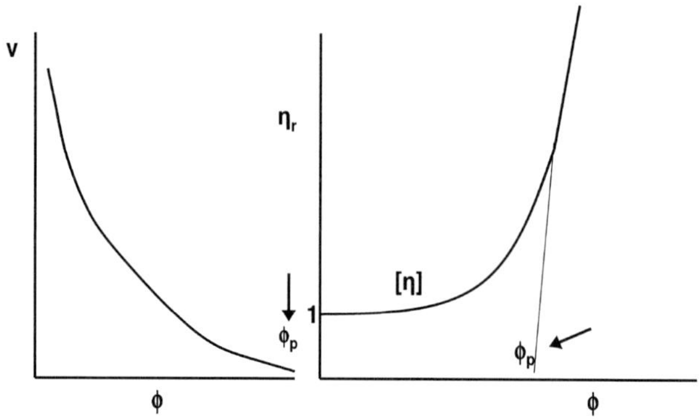

Figure 7.6 Variation of ν and η_r with ϕ.

polymer coils or rods begin to overlap. Under these conditions, the solutions become significantly non-Newtonian (viscoelastic; see Section 7.6.2) and they produce stresses that are sufficient to overcome the stress exerted by the particles. The settling of suspensions in these non-Newtonian fluids is not simple since one has to consider the non-Newtonian behavior of these polymer solutions. This problem has been addressed by Buscall *et al.* [19]. In order to describe adequately the settling of particles in non-Newtonian fluids, one needs to know how the viscosity of the medium changes with shear rate or shear stress. Most of these viscoelastic fluids show a gradual increase in viscosity with decrease in shear rate or shear stress, but below a critical stress or shear rate they show a Newtonian region with a limiting high viscosity that is denoted the residual (or zero shear) viscosity. This is illustrated in Figure 7.7, which shows the variation of the viscos-

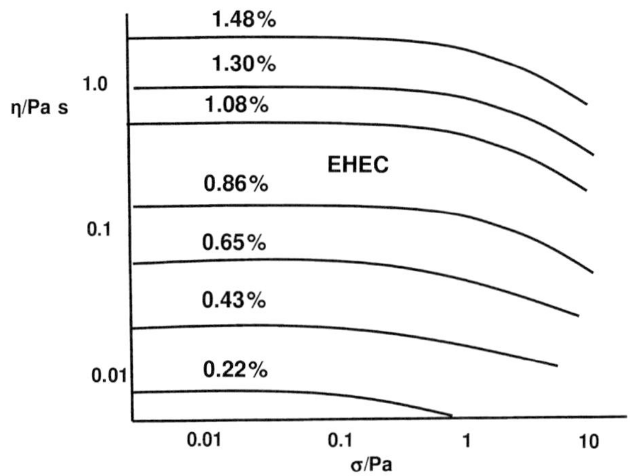

Figure 7.7 Constant stress (creep) measurements for polystyrene latex dispersions as a function of EHEC concentration.

ity with shear stress for a number of solutions of ethylhydroxyethylcellulose (EHEC) at various concentrations. It can be seen that the viscosity increases with decrease in stress and the limiting value, i.e. the residual viscosity $\eta(0)$, increases rapidly with increase in polymer concentration. The shear thinning behavior of these polymer solutions is clearly shown, since above a critical stress value the viscosity decreases rapidly with increase in shear stress. The limiting value of the viscosity is reached at low stresses (<0.2 Pa).

It is now important to calculate the stress exerted by the particles. This stress (see below) is equal to $a\Delta\rho g/3$. For polystyrene latex particles with radius 1.55 μm and density 1.05 g cm^{-3}, this stress is equal to 1.6×10^{-4} Pa. Such stress is lower than the critical stress for most EHEC solutions. In this case, one would expect a correlation between the settling velocity and the zero shear viscosity. This is illustrated in Figure 7.8, where v/R^2 is plotted versus $\eta(0)$. As is clear, a linear relationship between $\log(v/R^2)$ and $\log \eta(0)$ is obtained, with a slope of -1, over three decades of viscosity. This indicates that the settling rate is proportional to $[\eta(0)]^{-1}$. Thus, the settling rate of isolated spheres in non-Newtonian (pseudoplastic) polymer solutions is determined by the zero shear viscosity in which the particles are suspended. As we will see in Section 7.6.2, determination of the zero shear viscosity is not straightforward and requires the use of constant stress rheometers.

The above correlation applies to the simple case of relatively dilute suspensions. For more concentrated suspensions, other parameters should be taken into consideration, such as the bulk (elastic) modulus. It is also clear that the stress exerted by the particles depends not only on the particle size but also on the density difference between the particle and the medium. Many suspension concentrates have particles with radii up to 10 μm and density differences of more than 1 g cm^{-3}. However, the stress exerted by such particles will seldom exceed 10^{-2} Pa and most polymer solutions will reach their limiting viscosity value at higher stresses than this value. Therefore, in most cases the correlation between settling velocity and zero shear viscosity is justified, at least for relatively dilute systems. For more concentrated suspensions, an elastic network is produced in the system, which encompasses both the suspension particles and the polymer

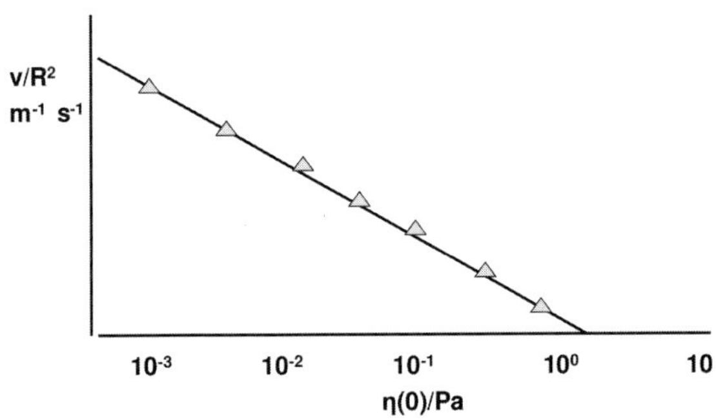

Figure 7.8 Sedimentation rate versus $\eta(0)$.

chains. In this case, settling of individual particles may be prevented. However, in this event the elastic network may collapse under its own weight and some liquid is squeezed out from between the particles.

This is manifested in a clear liquid layer at the top of the suspension, a phenomenon usually referred to as syneresis. If such separation is not significant, it may not cause any problem on application since on shaking the container the whole system redisperses. However, significant separation is not acceptable since it becomes difficult to homogenize the system. In addition, such extensive separation is cosmetically unacceptable and the formulation rheology should be controlled to reduce such separation to a minimum.

7.4.5 Prevention of Settling and Claying

Several methods are applied in practice to control the settling and prevent the formation of dilatant clays; these methods are summarized below.

7.4.5.1 Balance of the Density of Disperse Phase and Medium

This is obviously the simplest method for retarding settling, since it is clear from equation 16 that if $\rho = \rho_0$, then $v_0 = 0$. However, this method is of limited application and can only be applied to systems where the difference in density between the particle and the medium is not too large. For example, with many organic solids with densities between 1.1 and 1.3 g cm^{-3} suspended in water, some soluble substances such as sugar or electrolytes may be added to the continuous phase to increase the density of the medium to a level that is equal to that of the particles. However, one should be careful that the added substance does not cause any flocculation for the particles. This is particularly the case when using electrolytes, where one should avoid any 'salting-out' materials, which cause the medium to be a poor solvent for the stabilizing chains. It should also be mentioned that density matching can only be achieved at one temperature. Liquids usually have larger thermal expansion coefficients than solids and if, say, the density is matched at room temperature, settling may occur at higher temperatures. Therefore, one has to be careful when applying the density matching method, particularly if the formulation is subjected to large temperature changes.

7.4.5.2 Use of High Molecular Weight Polymers ('Thickeners')

High molecular weight materials such as natural gums, hydroxyethylcellulose and synthetic polymers such as poly(ethylene oxide) may be used to reduce the settling of suspension concentrates. The most commonly used material in agrochemical formulations is xanthan gum (produced by converting waste sugar into a high molecular weight material using a microorganism and sold under the trade names Kelzan and Rhodopol), which is effective at relatively low concentrations (of the order of 0.1–0.2% depending on the formulation). As mentioned above, these high molecular weight materials produce viscoelastic solutions above a critical concentration. This viscoelasticity produces sufficient residual viscosity to stop the settling of individual particles. The solutions also give enough elasticity to overcome separation of the suspension. However, one cannot rule out the interaction of these polymers with the suspension particles, which may result in 'bridging', hence the mechanism by which such molecules reduce settling and prevent the formation of clays may be complex. To arrive at the optimum concentration and molecular weight of the polymer necessary for prevention of settling and claying, one should study the rheological characteristics of the formulation as a function of the variables of the system such as the volume fraction, concentration and molecular weight of the polymer and temperature.

7.4.5.3 Use of 'Inert' Fine Particles

It has long been known that fine inorganic materials such as swellable clays and finely divided oxides (silica or alumina), when added to the dispersion medium of coarser suspensions, can eliminate claying or caking. These fine inorganic materials form a 'three-dimensional' network in the continuous medium which, by virtue of its elasticity, prevents sedimentation and claying. With swellable clays such as sodium montmorillonite, the gel arises from the interaction of the plate-like particles in the medium. The plate-like particles of sodium montmorillonite consist of an octahedral alumina sheet sandwiched between two tetrahedral silica sheets [21]. In the tetrahedral sheets, tetravalent Si may be replaced by trivalent Al, whereas in the octahedral sheet there may be replacement of trivalent Al with divalent Mg, Fe, Cr or Zn. This replacement is usually referred to as isomorphic substitution [22], i.e. an atom of higher valency is replaced by one of lower valency. This results in a deficit of positive charges or an excess of negative charges. Thus, the faces of the clay platelets become negatively charged and these negative charges are compensated by counterions such as Na^+ or Ca^{2+}. As a result, a double layer is produced with a constant charge (that is independent of the pH of the solution). However, at the edges of the platelets, some disruption of the bonds occurs, resulting in the formation of an oxide-like layer, e.g. –Al–OH, which undergoes dissociation giving a negative (–Al–O^-) or positive (–Al–OH_2^+) ion depending on the pH of the solution. An isoelectric point may be identified for the edges (usually between pH 7 and 9). This means that the double layer at the edges is different from that at the faces and the surface charges can be positive or negative depending on the pH of the solution. For that reason, van Olphen [21] suggested an edge-to-face association of clay platelets (which he termed the 'house of cards' structure), and this was assumed to be the driving force for gelation of swellable clays. However, Norrish [22] suggested that clay gelation is caused simply by the interaction of the expanded double layers. This is particularly the case in dilute electrolyte solutions where the double layer thickness can be several orders of magnitude higher than the particle dimensions.

With oxides, such as finely divided silica, gel formation is caused by the formation of chain aggregates, which interact forming a three-dimensional network that is elastic in nature. Clearly, the formation of such networks depends on the nature and particle size of the silica particles. For effective gelation, one should choose silicas with very small particles and highly solvated surfaces.

7.4.5.4 Use of Mixtures of Polymers and Finely Divided Solids

Mixtures of polymers such as hydroxyethylcellulose or xanthan gum with finely divided solids such as sodium montmorillonite or silica offer one of the most robust antisettling systems. By optimizing the ratio of the polymer to the solid particles, one can arrive at the right viscosity and elasticity to reduce settling and separation. Such systems are more shear thinning than the polymer solutions and hence they are more easily dispersed in water on application. The most likely mechanism by which these mixtures produce a viscoelastic network is probably through bridging or depletion flocculation. The polymer–particulate mixtures also show less temperature dependence of viscosity and elasticity than the polymer solutions and hence they ensure long-term physical stability at high temperatures.

7.4.6 Controlled Flocculation

For systems where the stabilizing mechanism is electrostatic in nature, for example those stabilized by surfactants or polyelectrolytes, the energy–distance curve (Figure 7.3a) shows a secondary minimum at larger particle separations. Such a minimum can be fairly deep (a few tens of kT units), particularly for large ($>1\,\mu m$) and asymmetric particles. The depth of the minimum also depends on electrolyte concentration. Thus, by adding small amounts of electrolyte, weak flocculation may occur. This weak flocculation may produce a gel network (self-structured systems) which has sufficient elasticity to reduce settling and eliminate claying. This has been demonstrated by Tadros [4] for ethirimol suspensions stabilized with phenol formaldehyde sulfonated condensate (a polyelectrolyte with modest molecular weight). With increasing NaCl concentration, the depth of the secondary minimum increases, reaching $\sim 50kT$ at the highest electrolyte concentration. By using electrolytes of higher valency such as $CaCl_2$ or $AlCl_3$, such deep minima are produced at much lower electrolyte concentrations. Therefore, by controlling the electrolyte concentration and valency, one can reach a sufficiently deep secondary minimum for producing a gel with enough elasticity to reduce settling and eliminate claying.

For systems stabilized by nonionic surfactants or macromolecules, the energy–distance curve also shows a minimum (Figure 7.3b), the depth of which depends on particle size, Hamaker constant and the thickness of the adsorbed layer, which in turn depends on the molecular weight of the polymer. As the molecular weight of the polymer is reduced below a certain value, the depth of the minimum becomes sufficiently high for weak flocculation to occur. The gel network structure of these weakly flocculated suspension prevents sedimentation.

7.4.7 Depletion Flocculation

The addition of 'free' (non-adsorbing) polymer can induce weak flocculation of the suspension, when the concentration or volume fraction of the free polymer

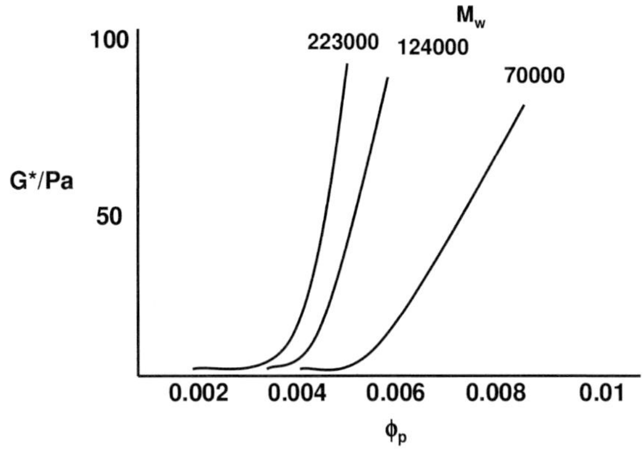

Figure 7.9 Variation of g^* with ϕ_p for an ethirimol suspension flocculated by HEC.

(ϕ_p) exceeds a critical value that is denoted $\phi_p{}^+$. The first quantitative analysis of the phenomenon was reported by Asakura and Oosawa [12]). They showed that when two particles approach to a distance of separation that is smaller than the diameter of the free coil, exclusion of the polymer molecules from the interstices between the particles takes place, leading to the formation a polymer-free zone (depletion zone). This is illustrated in Figure 7.4, which shows the situation below and above $\phi_p{}^+$. As a result of this process, an attractive force, associated with the lower osmotic pressure in the region between the particles, is produced. This weak flocculation process can be applied to prevent sedimentation and formation of clays. This has been illustrated by using ethirimol suspensions stabilized by a graft copolymer containing poly(ethylene oxide) (PEO) side chains (with $M = 750$) to which free hydroxyethylcellulose (HEC) of various molecular weights was added. Above a critical volume fraction of the free polymer (which decreased with increase in molecular weight), weak flocculation occurred, which increased with further increase in free polymer concentration.

The above structure may have a sufficient yield value or modulus, G^*, that prevents particle sedimentation. This is illustrated in Figure 7.9, which shows the variation of G^* with ϕ_p of HEC (with three different molecular weights) for ethirimol suspensions. G^* shows a rapid increase with increase in ϕ_p when the latter is higher than $\phi_p{}^+$. As predicted, the higher the molecular weight of HEC, the lower is the value of $\phi_p{}^+$.

7.5
Characterization of Suspension Concentrates and Assessment of Their Long-term Physical Stability

For the full assessment of the properties of suspension concentrates, three main types of investigations are needed: (1) fundamental investigation of the system at a molecular level, (2) investigations into the state of the suspension on standing and (3) bulk properties of the suspension.

All theses investigations require a number of sophisticated techniques, such as zeta potential determinations, measurement of surfactant and polymer adsorption and their conformation at the solid/liquid interface, measurement of the rate of flocculation and crystal growth and several rheological measurements. Apart from these practical methods, which are present in most industrial laboratories, more fundamental information can be obtained using modern sophisticated techniques such as small-angle X-ray and neutron scattering measurements, ultrasonic absorption techniques, etc. Several other modern techniques are also now available for investigation of the state of suspensions, e.g. freeze fracture and electron microscopy, atomic force microscopy, scanning tunneling microscopy and confocal laser microscopy.

In all of the above methods, care should be taken in sampling the suspension, which should cause as little disturbance as possible to the 'structure' to be investigated. For example, when one investigates the flocculation of a concentrated suspension, dilution of the system for microscopic investigation may lead to breakdown of the flocs and a false assessment will be obtained. The same applies when one investigates the rheology of a concentrated suspension, since transfer of the system from its container to the rheometer may lead to breakdown of the structure.

For the above reasons, one must establish well-defined procedures for every technique and this requires a great deal of skill and experience. It is advisable in all cases to develop standard operating procedures for the above investigations.

7.5.1 Assessment of the Structure of the Solid/Liquid Interface

7.5.1.1 Double Layer Investigations

The main procedure that may be applied to investigate the charge at the solid/liquid interface (e.g. when using ionic surfactants or polyelectrolytes), which is important in the assessment of electrostatic stabilization, is to measure the zeta potential. As discussed in Chapter 3, the zeta potential is obtained from the measured electrophoretic mobility provided that information is available on particle size and electrolyte concentration. In cases when the above information is not available (as is the case with many practical systems), one should use the electrophoretic mobility for relative comparison between various systems; the assumption can be made that the higher the mobility, the higher is the surface charge and the more likely the system is to be stable against flocculation, if the charge is the main stabilizing factor.

Clearly, for systems stabilized by nonionic surfactants and polymers, electrophoretic mobility measurements are less informative. However, zeta potential measurements can be used qualitatively to obtain information on the adsorbed layer thickness for nonionic surfactants and polymers. When a nonionic surfactant or polymer adsorbs at the solid/liquid interface, a shift in the shear plane occurs and this results in a reduction in the zeta potential. If the zeta potential of the particles is measured in the presence and absence of nonionic surfactant or polymer, then the adsorbed layer thickness can be roughly estimated from the reduction in zeta potential.

7.5.1.2 Surfactant and Polymer Adsorption

A representative sample of the solid with known mass m and surface area A per gram is equilibrated with a surfactant or polymer with concentration C_1. After equilibrium has been reached (at a given constant temperature), the solid is removed by centrifugation and the equilibrium concentration C_2 is determined analytically.

The amount of adsorption Γ in mol m^{-2} is given by

$$\Gamma = \frac{C_1 - C_2}{mA} = \frac{\Delta C}{mA} \tag{21}$$

In most cases (particularly with surfactants), a plot of Γ versus C_2 gives a Langmuir-type isotherm. The data can be fitted using the Langmuir equation:

$$\Gamma = \frac{\Gamma_\infty b C_2}{1 + bC_2} \tag{22}$$

where b is a constant that is related to the free energy of adsorption:

$$b \propto -\frac{\Delta G_{\text{ads}}}{RT} \tag{23}$$

Most polymers (particularly those with high molecular weight) give a high affinity isotherm, as discussed in Chapter 3.

7.5.2 Assessment of the State of the Dispersion

7.5.2.1 Measurement of Rate of Flocculation

Two general techniques may be applied for measuring the rate of flocculation of suspensions, both of which can only be applied for dilute systems. The first method is based on measuring the scattering of light by the particles. For monodisperse particles with a radius that is less than $\lambda/20$ (where λ is the wavelength of light), one can apply the Rayleigh equation, where the turbidity τ_0 is given by

$$\tau_0 = A' n_0 V_1^2 \tag{24}$$

where A' is an optical constant (which is related to the refractive index of the particle and medium and the wavelength of light) and n_0 is the number of particles, each with a volume V_1.

By combining the Rayleigh theory with the Smoluchowski–Fuchs [23, 24] theory of flocculation kinetics, one can obtain the following expression for the variation of turbidity with time:

$$\tau = A' n_0 V_1^2 (1 + 2n_0 kt) \tag{25}$$

where k is the rate constant of flocculation.

The second method for obtaining the rate constant of flocculation is by direct particle counting as a function of time. For this purpose, optical microscopy or image analysis may be used, provided that the particle size is within the resolution limit of the microscope. Alternatively, the particle number may be determined using electronic devices such as the Coulter Counter or the flow ultramicroscope.

The rate constant of flocculation is determined by plotting $1/n$ versus t, where n is the number of particles after time t, i.e.

$$\frac{1}{n} = \frac{1}{n_0} + kt \tag{26}$$

The rate constant k of slow flocculation is usually related to the rapid rate constant k_0 (the Smoluchowski rate) by the stability ratio W [25]:

$$W = \frac{k}{k_0} \tag{27}$$

One usually plots log W versus log C (where C is the electrolyte concentration) to obtain the critical coagulation concentration (ccc), which is the point at which $\log W = 0$.

7.5.2.2 Measurement of Incipient Flocculation

This can be done for sterically stabilized suspensions when the medium for the chains becomes a θ-solvent. This occurs, for example, on heating an aqueous suspension stabilized with poly(ethylene oxide) (PEO) or poly(vinyl alcohol) chains. Above a certain temperature (the θ-temperature) that depends on electrolyte concentration, flocculation of the suspension occurs. The temperature at which this occurs is defined as the critical flocculation temperature (CFT).

This process of incipient flocculation can be followed by measuring the turbidity of the suspension as a function of temperature. Above the CFT, the turbidity of the suspension rises very sharply.

For the above purpose, the cell in the spectrophotometer that is used to measure the turbidity is placed in a metal block that is connected to a temperature programming unit (which allows one to increase the temperature rise at a controlled rate).

7.5.2.3 Measurement of Crystal Growth (Ostwald Ripening)

As discussed in Chapter 6, Ostwald ripening is the result of the difference in solubility S between small and large particles. The smaller particles have higher solubility than the larger particles:

$$S \propto \frac{2\sigma}{r} \tag{28}$$

where σ is the solid/liquid interfacial tension and r is the particle radius. For two particles with radii r_1 and r_2:

$$\frac{RT}{M} \ln\left(\frac{S_1}{S_2}\right) = \left(\frac{2\sigma}{\rho}\right)\left(\frac{1}{r_1} - \frac{1}{r_2}\right) \tag{29}$$

where R is the gas constant, T is the absolute temperature, M is the molecular weight and ρ is the density of the particles.

To obtain a measure of the rate of crystal growth, the particle size distribution of the suspension is followed as a function of time, using either a Coulter Counter, a Mastersizer or an optical disc centrifuge. One usually plots the cube of the average radius versus time, which gives a straight line, from the slope of which the rate of crystal growth can be determined.

7.6 Bulk Properties of Suspension Concentrates

7.6.1 Equilibrium Sediment Volume (or Height) and Redispersion

For a 'structured' suspension, obtained by 'controlled' flocculation or addition of 'thickeners' (such as polysaccharides, clays or oxides), the 'flocs' sediment at a rate depending on their size and porosity of the aggregated mass. After this initial sedimentation, compaction and rearrangement of the floc structure occur, a phenomenon referred to as consolidation.

Normally in sediment volume measurements, one compares the initial volume V_0 (or height H_0) with the ultimately reached value V (or H). A colloidally stable suspension gives a 'close-packed' structure with relatively small sediment volume (dilatant sediment referred to as clay). A weakly 'flocculated' or 'structured' suspension gives a more open sediment and hence a higher sediment volume. Thus, by comparing the relative sediment volume V/V_0 or height H/H_0, one can distinguish between a clayed and flocculated suspension.

7.6.2 Rheological Measurements

Three different rheological measurements may be applied: (1) steady-state shear stress–shear rate measurements (using a controlled shear rate instrument); (2) constant stress (creep) measurements (carried out using a constant stress instrument); and (3) dynamic (oscillatory) measurements (preferably carried out using a constant strain instrument).

The above rheological techniques can be used to assess sedimentation and flocculation of suspensions and this is discussed in detail below.

7.6.2.1 Assessment of Sedimentation

As discussed before, the rate of sedimentation decreases with increase in the volume fraction of the disperse phase, ϕ, and ultimately it approaches zero at a critical volume fraction ϕ_p (the maximum packing fraction). However, at $\phi \approx \phi_p$, the viscosity of the system approaches ∞. Therefore, for most practical emulsions, the system is prepared at ϕ values below ϕ_p and then 'thickeners' are added to reduce sedimentation. These 'thickeners' are usually high molecular weight polymers (such as xanthan gum, hydroxyethylcellulose or associative thickeners), finely divided inert solids (such as silica or swelling clays) or a combination of the two.

In all cases, a 'gel' network is produced in the continuous phase which is shear thinning (i.e. its viscosity decreases with increase in shear rate) and viscoelastic (i.e. it has a viscous and elastic components of the modulus). If the viscosity of the elastic network, at shear stresses (or shear rates) comparable to those exerted by the particles, exceeds a certain value, then sedimentation is completely eliminated.

The shear stress, σ_p, exerted by a particle (force/area) can be simply calculated:

$$\sigma_p = \frac{\frac{4}{3}\pi R^3 \Delta\rho g}{4\pi R^2} = \frac{\Delta\rho R g}{3} \tag{30}$$

For a 10 μm radius particle with density difference $\Delta\rho = 0.2$, σ_p is equal to

$$\sigma_p = \frac{0.2 \times 10^3 \times 10 \times 10^{-6} \times 9.8}{3} \approx 6 \times 10^{-3} \tag{31}$$

For smaller particles, smaller stresses are exerted.

Thus, to predict sedimentation, one has to measure the viscosity at very low stresses (or shear rates). These measurements can be carried out using a constant stress rheometer (Carrimed, Bohlin, Rheometrics or Physica). A constant stress σ (using, for example, a drag cup motor that can apply very small torques and using an air bearing system to reduce the frictional torque) is applied on the system (which may be placed in the gap between two concentric cylinders or a cone–plate geometry) and the deformation (strain γ or compliance $J = \gamma/\sigma = \text{Pa}^{-1}$) is followed as a function of time [26, 27].

For a viscoelastic system, the compliance shows a rapid elastic response J_0 at $t \rightarrow 0$ [instantaneous compliance $J_0 = 1/G_0$, where G_0 is the instantaneous modulus that is a measure of the elastic (i.e. 'solid-like') component]. At $t > 0$, J increases slowly with time and this corresponds to the retarded response ('bonds' are broken and re-formed but not at the same rate). Above a certain time (that depends on the system), the compliance shows a linear increase with time (i.e.

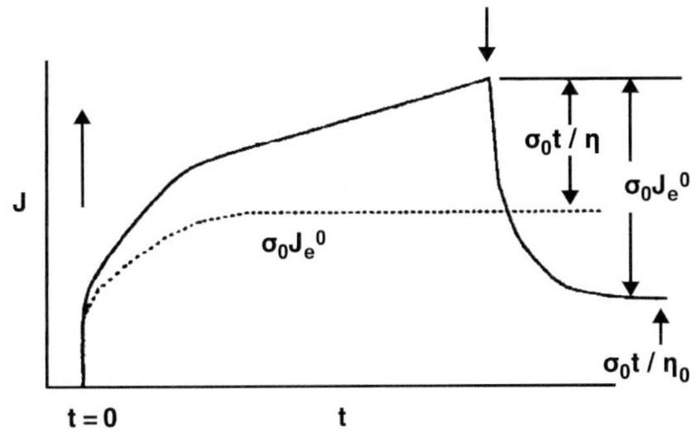

Figure 7.10 Creep curve for a viscoelastic liquid. Creep is the sum of a constant value $J_e\sigma_0$ (elastic part) and a viscous contribution $\sigma_0 t/\eta_0$.

the system reaches a steady state with constant shear rate). If after the steady state has been reached the stress is removed, elastic recovery occurs and the strain changes sign. The above behavior (usually referred to as 'creep') is represented schematically in Figure 7.10.

The slope of the linear part of the creep curve gives the value of the viscosity at the applied stress, η_σ:

$$\frac{J}{t} = \frac{\mathrm{Pa}^{-1}}{\mathrm{s}} = \frac{1}{\mathrm{Pa\ s}} = \eta_\sigma \tag{32}$$

The recovery curve will only give the elastic component, which if superimposed on the ascending part of the curve will give the viscous component. Therefore, one measures creep curves as a function of the applied stress (starting from a very small stress of the order of 0.01 Pa). This is illustrated in Figure 7.11.

The viscosity η_σ (which is equal to the reciprocal of the slope of the straight portion of the creep curve) is plotted as a function of the applied stress. This is shown schematically in Figure 7.12.

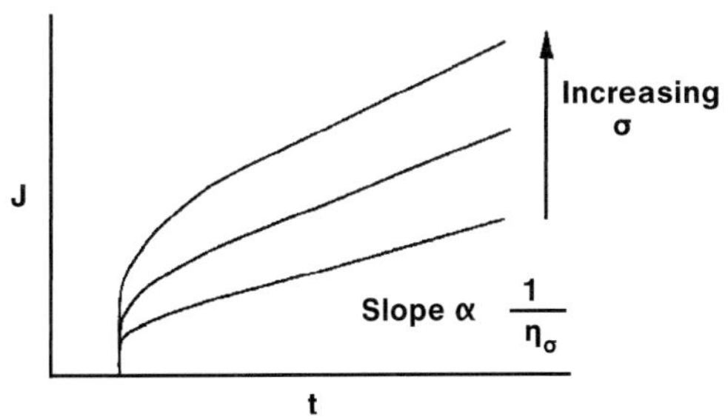

Figure 7.11 Creep curves at increasing applied stress. Creep measurements (constant stress) can be used to obtain the residual or zero shear viscosity.

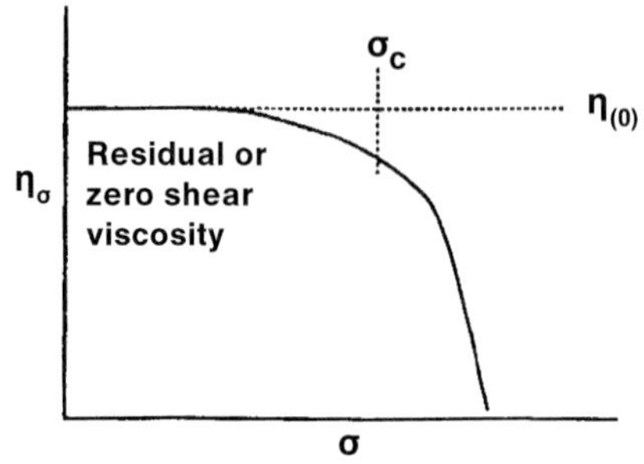

Figure 7.12 Variation of viscosity with applied stress. Critical stress is a useful parameter (related to yield stress) as it denotes the stress at which structure 'breaks down'.

Below a critical stress, σ_{cr}, the viscosity reaches a limiting value, $\eta_{(0)}$, namely the residual (or zero shear) viscosity. σ_{cr} may be denoted the 'true yield stress' of the emulsion, i.e. the stress above which the 'structure' of the system is broken down. Above σ_{cr}, η_σ decreases rapidly with further increase in the shear stress (the shear thinning regime). It reaches another Newtonian value η_∞, which is the high shear limiting viscosity.

$\eta_{(0)}$ could be several orders of magnitudes (10^4–10^8) higher than η_∞. Usually one obtains a good correlation between the rate of sedimentation v and the residual viscosity $\eta_{(0)}$ [19]. This is illustrated in Figure 7.8.

Above a certain value of $\eta_{(0)}$, v becomes equal to zero. Clearly, to minimize sedimentation one has to increase $\eta_{(0)}$; an acceptable level for the high shear viscosity η_∞ must be achieved, depending on the application. In some cases, a high $\eta_{(0)}$ may be accompanied by a high η_∞ (which may not be acceptable for application, for example if spontaneous dispersion on dilution is required). If this is the case, the formulation chemist should look for an alternative thickener.

Another problem encountered with many suspensions is that of 'syneresis', i.e. the appearance of a clear liquid film at the top of the suspension. 'Syneresis' occurs with most 'flocculated' and/or 'structured' (i.e. those containing a thickener in the continuous phase) suspensions. 'Syneresis' may be predicted from measurement of the yield value (using steady-state measurements of shear stress as a function of shear rate) as a function of time or using oscillatory techniques (where the storage and loss modulus are measured as a function of strain amplitude and frequency of oscillation). The above techniques are discussed in detail below.

It is sufficient to state in this section that when a network of the suspension particles (either alone or combined with the thickener) is produced, the gravity force will cause some contraction of the network (which behaves as a porous plug), thus causing some separation of the continuous phase which is entrapped between the droplets in the network.

7.6.2.2 Assessment of Flocculation

As mentioned before, flocculation of suspensions is the result of the long-range van der Waals attraction. Flocculation can be weak (and reversible) or strong, depending on the magnitude of the net attractive forces. Weak flocculation may result in a reversible time dependence of the viscosity, i.e. on shearing the suspension at a given shear rate, the viscosity decreases and on standing the viscosity recovers to its original value. This phenomenon is referred to as thixotropy (sol ↔ gel transformation).

Rheological techniques are most convenient to assess suspension flocculation without the need for any dilution (which in most cases results in breakdown of the floc structure). In steady-state measurements, the suspension is carefully placed in the gap between concentric cylinders or cone-and-plate platens. For the concentric cylinder geometry, the gap width should be at least 10 times larger than the largest particle size (a gap width that is greater than 1 mm is usually used). For the cone-and-plate geometry, a cone angle of 4° or smaller is usually used.

A controlled rate instrument is usually used for the above measurements; the inner (or outer) cylinder, the cone (or the plate) is rotated at various angular velocities (which allows one to obtain the shear rate γ) and the torque is measured on the other element (this allows one to obtain the stress σ).

For a Newtonian system (such as the case of a dilute suspension, with a volume fraction $\phi < 0.1$), σ is related to γ by the equation

$$\sigma = \eta\dot{\gamma} \tag{33}$$

where η is the Newtonian viscosity (which is independent of the applied shear rate).

For most practical suspensions (with $\phi > 0.1$ and containing thickeners to reduce sedimentation), a plot of σ versus γ is not linear (i.e. the viscosity depends on the applied shear rate). The most common flow curve is shown in Figure 7.13

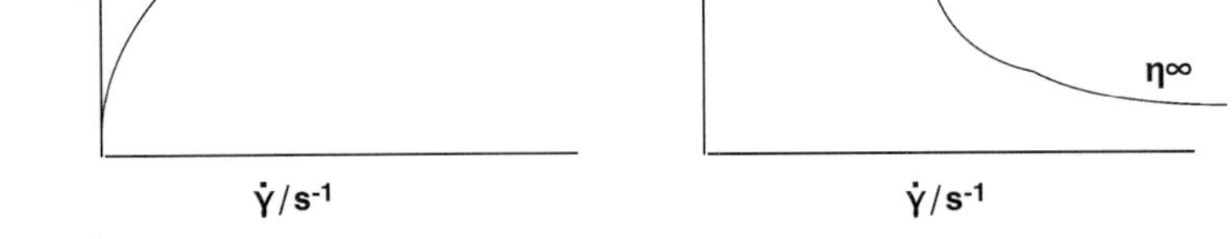

Figure 7.13 Shear stress and viscosity versus shear rate for a pseudoplastic system.

(usually described as a pseudoplastic or shear thinning system). In this case, the viscosity decreases with increase in shear rate, reaching a Newtonian value above a critical shear rate.

Several models may be applied to analyze the results in Figure 7.13.

(a) Power law model:

$$\sigma = k\dot{\gamma}^{n} \tag{34}$$

where k is the consistency index of the suspension and n is the power (shear thinning) index ($n < 1$); the lower the value of n, the more shear thinning the suspension is. This is usually the case with weakly flocculated suspensions or those to which a 'thickener' is added. By fitting the results in Figure 7.13 to equation 34 (this is usually in the software of the computer connected to the rheometer), one can obtain the viscosity of the suspension at a given shear rate:

$$\eta(\text{at a given shear rate}) = \frac{\sigma}{\gamma} = k\dot{\gamma}^{n-1} \tag{35}$$

(b) Bingham model:

$$\sigma = \sigma_{\beta} + \eta_{\text{pl}}\dot{\gamma} \tag{36}$$

where σ_{β} is the extrapolated yield value (obtained by extrapolation of the shear stress–shear rate curve to $\gamma = 0$). Again, this is provided in the software of the rheometer. η_{pl} is the slope of the linear portion of the σ–γ curve (usually referred to as the plastic viscosity).

Both σ_{β} and η_{pl} may be related to the flocculation of the suspension. At any given volume fraction of the emulsion and at a given particle size distribution, the higher the value of σ_{β} and η_{pl}, the more flocculated the suspension is. Therefore, if one stores a suspension at any given temperature and makes sure that the particle size distribution remains constant (i.e. no Ostwald ripening occurs), an increase in the above parameters indicates flocculation of the suspension on storage. Clearly, if Ostwald ripening occurs simultaneously, σ_{β} and η_{pl} may change in a complex manner with storage time. Ostwald ripening results in a shift of the particle size distribution to larger diameters; this has the effect of reducing σ_{β} and η_{pl}. If flocculation occurs simultaneously (having the effect of increasing these rheological parameters), the net effect may be an increase or decrease in the rheological parameters.

The above trend depends on the extent of flocculation relative to Ostwald ripening. Therefore, following σ_{β} and η_{pl} with storage time requires a knowledge of Ostwald ripening and/or coalescence. Only in the absence of this latter breakdown process can one use rheological measurements as a guide of assessment of flocculation.

(c) Herschel–Buckley model [28]:

In many cases, the shear stress–shear rate curve may not show a linear portion at high shear rates. In this case, the data may be fitted with a Hershel–Buckley model:

$$\sigma = \sigma_\beta + k\dot{\gamma}^n \tag{37}$$

(d) Caisson's model [29]:

This is another semi-empirical model that may be used to fit the data in Figure 7.14:

$$\sigma^{\frac{1}{2}} = \sigma_C^{\frac{1}{2}} + \eta_C^{\frac{1}{2}}\dot{\gamma}^{\frac{1}{2}} \tag{38}$$

Note that σ_β is not equal to C_D. Equation 38 shows that a plot of $\sigma^{1/2}$ versus $\gamma^{1/2}$ gives a straight line from which C_D and C_D can be evaluated.

In all the above analyses, the assumption was made that a steady state was reached. In other words, no time effects occurred during the duration of the flow experiment.

7.6.3 Time Effects During Flow – Thixotropy

Many suspensions (particularly those that are weakly flocculated or 'structured' to reduce sedimentation) show time effects during flow. At any given shear rate, the viscosity of the suspension continues to decrease with increasing time of shear; on stopping the shear, the viscosity recovers to its initial value. This reversible decrease in viscosity is referred to as thixotropy.

The most common procedure for studying thixotropy is to apply a sequence of shear stress–shear rate regimes within controlled periods. If the flow curve is carried out within a very short time (say increasing the rate from 0 to 500 s^{-1} in 30 s and then reducing it again from 500 to 0 s^{-1} within the same period), one finds that the descending curve is below the ascending curve.

The above behavior can be explained from consideration of the structure of the system. If, for example, the suspension is weakly flocculated, then on applying a shear force to the system, this flocculated structure is broken down (and this is the cause of the shear thinning behavior). On reducing the shear rate back to zero, the structure builds up only partly within the duration of the experiment (30 s).

The ascending and descending flow curves show hysteresis that is usually referred to as a 'thixotropic loop'. If the same experiment is now repeated within a longer time experiment (say 120 s for the ascending and 120 s for the descending curves), the hysteresis decreases, i.e. the 'thixotropic loop' becomes smaller.

The above approach may be used to investigate the state of flocculation of a suspension. Weakly flocculated suspensions usually show thixotropy and the change in thixotropy with applied time can be used as an indication of the strength of this weak flocculation.

The above analysis is only qualitative and one cannot use the results in a quantitative manner. This is due to the possible breakdown of the structure on transferring the suspension to the rheometer and also during the uncontrolled shear experiment.

A very important point that must be considered during rheological measurements is the possibility of 'slip' during the measurements. This is particularly the case with highly concentrated suspensions, where the flocculated system may form a 'plug' in the gap of the platens, leaving a thin liquid film at the walls of the concentric cylinder or cone-and-plate geometry. To reduce 'slip' one should use roughened walls for the platens.

Strongly flocculated suspensions usually show much less thixotropy than weakly flocculated systems. Again, one must be careful in drawing definite conclusions without other independent techniques (e.g. microscopy).

7.6.4
Constant Stress (Creep) Experiments

This method has been described in detail in Section 7.6.2.1. Basically, a constant stress σ is applied to the system and the compliance J (Pa^{-1}) is plotted as a function of time, as illustrated in Figure 7.11 (creep curve).

The above experiment is repeated several times, increasing the stress from the smallest possible value (that can be applied by the instrument) in small increments. A set of creep curves are produced at various applied stresses, as illustrated in Figure 7.12. From the slope of the linear portion of the creep curve (after the system has reached a steady state), the viscosity at each applied stress, η_σ, is calculated. A plot of η_σ versus σ (as illustrated in Figure 7.14) allows one to obtain the limiting (or zero shear) viscosity $\eta_{(0)}$ and the critical stress σ_{cr} (which may be identified with the 'true' yield stress of the system).

The values of $\eta_{(0)}$ and σ_{cr} may be used to assess the flocculation of the suspension on storage. If flocculation occurs on storage (without any Ostwald ripening), the values of $\eta_{(0)}$ and σ_{cr} may show a gradual increase with increase in storage time.

As discussed in the previous section (on steady-state measurements), the trend becomes complicated if Ostwald ripening occurs simultaneously (both have the effect of reducing $\eta_{(0)}$ and σ_{cr}).

The above measurements should be supplemented by particle size distribution measurements of the diluted suspension (making sure that no flocs are present after dilution) to assess the extent of Ostwald ripening.

Another complication may arise from the nature of the flocculation. If the latter occurs in an irregular way (producing strong and tight flocs), $\eta_{(0)}$ may increase, while σ_{cr} may show some decrease, and this complicates the analysis of the results.

In spite of the above complications, constant stress measurements may provide valuable information on the state of the suspension on storage. Carrying out creep experiments and ensuring that a steady state is reached can be time consuming. One usually carries out a stress sweep experiment, where the stress is gradually increased (within a predetermined time period to ensure that one is not too far from reaching the steady state) and plots of η_σ versus σ are established.

The above experiments are carried out at various storage times (say every 2 weeks) and temperatures. From the changes in $\eta_{(0)}$ and σ_{cr} with storage time and temperature, one can obtain information on the degree and the rate of flocculation of the system.

Clearly, interpretation of the rheological results requires expert knowledge of rheology and measurement of the particle size distribution as a function of time.

One major problem in carrying the above experiments is sample preparation. When a flocculated emulsion is removed from the container, care should be taken not to cause much disturbance to that structure (minimum shear should be applied on transferring the emulsion to the rheometer). It is also advisable to use separate containers for assessment of the flocculation. A relatively large sample is prepared and this is then transferred to a number of separate containers.

7.6.5 Dynamic (Oscillatory) Measurements

This is by far the most commonly used method to obtain information on the flocculation of a suspension. A strain is applied in a sinusoidal manner, with an amplitude γ_0 and a frequency ν (cycles s^{-1} or Hz) or ω (rad s^{-1}). In a viscoelastic system (such as the case with a flocculated suspension), the stress oscillates with the same frequency, but out of phase from the strain [30–32]. From measurement of the time shift between strain and stress amplitudes (Δt), one can obtain the phase angle shift δ:

$$\delta = \Delta t \omega \tag{39}$$

A schematic representation of the variation of strain and stress with t is shown in Figure 7.14.

From the amplitudes of stress and strain and the phase angle shift, one can obtain the various viscoelastic parameters: the complex modulus G^*, the storage modulus (the elastic component of the complex modulus) G', the loss modulus (the viscous component of the complex modulus) G'', $\tan\delta$ and the dynamic viscosity η':

$$\textit{Complex modulus} \quad |G^*| = \frac{\sigma_0}{\gamma_0} \tag{40}$$

$$\textit{Storage modulus} \quad G' = |G^*| \cos\delta \tag{41}$$

$$\textit{Loss modulus} \quad G'' = |G^*| \sin\delta \tag{42}$$

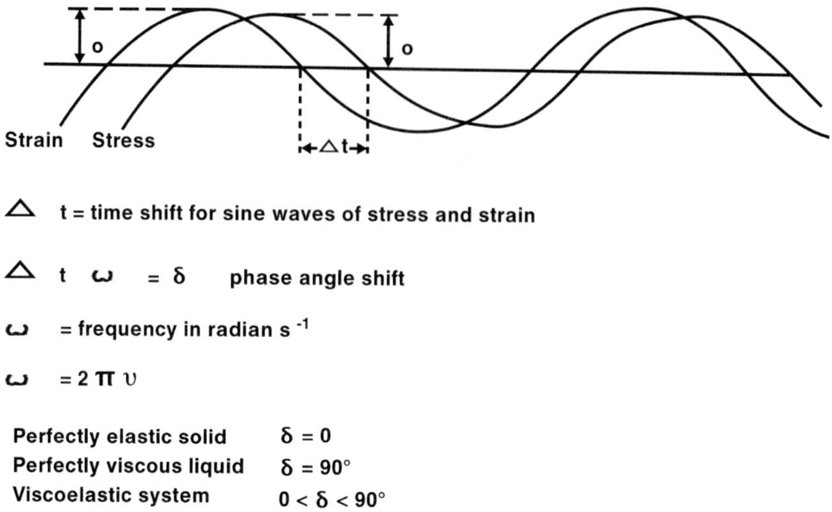

Figure 7.14 Stress–strain relationship for a viscoelastic system.

$$\tan\delta = \frac{G''}{G'} \tag{43}$$

$$\textit{Dynamic viscosity}\quad \eta' = \frac{G''}{\omega} \tag{44}$$

where G′ is a measure of the energy stored in a cycle of oscillation, G″ is a measure of the energy dissipated as viscous flow in a cycle of oscillation and $\tan\delta$ is a measure of the relative magnitudes of the viscous and elastic components. Clearly, the smaller the value of $\tan\delta$, the more elastic the system is, and vice versa.

The dynamic viscosity, η', shows a decrease with increase in frequency ω; η' reaches a limiting value as $\omega \rightarrow 0$ – the value of η' in this limit is identical with the residual (or zero shear) viscosity $\eta_{(0)}$. This is referred to as the Cox–Mertz rule.

In oscillatory measurements, one carries out two sets of experiments:

(a) Strain sweep measurements. In this case, the oscillation is fixed (say at 0.1 or 1 Hz) and the viscoelastic parameters are measured as a function of strain amplitude. This allows one to obtain the linear viscoelastic region. In this region, all moduli are independent of the applied strain amplitude and become a function only of time or frequency. This is illustrated in Figure 7.15, which shows a schematic representation of the variation of G^*, G' and G'' with strain amplitude (at a fixed frequency).

It can be seen that G^*, G' and G'' remain virtually constant up to a critical strain value, γ_{cr}. This region is the linear viscoelastic region. Above γ_{cr}, G^* and

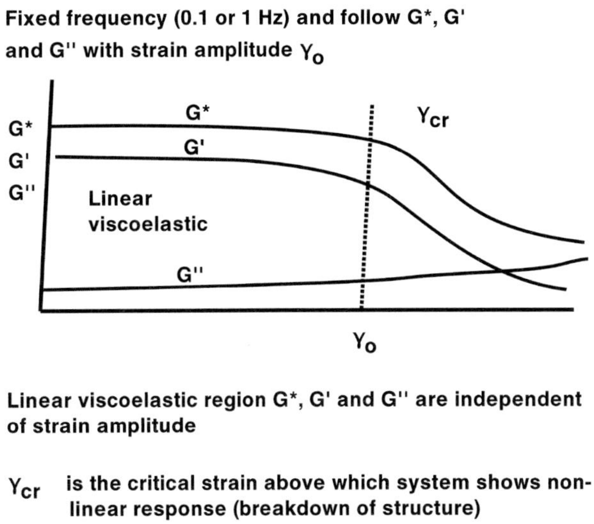

Figure 7.15 Strain sweep results.

G' starts to fall, whereas G'' starts to increase. This is the nonlinear region. The value of γ_{cr} may be identified with the minimum strain above which the 'structure' of the suspension starts to break down (for example, breakdown of flocs into smaller units and/or breakdown of a 'structuring' agent).

From γ_{cr} and G', one can obtain the cohesive energy E_c (J m^{-3}) of the flocculated structure:

$$E_c = \int_0^{\gamma_{cr}} \sigma\,d\gamma = \int_0^{\gamma_{cr}} G'\gamma\,d\gamma = \frac{1}{2}G'{\gamma_{cr}}^2 \tag{45}$$

E_c may be used in a quantitative manner as a measure of the extent and strength of the flocculated structure in a suspension. The higher the value of E_c, the more flocculated the structure is. Clearly, E_c depends on the volume fraction of the suspension in addition to the particle size distribution (which determines the number of contact points in a floc). Therefore, for quantitative comparison between various systems, one has to make sure that the volume fraction of the disperse particles is the same and the suspensions have very similar particle size distributions.

E_c also depends on the strength of the flocculated structure, i.e. the energy of attraction between the particles. This depends on whether the flocculation is in the primary or secondary minimum. Flocculation in the primary minimum is associated with a large attractive energy and this leads to higher values of E_c compared with the values obtained for secondary minimum flocculation. For a weakly flocculated suspension, such as the case with secondary minimum flocculation of an electrostatically stabilized suspension, the deeper the secondary minimum, the higher is the value of E_c (at any given volume fraction and particle size distribution of the suspension).

With a sterically stabilized suspension, weak flocculation can also occur when the thickness of the adsorbed layer decreases. Again, the value of E_c can be used as a measure of the flocculation: the higher the value of E_c, the stronger is the flocculation. If incipient flocculation occurs (on reducing the solvency of the medium for the change to worse than θ-condition), a much deeper minimum is observed and this is accompanied by a much larger increase in E_c.

To apply the above analysis, one must have an independent method for assessing the nature of the flocculation. Rheology is a bulk property that can give information on the interdroplet interaction (whether repulsive or attractive) and to apply it in a quantitative manner one must know the nature of these interaction forces. However, rheology can be used in a qualitative manner to follow the change in the suspension on storage. Provided that the system does not undergo any Ostwald ripening, the change in the moduli with time and in particular the change in the linear viscoelastic region may be used as an indication of flocculation. Strong flocculation is usually accompanied by a rapid increase in G' and this may be accompanied by a decrease in the critical strain above which the 'structure' breaks down. This may be used as an indication of the formation of 'irregular' flocs which become sensitive to the applied strain. The floc structure will entrap a large amount of the continuous phase, and this leads to an apparent increase in the volume fraction of the suspension and hence an increase in G'.

(b) Oscillatory sweep measurements. In this case, the strain amplitude is kept constant in the linear viscoelastic region (one usually takes a point far from γ_{cr} but not too low, i.e. in the mid-point of the linear viscoelastic region) and measurements are carried out as a function of frequency. This is represented schematically in Figure 7.16 for a viscoelastic liquid system.

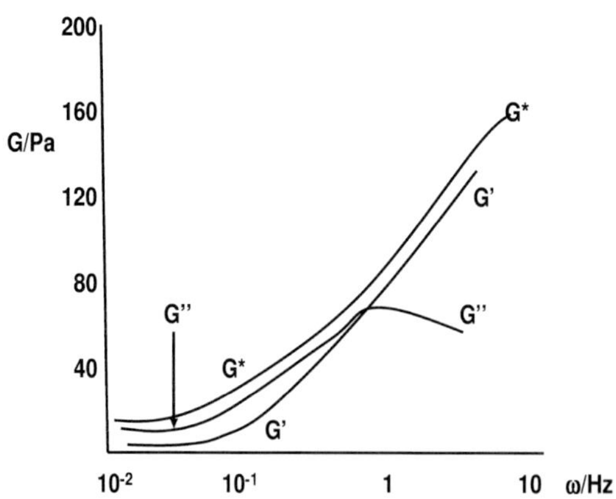

Figure 7.16 Schematic representation of oscillatory measurements for a viscoelastic liquid.

Both G^* and G' increase with increase in frequency and ultimately, above a certain frequency, they reach a limiting value and show little dependence on frequency. G'' is higher than G' in the low-frequency regime; it also increases with increase in frequency and at a certain characteristic frequency ω^* (which depends on the system) it becomes equal to G' (usually referred to as the cross-over point), after which it reaches a maximum and then shows a reduction with further increase in frequency.

In the low-frequency regime, i.e. below ω^*, $G'' > G'$; this regime corresponds to longer times (remember that the time is the reciprocal of frequency) and under these conditions the response is more viscous than elastic. In the high-frequency regime, i.e. above ω^*, $G' > G''$; this regime corresponds to short times and under these conditions the response is more elastic than viscous.

At sufficiently high frequency, G'' approaches zero and G' becomes nearly equal to G^*; this corresponds to very short time scales where the system behaves as a near-elastic solid. Very little energy dissipation occurs at such high frequency.

The characteristic frequency ω^* can be used to calculate the relaxation time of the system, t^*:

$$t^* = \frac{1}{\omega^*} \tag{46}$$

The relaxation time may be used as a guide for the state of the suspension. For a colloidally stable suspension (at a given particle size distribution), t^* increases with increase in the volume fraction of the oil phase, ϕ. In other words, the cross-over point shifts to lower frequency with increase in ϕ. For a given suspension, t^* increases with increase in flocculation provided that the particle size distribution remains the same (i.e. no Ostwald ripening).

The value of G' also increases with increase in flocculation, since aggregation of particles usually results in liquid entrapment and the effective volume fraction of the suspension shows an apparent increase. With flocculation, the net attraction between the droplets also increases and this results in an increase in G'. G' is determined by the number of contacts between the particles and the strength of each contact (which is determined by the attractive energy.

It should be mentioned that in practice one may not obtain the full curve, due to the frequency limit of the instrument, and also measurement at low frequency is time consuming. Usually one obtains part of the frequency dependence of G' and G''. In most cases, one has a more elastic than viscous system. Most suspension systems used in practice are weakly flocculated and they also contain 'thickeners' or 'structuring' agents to reduce sedimentation and to acquire the right rheological characteristics for application.

The exact values of G' and G'' required depend on the system and its application. In most cases, a compromise has to be made between acquiring the right rheological characteristics for application and the optimum rheological parameters for long-term physical stability.

Application of rheological measurements to achieve the above conditions requires a great deal of skill and understanding of the factors that affect rheology.

References

1 Th.F. Tadros, *Applied Surfactants*, Wiley-VCH, Weinheim (2005).
2 Th.F. Tadros, *Adv. Colloid Interface Sci.*, *12*, 141 (1980).
3 Th.F. Tadros(ed.), *Solid/Liquid Dispersions*, Academic Press, London (1987).
4 Th.F. Tadros, *Surfactants in Agrochemicals*, Marcel Dekker, New York (1994).
5 G.D. Parfitt, Fundamental aspects of dispersions, in *Dispersion of Powders in Liquids*, ed. G.D. Parfitt, Applied Science, London (1973).
6 E.K. Rideal, *Philos. Mag.*, *44*, 1152 (1922).
7 E.D. Washburn, *Phys. Rev.*, *17*, 273 (1921).
8 P.A. Rehbinder, *Colloid J. USSR*, *20*, 493 (1958).
9 P.A. Rehbinder and V.I. Likhtman, in *Proceedings of International Conference Surface Activity*, Vol. 3, Butterworth, London, p. 157 (1957).
10 E.D. Schukin and P.A. Rehbinder, *Colloid J. USSR*, *20*, 601 (1958).
11 C.A. Smolders, *Recl. Trav. Chim. Pays-Bas*, *80*, 650 (1961).
12 A. Asakura and F. Oosawa, *J. Chem. Phys.*, *22*, 1235 (1954); *J. Polym. Sci.*, *93*, 183 (1958).
13 E.V. Khaminski, *Crystallization from Solution*, Consultants Bureau, New York (1969).
14 G.H. Nancollas and N. Purdie, *Q. Rev. Chem. Soc.*, *18*, 1 (1964).
15 A.L. Smith (ed.), *Particle Growth in Suspensions*, Academic Press, London (1973).
16 A.D. Maude and R.L. Whitmore, *Br. J. Appl. Phys.*, *9*, 477 (1958).
17 G.K. Bachelor, *J. Fluid Mech.*, *52*, 245 (1972).
18 C.C. Reed and J.L. Anderson, *AIChE J.*, *26*, 814 (1980).
19 R. Buscall, J.W. Goodwin, R.H. Ottewill and Th.F. Tadros, *J. Colloid Interface Sci.*, *85*, 78 (1982).
20 I.M. Krieger, *Adv. Colloid Interface Sci.*, *3*, 45 (1971).
21 H. van Olphen, *Clay Colloid Chemistry*, John Wiley and Sons, New York (1963).
22 K. Norrish, *Discuss. Faraday Soc.*, *18*, 120 (1954).
23 M.V. von Smoluchowski, *Handbuch der Electricität und des Magnetismus*, Vol. II, Barth, Leipzig (1914).
24 N. Fuchs, *Z. Phys.*, *89*, 736 (1936).
25 H. Reerink and J.Th.G. Overbeek, *Discuss. Faraday Soc.*, *18*, 74 (1954).
26 Th.F. Tadros, *Adv. Colloid Interface Sci.*, *46*, 1 (1993).
27 Th.F. Tadros, *Adv. Colloid Interface Sci.*, *68*, 91 (1996).
28 W.H. Herschel and R. Buckley, *Proc. Am. Soc. Test Mater.*, *26*, 621 (1926); Kolloid Z., 39, 291 (1926).
29 N. Casson, in *Rheology of Disperse Systems*, ed. C.C. Mill, Pergamon Press, Oxford, pp. 84–104 (1959).
30 J.D. Ferry, *Viscoelastic Properties of Polymers*, John Wiley & Sons, New York (1980).
31 R.W. Whorlow, *Rheological Techniques*, Ellis Horwood, Chichester (1980).
32 J.W. Goodwin and R. Hughes, *Rheology for Chemists*, Royal Society of Chemistry, Cambridge (2000).

8
Oil-based Suspension Concentrates

8.1
Introduction

Oil-based suspensions are currently used for the formulation of many agrochemicals, in particular those which are chemically unstable in aqueous media. These suspensions allow one to use oils (such as methyl oleate) which may enhance the biological efficacy of the active ingredient. In addition, one may incorporate water-insoluble adjuvants in the formulation. The most important criterion for the oil used is to have minimum solubility of the active ingredient, otherwise Ostwald ripening or crystal growth will occur on storage.

The oil-based suspension has to be diluted into water to produce an oil-in-water emulsion. A self-emulsifiable system has to be produced and this requires the presence of the appropriate surfactants for self-emulsification. The surfactants used for self-emulsification should not interfere with the dispersing agent that is used to stabilize the suspension particles in the nonaqueous media. Displacement of the dispersing agent with the emulsifiers can lead to flocculation of the suspension.

To prevent sedimentation of the particles (since the density of the active ingredient is higher than that of the oil in which it is dispersed), an appropriate rheology modifier (antisettling agent) that is effective in the nonaqueous medium must be incorporated in the suspension. This rheology modifier should not interfere with the self-emulsification process of the oil-based suspension.

Two main types of nonaqueous suspensions may be distinguished: (1) suspensions in polar media such as alcohol, glycols, glycerol, esters; these media have a relative permittivity $\varepsilon_r > 10$ In this case double layer repulsion plays an important role, in particular when using ionic dispersing agents; (2) suspensions in nonpolar media, $\varepsilon_r < 10$, such as hydrocarbons (paraffinic or aromatic oils) which can have a relative permittivity as low as 2. In this case charge separation and double layer repulsion are not effective and hence one has to depend on the use of dispersants that produce steric stabilization.

Colloids and Interface Science Series, Vol. 5
Colloids in Agrochemicals. Tharwat F. Tadros

ISBN: 978-3-527-31465-2

8.2
Stability of Suspensions in Polar Media

This follows the same principles as aqueous suspensions, but one must take into account the 'incomplete' dissociation of the ionic species which is the case when $\varepsilon_r < 40$. This could result in a low surface charge and hence a low zeta potential. However, the latter may be sufficient to produce an effective energy barrier that prevents any flocculation as explained by the Deryaguin–Landau–Verwey–Overbeek (DLVO) theory [1, 2].

Due to the lack of complete dissociation, one cannot use the electrolyte concentration for calculation of the thickness of the double layer. In this case one has to obtain the effective electrolyte concentration that must take into account the incomplete dissociation. The effective ionic concentration C can be determined from conductivity measurements and this allows one to calculate the double layer thickness $(1/\kappa)$.

$$\frac{1}{\kappa} = \left(\frac{\varepsilon_r \varepsilon_0 kT}{2e^2 N_A C \times 10^3}\right)^{\frac{1}{2}} \tag{1}$$

where ε_0 is the permittivity of free space, k is the Boltzmann constant, T is the absolute temperature and N_A is Avogadro's number.

Since the effective ionic concentration in a polar medium is less than that in water, the double layer will be more extended [3]. For polar substances in polar media, such as TiO_2 in glycol, the double layer charge may be sufficient for effective repulsion to occur and in this case a stable dispersion may be produced. For hydrophobic particles such as the case with agrochemicals stabilized with ionic surfactants or polyelectrolytes, the double layer charge is also effective for stabilization of the suspension.

In the above case, one may describe stability using the classical theory of colloid stability due to DLVO theory [1, 2]. For systems, where $\kappa a > 1$ (where a is the particle radius) and weak interaction, the electrostatic repulsion, G_{el}, is given by the following expression:

$$G_{el} = 4\pi\varepsilon_r\varepsilon_0 a \psi_0{}^2 \ln[1 + \exp(-\kappa h)] \tag{2}$$

where h is the surface-to-surface separation and ψ_0 is the surface potential which can be replaced by the measured zeta potential. G_{el} decreases exponentially with increase in h and the rate of decay increases with increase in electrolyte concentration.

The van der Waals attraction is simply given by the following expression [4]:

$$G_A = -\frac{Aa}{12h} \tag{3}$$

where A is the effective Hamaker constant:

$$A = (A_{11}^{\frac{1}{2}} - A_{22}^{\frac{1}{2}})^2 \tag{4}$$

where A_{11} is the Hamaker constant of the particles and A_{22} is the Hamaker constant of the medium. G_A increases with decrease in h and at very short distances it reaches very high values.

The basis of the DLVO theory is to sum G_{el} and G_A at all distances:

$$G_T = G_{el} + G_A \tag{5}$$

The general form of the G_T–h curve is illustrated schematically in Figure 8.1, which shows the case at low electrolyte concentrations. The manner in which G_T varies with interparticle distance is determined by the way by which G_{el} and G_A vary with h. G_{el} decays exponentially with h and approaches zero at large separation distances. G_A decays with h as an inverse power law and a residual attraction remains at large values of h. The G_T–h curve shows two minima (at long and short distances of separation) and one maximum (at intermediate distance), as illustrated in Figure 8.1.

At large h, $G_A > G_{el}$, resulting in a shallow secondary minimum. At small h, $G_A \gg G_{el}$, resulting in a deep primary minimum. At intermediate h, $G_{el} > G_A$, resulting in an energy maximum G_{max}. Three cases can be distinguished: (1) G_{max} large (>25 kT), resulting in a colloidally stable suspension; (2) G_{max} small or absent, resulting in an unstable suspension (coagulated); and (3) G_{max} intermediate and G_{sec} deep (1–10 kT), resulting in weak (reversible) flocculation.

It should be mentioned that for hydrophobic particles in polar solvents, stabilized by ionic surfactants, the surface charge may be very low and hence electrostatic stabilization cannot be achieved. For this case, it is essential to use a dispersant that adsorbs strongly on the particle surface (with an anchor B chain).

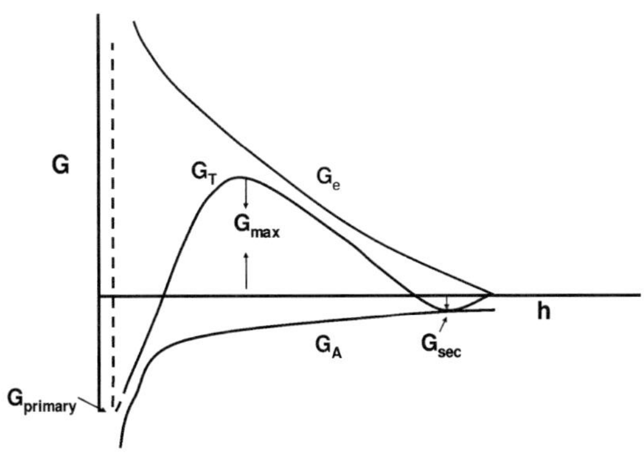

Figure 8.1 Energy–distance curves according to the DLVO theory.

This dispersant should contain stabilizing chains (A chains) that are strongly solvated by the medium. The use of block and graft copolymers for the stabilization of particles in polar media follows the same principles that are applied for stabilization in nonpolar media. This stabilizing mechanism is referred to as steric stabilization, which will be discussed below.

The choice of dispersants for stabilization of suspensions in polar media is not an easy task. The 'anchor' chain B needs to be insoluble in the polar medium and has high affinity to the surface. Since most polymer chains have some solubility in nonaqueous polar media, it is very difficult to have strong adsorption to the particle surface. The stabilizing chains A have to be highly soluble in the polar solvent and strongly solvated by its molecules. This is not as difficult as the choice of the B chains. For these reasons, it is difficult to find a commercially available dispersant for stabilization of hydrophobic particles in polar media.

8.3
Stability of Suspensions in Nonpolar Media

For nonpolar media such as hydrocarbon oils (e.g. paraffinic oils) or some modified vegetable oils such as methyl oleate that are commonly used for the formulation of oil-based suspensions, the relative permittivity is lower than 10 and can reach values as low as 2. In this case, the double layer is very extended, since the effective ionic concentration is very small ($\sim 10^{-10}$ mol dm^{-3}) and $1/\kappa$ can be as high as 10 µm. In this case, the double layer repulsion plays a very minor role. Effective stabilization requires the presence of an adsorbed layer of an effective dispersant with particular properties (sometimes referred to as 'protective colloid').

The most effective dispersants are polymeric surfactants, which may be classified into two main categories [5]: (1) homopolymers and (2) block and graft copolymers. The homopolymers adsorb as random coils with tail–rain–loop configurations. In most cases, there is no specific interaction between the homopolymer and the particle surface and it can seldom provide effective stabilization. The block and graft copolymers can provide effective stabilization, provided that they satisfy the following criteria: (1) strong adsorption of the dispersant to the particle surface; this can be provided by a block B that is chosen to be insoluble in the nonaqueous medium and has some affinity to the particle surface; in cases where the affinity to the surface is not strong, one can rely on 'rejection anchoring', where the insoluble B chain is rejected towards the surface as a result of its insolubility in the nonaqueous medium; (2) strongly solvated A blocks that provide effective steric stabilization as a result of their unfavorable mixing and loss of entropy when the particles approach each other in the suspension; (3) a reasonably thick adsorbed layer to prevent any strong flocculation.

Based on the above principles, various dispersants have been designed for suspensions in nonaqueous media. One of the most effective stabilizing chains in nonaqueous media is poly(12-hydroxystearic acid) (PHS), which has a molar mass in the range 1000–2000 Da. This chain is strongly solvated in most hydrocarbon oils (paraffinic or aromatic oils). It is also strongly solvated in many esters such as methyl oleate, which is commonly used for oil-based suspensions. For the B chain, one can choose a polar chain, such as polyethylenimine or poyvinylpyrrolidone, which is insoluble in most oils. The B chain could also be polystyrene or poly(methyl methacrylate), which is insoluble in aliphatic hydrocarbons and may have some affinity to the hydrophobic agrochemical particle.

8.3.1
Characterization of the Adsorbed Polymer Layer

For full characterization of the adsorbed polymer layer, one needs to determine the following parameters [6]: (1) amount of polymer adsorbed per unit area, Γ; this requires determination of the adsorption isotherms at various temperatures; the influence of solvency for the stabilizing chain should be studied; (2) number of segments in direct contact with the surface, p (in trains); this can be obtained using pulse gradient NMR [6]; (3) the adsorption energy per segment χ^s; for polymer adsorption to occur; a minimum value for χ^s is required to overcome the entropy loss when the polymer adsorbs at the surface; the adsorption energy can be determined using microcalorimetry; (4) extension of the layer from the surface, i.e. segment density distribution $\rho(z)$ or the adsorbed layer thickness δ; the most convenient method for determination of δ is to use dynamic light scattering, usually referred to as photon correlation spectroscopy (PCS) [7]; for this purpose, small spherical particles such as poly(methyl methacrylate) can be used.

8.4
Theory of Steric Stabilization

When two particles with adsorbed polymer layers, each with thickness δ, approach to a distance of separation h that is smaller than twice the adsorbed layer thickness 2δ, the layers will either overlap with each other or they become compressed [5]. This is illustrated in Figure 8.2.

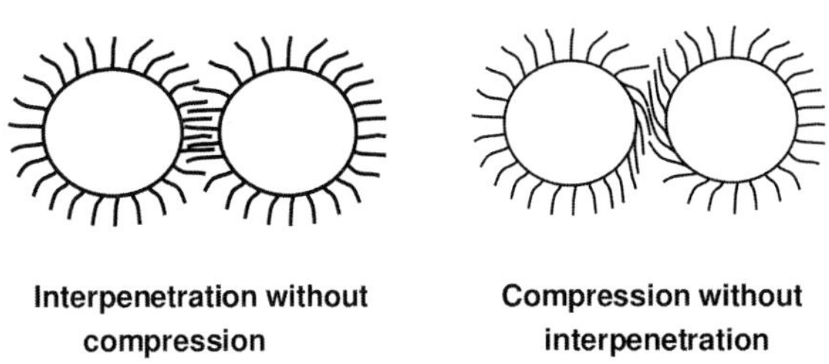

Figure 8.2 Schematic representation of the interaction between particles containing adsorbed polymer layers.

In both cases, there will be an increase in the local segment density of the polymer chains in the interaction region. Provided that the dangling chains (the A chains in A–B or A–B–A block or BA_n graft copolymers) are in a good solvent, this local increase in segment density in the interaction zone will result in strong repulsion as a result of two main effects: (1) increase in the osmotic pressure in the overlap region as a result of the unfavorable mixing of the polymer chains, when these are in good solvent conditions [8]; this is referred to as osmotic repulsion or mixing interaction and it is described by a free energy of interaction G_{mix}; (2) reduction in the configurational entropy of the chains in the interaction zone; this entropy reduction results from the decrease in the volume available for the chains when these are either overlapped or compressed; this is referred to as volume restriction interaction, entropic or elastic interaction and it is described by a free energy of interaction G_{el}.

Combination of G_{mix} and G_{el} is usually referred to as the steric interaction free energy, G_s, i.e.

$$G_s = G_{mix} + G_{el} \tag{6}$$

The sign of G_{mix} depends on the solvency of the medium for the chains. If in a good solvent, i.e. the Flory–Huggins interaction parameter χ is less than 0.5, then G_{mix} is positive and the mixing interaction leads to repulsion. In contrast, if $\chi > 0.5$ (i.e. the chains are in a poor solvent condition), G_{mix} is negative and the mixing interaction becomes attractive. G_{el} is always positive and hence in some cases one can produce stable dispersions in a relatively poor solvent (enhanced steric stabilization).

In the overlap region, the chemical potential of the polymer chains is now higher than in the rest of the layer (with no overlap). This amounts to an increase in the osmotic pressure in the overlap region and, as a result, solvent will diffuse from the bulk to the overlap region, thus separating the particles and hence a strong repulsive energy arises from this effect. This repulsive energy can be calculated by considering the free energy of mixing of two polymer solutions, as treated by, for example, Flory and Krigbaum [9]. The free energy of mixing is given by two terms: (1) an entropy term that depends on the volume fraction of polymer and solvent and (2) an energy term that is determined by the Flory–Huggins interaction parameter χ.

Using the above theory, one can derive an expression for the free energy of mixing of two polymer layers (assuming a uniform segment density distribution in each layer) surrounding two spherical particles as a function of the separation distance h between the particles.

The expression for G_{mix} is

$$\frac{G_{mix}}{kT} = \left(\frac{2V_2^2}{V_1}\right)\nu_2^2\left(\frac{1}{2}-\chi\right)\left(\delta-\frac{h}{2}\right)^2\left(3R+2\delta+\frac{h}{2}\right) \tag{7}$$

where k is the Boltzmann constant, T is the absolute temperature, V_2 is the molar volume of the polymer, V_1 is the molar volume of the solvent and v_2 is the number of polymer chains per unit area.

The sign of G_{mix} depends on the value of the Flory–Huggins interaction parameter χ: if $\chi < 0.5$, G_{mix} is positive and the interaction is repulsive; if $\chi > 0.5$, G_{mix} is negative and the interaction is attractive. The condition $\chi = 0.5$, i.e. $G_{mix} = 0$, is referred to as the θ-condition.

The elastic interaction arises from the decrease in configurational entropy of the chains on the approach of a second particle [10]. As a result of this approach, the volume available for the chains becomes restricted, resulting in a decrease in the number of configurations.

For two flat plates, G_{el} is given by the following expression (10),

$$\frac{G_{el}}{kT} = 2v_2 \ln\left[\frac{\Omega(h)}{\Omega(\infty)}\right] = 2v_2 R_{el}(h) \tag{8}$$

where $R_{el}(h)$ is a geometric function whose form depends on the segment density distribution.

It should be stressed that G_{el} is always positive and could play a major role in steric stabilization. It becomes very strong when the separation distance between the particles becomes comparable to the adsorbed layer thickness δ.

Combination of G_{mix} and G_{el} with G_A gives the total energy of interaction G_T (assuming there is no contribution from any residual electrostatic interaction), i.e.

$$G_T = G_{mix} + G_{el} + G_A \tag{9}$$

A schematic representation of the variation of G_{mix}, G_{el}, G_A and G_T with surface–surface separation distance h is shown in Figure 8.3.

G_{mix} increases very sharply with decrease in h, when $h < 2\delta$. G_{el} increases very sharply with decrease in h, when $h < \delta$. G_T versus h shows a minimum, G_{min}, at separation distances comparable to 2δ. When $h < 2\delta$, G_T shows a rapid increase with decrease in h.

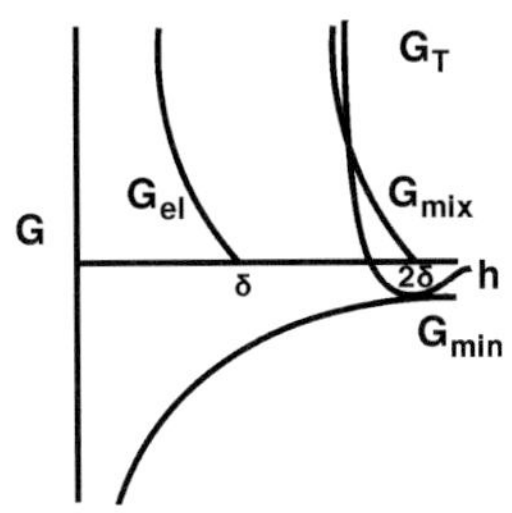

Figure 8.3 Energy–distance curves for sterically stabilized systems.

Unlike the G_T–h curve predicted by the DLVO theory (which shows two minima and one energy maximum), the G_T–h curves for systems that are sterically stabilized show only one minimum, G_{min}, followed by sharp increase in G_T with decrease in h (when $h < 2\delta$). The depth of the minimum depends on the Hamaker constant A, the particle radius R and adsorbed layer thickness δ. G_{min} increases with increase in A and R. At a given A and R, G_{min} increases with decrease in δ (i.e. with decrease in the molecular weight, M_w, of the stabilizer). The larger the value of δ/R, the smaller is the value of G_{min}. In this case, the system may approach thermodynamic stability, as is the case with nano-dispersions.

8.4.1
Criteria for Effective Steric Stabilization

The criteria are as follows:

1. Full coverage of the particles by the polymer. The amount of polymer added should correspond to the plateau of the adsorption isotherm. Any bare patches may lead to flocculation as a result of van der Waals attraction and/or bridging.
2. Strong adsorption of the polymer to the particle surface. A strong B 'anchor' chain: this requires the chain to be insoluble in the medium and has some affinity to the surface. Lack of a strong 'anchor' may lead to chain displacement on approach of the particles in a suspension. This is particularly important for concentrated suspensions.
3. The stabilizing chain(s) A should be in good solvent conditions. The Flory–Huggins interaction parameter χ should be less than 0.5 under all storage conditions, e.g. temperature changes.
4. An optimum layer thickness δ to ensure that G_{min} is not deep. This is particularly the case with concentrated suspensions where weak flocculation may occur at relatively small G_{min}. In most cases, an adsorbed layer thickness in the range 5–10 nm is sufficient. This is particularly the case with graft copolymers where the A chains are stretched ('comb-like' or 'brush' structure).

8.5
Settling of Suspensions

8.5.1
Prevention of Formation of Dilatant Sediments

This follows the same principles as described for aqueous suspension concentrates (see Chapter 7) and only a summary is given here.

Settling (or sedimentation) results from gravity forces [11]. When $\rho_{particles} > \rho_{medium}$, the gravity force $(4/3)\pi a^3 \Delta\rho g L$ (where g is the acceleration due to gravity and L is the height of the container) will overcome the thermal (Brownian) motion, leading to particle sedimentation.

Only when $kT > (4/3)\pi a^3 \Delta\rho g L$ does sedimentation become insignificant. This condition is only satisfied with very small particles (<0.1 μm) and $\Delta\rho < 0.1$. Thus, with most colloidally stable suspensions, sedimentation is the rule rather than the exception. The particles may sediment individually to the bottom of the container and the repulsive forces necessary to maintain stability allow them to move past each other, resulting in the formation of a very compact sediment. The compact sediment, which is technically referred to as 'clay' or 'cake', is very difficult to redisperse by shaking. Such clays are dilatant (shear thickening) and must be avoided.

For an infinitely dilute suspension of non-interacting particles, the sedimentation velocity v_0 is given by Stokes' law:

$$v_0 = \frac{2}{9}\frac{a^2 \Delta\rho g}{\eta} \tag{10}$$

For 1 μm particles with $\Delta\rho = 0.2$ and $\eta = 10^{-3}$ Pa s (low-viscosity oil), $v_0 = 4.4 \times 10^{-7}$ m s^{-1}, whereas for 10 μm particles $v_0 = 4.4 \times 10^{-5}$ m s^{-1}. In a 100 cm container, the 1 μm particles will settle in about 3 days and the 10 μm particles will settle in about 40 min.

For more concentrated suspensions, the sedimentation velocity v is reduced below the Stokes velocity v_0 as a result of interparticle interaction. For suspensions with volume fraction $\phi \approx 0.1$, v is related to v_0 by the equation [12]

$$v = v_0(1 - k\phi) \tag{11}$$

where $k = 6.55$, a constant that accounts for hydrodynamic interaction.

For suspensions with $\phi > 0.1$, v becomes a complex function of ϕ and only empirical equations are used to describe the sedimentation velocity [11]. When v is plotted against ϕ, an exponential reduction in v with increase in ϕ is found and eventually v approaches zero at some limiting volume fraction [13], the so-called maximum packing fraction ϕ_p. At this maximum packing fraction, the relative viscosity of the suspension approaches infinity.

In most practical suspensions, a volume fraction in the range 0.1–0.5 is used and hence one should use a suspending agent to prevent sedimentation and formation of dilatant clays. The suspending agent (sometimes referred to as an anti-settling agent) should provide a restoring force to overcome the gravity.

It is useful to calculate the stress exerted by a particle σ_p in a suspension [14]:

$$\sigma_p = \frac{\text{gravity force}}{\text{area}} = \frac{\frac{4}{3}\pi a^3 \Delta\rho g}{4\pi a^2} = \frac{a\Delta\rho g}{3} \tag{12}$$

For 10 μm particles with $\Delta\rho = 1$, $\sigma_p = 3.3 \times 10^{-2}$ Pa. This clearly illustrates the need to measure the viscosity at such low stresses (using a constant stress rheometer).

8.5.2 Examples of Suspending Agents That Can Be Applied for Prevention of Settling in Non-aqueous Suspensions

The main criteria for an effective suspending agent are as follows:

1. It should produce a 'three-dimensional' gel network in the continuous phase with optimum rheological characteristics.
2. It should have a very high viscosity at low shear rates. A residual (zero shear) viscosity in excess of 100 Pa s is desirable. The residual viscosity can be measured using constant stress (creep) measurements.
3. It should have sufficient 'bulk' modulus to prevent separation of the suspension and syneresis. The bulk modulus is related to the shear modulus, which can be measured using dynamic (oscillatory) techniques.
4. The suspending agent should produce a shear thinning system such that on application of the suspension, the high shear viscosity is not too high.

Several suspending systems can be used and these are summarized below:

1. Hydrophobically modified clays (bentones). Clays such as sodium montmorillonite can be made hydrophobic by addition of long-chain alkylammonium surfactants (e.g. dodecyl- or cetyltrimethylammonium chloride). The alkylammonium cation exchanges with the Na^+ ions producing a hydrophobic surface and the bentone particles can be dispersed in the nonaqueous medium. The exchange is carried out in an optimum manner such that some hydrophilic sites remain on the bentone particles. On addition of a polar solvent such as propylene carbonate or alcohol, a gel is produced. The most likely mechanism of gel formation in this case is hydrogen bonding by the polar molecules between the polar sites on the bentone particles.
2. Fumed silica. These are commercially available under the trade names Aerosil and Cabosil. They are produced by reaction of silicon tetrachloride with steam. Fine particles (primary particle size $<0.02\,\mu m$) are produced with a surface containing silanol groups that are separated by siloxane bonds. When dispersing the fumed silica powder in nonaqueous media, a gel is produced by hydrogen bonding between the silanol groups. Chain aggregates are produced and, by controlling the dispersion procedure, adequate rheological characteristics are produced.
3. Trihydroxystearin (Thixin R). This product is derived from castor oil and it needs both shear and heat for full activation to produce a gel.
4. Aluminum magnesium hydroxide stearate. This has been used to gel a number of cosmetic oils and it can provide suspending properties. It could also be applied for oil-based suspensions of agrochemicals. However, one should be careful in applying this system since in many cases the gel strength is too high for adequate dispersion on application.
5. Use of high molecular weight polymers. Polyethylenes and copolymers of polyethylene can be used to gel mineral oils and several aliphatic solvents.

Some of these materials need high incorporation temperatures and one must bear in mind that the cooling procedures affect the final appearance and rheological characteristics of the gel.

The above systems are the most commonly used rheology modifiers in most nonaqueous suspensions. However, two other methods may be applied for the reduction of settling of nonaqueous suspensions:

1. Controlled flocculation (self-structured systems). By reducing the adsorbed layer thickness, one can increase the depth of the minimum, G_{min}, to such an extent that weak flocculation may produce a 'three-dimensional' gel structure that is sufficient to reduce sedimentation and formation of dilatant clays. The adsorbed layer thickness may be reduced by reducing the molecular weight of the stabilizer or by the addition of a small amount of a non-solvent. This is illustrated in Figure 8.4, where the energy–distance curve is plotted as a function of the ratio of adsorbed layer thickness to particle radius. This procedure is particularly useful with large and asymmetric particles.
2. Depletion flocculation [15]. This is obtained by the addition of 'free' (non-adsorbing) polymer to a sterically stabilized suspension. Above a critical volume fraction, ϕ_p^+, of the free polymer, weak flocculation occurs. Above ϕ_p^+, the polymer chains are 'squeezed out' from between the particles, leaving a polymer-free zone in the interstices. The higher osmotic pressure outside the particle surfaces causes weak flocculation. This is shown schematically in Figure 8.5.

The critical polymer volume fraction above which flocculation occurs decreases with increase in the molecular weight of the free polymer. Therefore, high molecular weight oil-soluble polymers should be used in order to reduce the amount required for flocculation. The amount of free polymer required for flocculation also decreases with increase in the volume fraction of the suspension. Therefore, depletion flocculation is more applicable for concentrated suspensions.

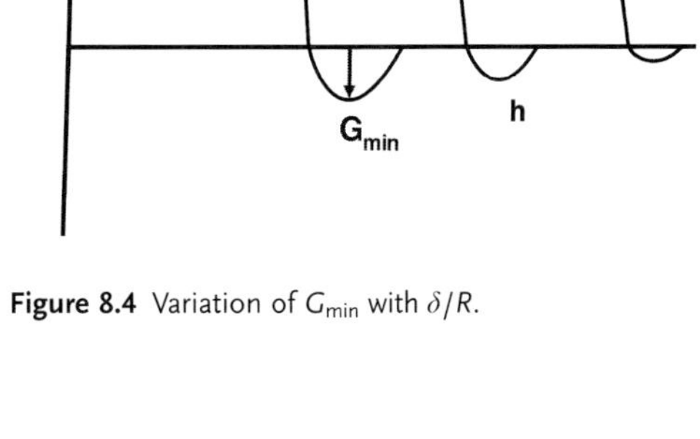

Figure 8.4 Variation of G_{min} with δ/R.

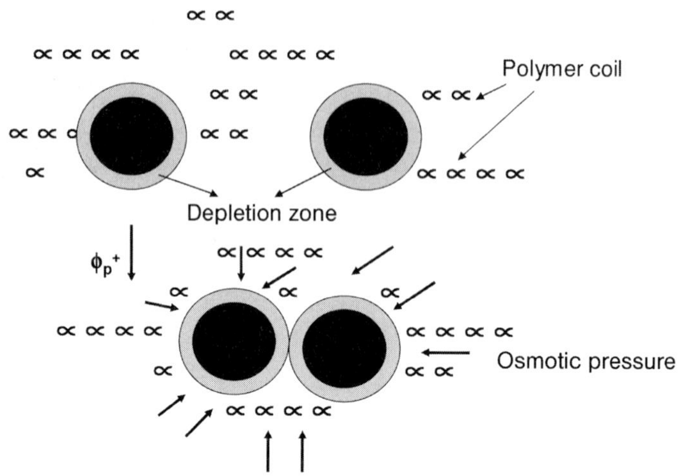

Figure 8.5 Schematic representation of depletion flocculation.

8.6 Emulsification of Oil-based Suspensions

The nonaqueous suspension concentrate containing the active ingredient particles and any antisettling system is emulsified into the spray tank before application. Here the same principles applied for self-emulsification of emulsifiable concentrates (ECs) are applied, as discussed below. Alternatively, the nonaqueous suspension concentrate is emulsified into an aqueous solution containing another water-soluble active ingredient (normally an electrolyte such as glyphosate) to produce a combined mixture for two active ingredients. In the first case, spontaneous emulsification is necessary since only gentle agitation in the spray tank is possible. In the second case, one may apply a normal emulsification procedure if a combined formulation is required. In some cases, spontaneous emulsification into an aqueous electrolyte solution may also be required.

The above oil-based suspension concentrates require careful formulation to achieve the required properties: (1) a stable nonaqueous suspension with adequate weak flocculation (to produce a shear thinning system) and no settling; (2) the antisettling agent used in the formulation should be water dispersible (e.g. fumed silica, Aerosil); (3) the emulsifying system used should not interfere with the stabilizing polymeric surfactant used for preparation of the nonaqueous suspension; (4) the viscosity of the formulation at intermediate shear rates should be low enough to ensure spontaneity of emulsification.

8.6.1 Mechanism of Spontaneous Emulsification and the Role of Mixed Surfactant Film

The first demonstration of spontaneous emulsification was given by Gad [16], who observed that when a solution of lauric acid in oil is carefully placed in an

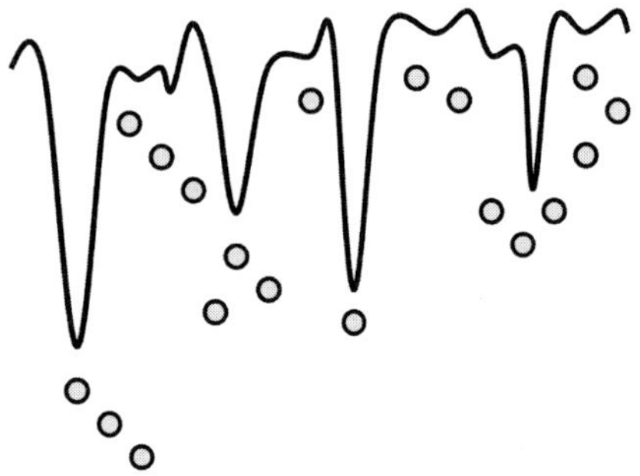

Figure 8.6 Schematic representation of interfacial turbulence.

aqueous alkaline solution, an emulsion forms at the interface. It is clear from this experiment that a mixture of lauric acid and sodium laurate is produced and this illustrates the role of the mixed film (which produces an ultralow interfacial tension). When the right conditions are produced, spontaneous emulsification occurs with minimum agitation. In this experiment, the oil used was Newtonian and it had a relatively low viscosity.

Three mechanisms may be established to explain the process of spontaneous emulsification.

(a) Interfacial turbulence (Figure 8.6). Interfacial turbulence occurs as a result of mass transfer from one phase to the other. The interface shows unsteady motions; streams of one phase are ejected and penetrate into the second phase, shedding small droplets. Localized reduction in interfacial tension is caused by non-uniform adsorption of the surfactant at the O/W interface or by mass transfer of surfactant molecules across the interface. When the two phases are not in equilibrium, convection currents may be formed transferring liquid rich in surfactant towards areas of liquid deficient in surfactant. These convection currents lead to local fluctuations in the interfacial tension causing oscillation of the interface (turbulence).

Such disturbances may amplify themselves, leading to violent interfacial perturbations and eventual disintegration of the interface, when liquid droplets of one phase are 'thrown' into the other phase. This mechanism requires the presence of two surfactant molecules or a surfactant plus alcohol. This facilitates the mass transfer and induces interfacial tension gradients. Several 'phases' may be produced at the O/W interface, as will be discussed later.

(b) Diffusion and stranding (Figure 8.7). This is best illustrated by placing an ethanol–toluene mixture (containing say 10% alcohol) on to water. The aqueous layer eventually becomes turbid as a result of the presence of toluene droplets. In this case interfacial turbulence does not occur although spontaneous emulsification takes place. The alcohol molecules diffuse into the aqueous phase, carrying some toluene molecules in a saturated three-component subphase. At some

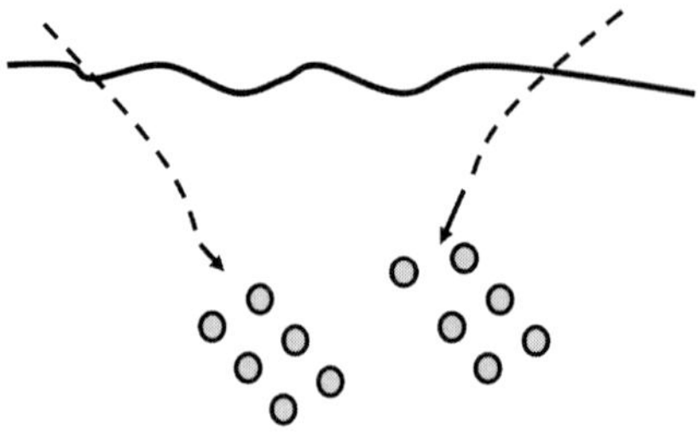

Figure 8.7 Schematic representation of diffusion and stranding.

distance from the interface, the alcohol becomes sufficiently diluted into water and the toluene droplets precipitate as droplets in the aqueous phase.

The above mechanism requires the presence of a third component (sometimes referred to as the co-surfactant) that increases the miscibility of the two previously immiscible phases (toluene and water). Many emulsifiable concentrates contain a high proportion of a polar solvent such as alcohol or ketone which facilitates the spontaneous emulsification. Its application for emulsification of nonaqueous suspension concentrates needs to be tested (addition of a polar solvent may enhance crystal growth).

(c) Production of ultralow (or transiently negative) interfacial tension (Figure 8.8). This is the same mechanism as for the production of microemulsions. The ultralow interfacial tension is produced by a combination of a surfactant and co-surfactant. This is illustrated in Figure 8.9.

Addition of surfactant to the aqueous or oil phase causes a gradual lowering of the interfacial tension γ, reaching a limiting value at the critical micelle concentration (cmc). Any further increase in C above the cmc causes little or no further decrease in γ. The limiting value of γ reached with most single surfactants is seldom lower than 0.1 mN m^{-1}, which is not sufficiently low for microemulsion formation. If a surfactant mixture, with one predominantly water soluble, such as sodium dodecyl sulfate (SDS), and the other predominantly oil soluble, usually referred to as the co-surfactant, such as pentanol or hexanol, is used the

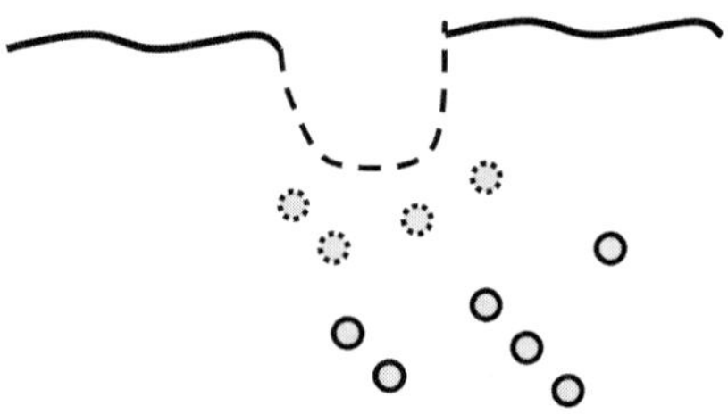

Figure 8.8 Schematic representation of production of ultralow interfacial tension.

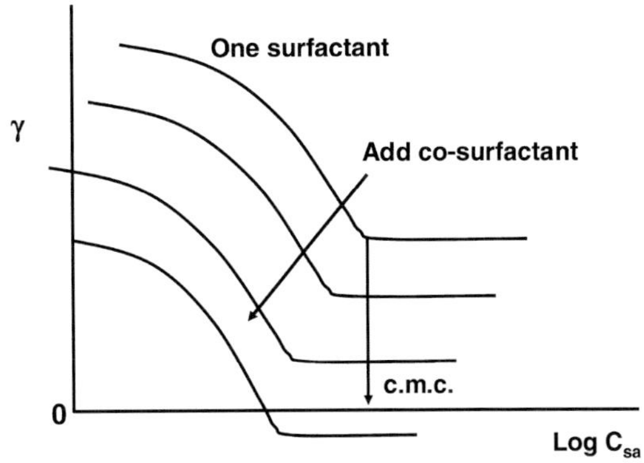

Figure 8.9 γ–log C curves for surfactant plus co-surfactant.

limiting γ value can reach very low values ($<10^{-2}$ mN m^{-1}) or even becomes transiently negative. When a transient negative interfacial tension is produced, the interface expands spontaneously (producing droplets) adsorbing all surfactant molecules until a small positive γ is reached [17].

The reason for the reduction in interfacial tension when using two molecules can be understood from a consideration of the Gibbs adsorption equation.

The reduction in γ is related to the adsorption of the surfactant molecules, which is referred to as the surface excess Γ (mol m^{-2}); for a multicomponent system the reduction in γ, i.e. dγ, is given by the summation of all surface excesses [17]:

$$d\gamma = -\sum \Gamma_i RT d \ln C_i \tag{13}$$

Thus, for two surfactants γ is reduced twice provided that the two molecules are adsorbed simultaneously at the interface. The two surfactant molecules should not interact with each other, otherwise they will lower their respective activities. This explains why the two molecules should vary in nature, one being predominantly water soluble and the other predominantly oil soluble.

Several other mechanisms have been proposed to explain the dynamics of spontaneous emulsification. Direct observation using phase contrast and polarizing microscopy showed that in some cases vesicles (closed bilayers) are produced in the oil phase near the interface with the water. These vesicles tend to 'explode', thereby pulverizing oil droplets into the aqueous phase. The above structures can be produced, for example, by using a mixture of a nonionic surfactant such as $C_{12}E_5$ and a long-chain alcohol such as $C_{12}H_{25}OH$ and an oil such as hexadecane. The oil–surfactant mixture is transparent, but on addition of a small amount of water it becomes turbid and vesicles can be observed under the microscope. The interfacial tension of the oil–surfactant mixture–water is also very low.

8.6.2
Polymeric Surfactants for Oil-based Suspensions and the Choice of Emulsifiers

When formulating a nonaqueous suspension concentrate that can be spontaneously emulsified into aqueous solutions, one should consider the following criteria. The polymeric emulsifiers should be chosen from the A–B or A–B–A block or BA_n graft copolymers as discussed above. The B chain should be chosen to be insoluble in the oil and should become strongly adsorbed on the particle surface (either by specific interaction or by rejection 'anchoring'). As discussed before, for nonpolar oils (hydrocarbons) the B chain could be a polar molecule such as polyethylenimine, which adsorbs on the particle by rejection anchoring. The A chain should be soluble in the oil and be highly solvated by its molecules – poly(hydroxystearic acid) or polyisobutylene is ideal.

The emulsifier system should be soluble in the oil phase and it should not cause desorption of the polymeric surfactant. A two-component emulsifier system is normally used, one predominantly water soluble (such as an ethoxylated surfactant) and the other oil soluble (such as a medium- or long-chain alcohol). In many cases calcium dodecylbenzenesulfonate may be used. The emulsifier system should lower the interfacial tension of O/W to very low values ($<0.1\,\text{mN m}^{-1}$).

The ultralow interfacial tension may be measured using the spinning drop technique. A drop of oil is injected into the aqueous phase, which is placed in a capillary tube that can be rotated at high speed to cause deformation of the spherical drop into a cylinder [17]. From the droplet profile (cylinder radius R) and speed of rotation ω one can calculate γ:

$$\gamma = \frac{\omega^2 \Delta\rho R^4}{4} \qquad (14)$$

where $\Delta\rho$ is the density difference between water and oil.

8.6.3
Emulsification into Aqueous Electrolyte Solutions

For the preparation of a combined formulation with one active ingredient suspended in the oil phase and another active ingredient soluble in water (salt), one can emulsify the nonaqueous suspension into the aqueous electrolyte solution using a polymeric surfactant with high HLB number, e.g. Synperonic PF127 [which contains ~55 units poly(propylene oxide) (PPO) and two poly(ethylene oxide) (PEO) chains with ~100 units each]. The PEO–PPO–PEO block copolymer is insoluble in the oil phase and hence it does not interfere with the polymeric surfactant used for the preparation of the nonaqueous suspension. Depending on the density of the active ingredient in the oil phase and the amount suspended, the density of the oil drops produced could be smaller or larger than that of the aqueous electrolyte. Hence the combined formulation could undergo creaming or settling on storage. It is essential, therefore, to include an antisettling system in the aqueous electrolyte phase, e.g. xanthan gum (Kelzan or Rhodopol) could be applied in this case.

In a formulation where spontaneous emulsification is required into the aqueous electrolyte solution, one should apply the same principles as discussed above, except in this case the emulsifier system should be more hydrophilic (higher HLB number) to prevent 'salting-out' of the emulsifier by the electrolyte. The nature and HLB number of the emulsifier system depend on the electrolyte concentration and nature and this could be checked independently using cloud point measurements. The resulting interfacial tension should be low.

8.7
Proper Choice of the Antisettling System

The antisettling system used for nonaqueous suspensions should be dispersible into water. Fumed silica (Aerosil 200) or microcrystalline cellulose are ideal systems for structuring nonaqueous suspensions. These systems produce a 'three-dimensional' gel network in the oil phase by hydrogen bonding between the particles forming chains and cross-chains. These gels produce sufficient 'yield stress' to prevent sedimentation of the coarse active ingredient particles. They also produce a very high viscosity at low shear rates, thus preventing sedimentation.

One of the main advantages of these hydrophilic particles is that they can partition into the aqueous phase and this produces two main effects. During partition into the aqueous phase and crossing the interface, they can enhance the interfacial tension gradients and hence promote the turbulence effect described above. This will facilitate self-emulsification. When these particles leave the oil phase, the yield value and viscosity of the suspension are lowered and this helps the self-emulsification process.

Hydrophilic polymers such as hydroxypropylcellulose may also be used either alone or in combination with silica or microcrystalline cellulose. The combined particulate–polymer system may be advantageous in reducing sedimentation and enhancing the self-emulsification of the oil. The optimum ratio between particles and polymer depends on the nature of the active ingredient suspended in the oil phase.

8.8
Rheological Characteristics of the Oil-based Suspensions

The rheological characteristics of the nonaqueous suspension (its yield value, zero shear viscosity and elastic modulus) need to be carefully adjusted to achieve the following criteria: (1) no settling or separation of the nonaqueous suspension on storage under all storage conditions and (2) ease of spontaneous emulsification with minimum agitation.

The above two criteria are not compatible and hence one has to control the rheology very carefully. For prevention of settling and separation, one needs a high 'yield stress', high zero shear viscosity and high modulus. High rheological parameters reduce the ease of spontaneous emulsification. For these reasons, one needs to prepare a highly shear thinning system. The suspension concentrate should have a high viscosity at low shear rates, but once the shear rate exceeds a certain value (say 10–100 s^{-1}) depending on the application, the viscosity decreases significantly. One can achieve the above effect by proper choice of the antisettling system.

By using an antisettling system that diffuses rapidly into the water phase, one can reduce the yield value, zero shear viscosity and modulus during the process of emulsification. Measurement of the rheological parameters is essential when formulating a nonaqueous suspension concentrate and this requires low-shear rheometers (constant stress and oscillatory techniques). Details of these techniques are available in several review articles [18, 19].

References

1 B.V. Deryaguin and L. Landau, *Acta Physicochim. USSR*, *14*, 633 (1941).
2 E.J.W. Verwey and J.Th.G. Overbeek, *Theory of Stability of Lyophobic Colloids*, Elsevier, Amsterdam (1948).
3 J. Lyklema, Structure of the solid/liquid interface and the electrical double layer, in *Solid/Liquid Dispersions*, ed. Th.F. Tadros, Academic Press, London (1987).
4 H.C. Hamaker, *Physica (Utrecht)*, *4*, 1058 (1937).
5 Th.F. Tadros, Polymer adsorption and dispersion stability, in *The Effect of Polymers on Dispersion Properties*, ed. Th.F. Tadros, Academic Press, London (1981).
6 G.J. Fleer, M.A. Cohen Stuart, J.M.H.M. Scheutjens, T. Cosgrove and B. Vincent, *Polymers at Interfaces*, Chapman and Hall, London (1993).
7 P.N. Pusey, in *Industrial Polymers: Characterization by Molecular Weight*, ed. J.H.S. Green and R. Dietz, Transcripta Books, London (1973).
8 D.H. Napper, *Polymeric Stabilization of Colloidal Dispersions*, Academic Press, London (1981).
9 P.J. Flory and W.R. Krigbaum, *J. Chem. Phys.*, *18*, 1086 (1950).
10 E.L. Mackor and J.H. van der Waals, *J. Colloid Sci.*, *7*, 535 (1951).
11 Th.F. Tadros, Settling suspensions, in *Solid/Liquid Dispersions*, ed. Th.F. Tadros, Academic Press, London (1987).
12 G.K. Bachelor, *J. Fluid Mech.*, *52*, 245 (1972).
13 I.M. Krieger, *Adv. Colloid Interface Sci.*, *3*, 45 (1971).
14 Th.F. Tadros, *Adv. Colloid Interface Sci.*, *12*, 141 (1980).
15 A. Asakura and F. Oosawa, *J. Chem. Phys.*, *22*, 1235 (1954); *J. Polym. Sci.*, *93*, 183 (1958).
16 Th.F. Tadros, *Surfactants in Agrochemicals*, Marcel Dekker, New York (1994).
17 Th.F. Tadros, *Applied Surfactants*, Wiley-VCH, Weinheim (2005).
18 Th.F. Tadros, *Adv. Colloid Interface Sci.*, *46*, 1 (1993).
19 Th.F. Tadros, *Adv. Colloid Interface Sci.*, *68*, 91 (1996).

9
Suspoemulsions

9.1
Introduction

Suspoemulsions are mixtures of suspensions and emulsions that are applied in many agrochemical formulations where two active ingredients are formulated with one as an aqueous suspension and the other as an oil-in-water emulsion. A schematic representation of suspoemulsions is given in Figure 9.1. Two main types can be distinguished: (1) a system where the solid particles and emulsion droplets remain as separate entities and (2) a system where the solid particles are dispersed in the oil droplets. The first system is the one that is commonly applied in agrochemical formulations.

As mentioned above, with suspoemulsions two active ingredients are formulated together, which offers convenience to the farmer and also may result in synergism in biological efficacy. A wider spectrum of disease control may be achieved particularly with many fungicides and herbicides. With many suspoemulsions, an adjuvant that enhances the biological efficacy is added.

The formulation of suspoemulsions is not an easy task; one may produce a stable suspension and emulsion separately, but when these mixed they become unstable due to the following interactions:

1. Homoflocculation of the suspension particles. This can occur if the dispersing agent used for preparation of the suspension is not strongly adsorbed and hence it becomes displaced by the emulsifier, which is more strongly adsorbed but not a good stabilizer for the suspension particles.
2. Emulsion coalescence. This can occur if the emulsifier is not strongly adsorbed at the O/W or W/O interface, resulting in its partial or complete displacement by the suspension dispersant, which is not a good emulsion stabilizer, and this results in coalescence of the emulsion droplets with ultimate separation of oil (for O/W) or water (for W/O).
3. Heteroflocculation between the oil droplets and suspension particles. The latter may be partially wetted by the oil and they may reside at the O/W interface (this is particularly the case if the oil droplets are much larger than the suspension particles). Heteroflocculation can also occur with suspension particles dispersed in a W/O emulsion.

Colloids and Interface Science Series, Vol. 5
Colloids in Agrochemicals. Tharwat F. Tadros

ISBN: 978-3-527-31465-2

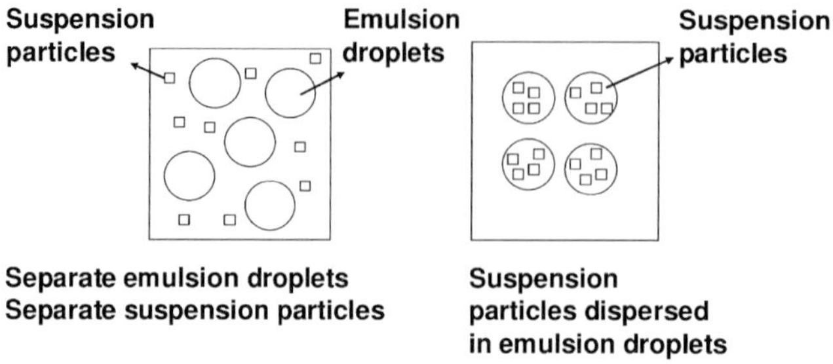

Figure 9.1 Schematic representation of suspoemulsions.

4. Phase transfer and crystallization. This occurs when the suspension particles have some solubility in the oil phase. The small suspension particles, which have higher solubility than the larger particles (due to curvature effects), may become dissolved in the oil phase and they become recrystallized on to the larger suspension particles (a form of an Ostwald ripening process). Large and sometimes needle-shaped crystals may be produced as a result of crystal habit modification (that sometimes occurs with Ostwald ripening).

One of the most useful methods to study interaction in suspoemulsions is rheology, in particular viscoelastic measurements [1, 2].

9.2 Systems Investigated

The following systems were investigated:

1. Practical agrochemical suspoemulsions. Chlorothalonil (density $\rho = 1.85\,g\,cm^{-3}$) and dichlobutrazol (density $\rho = 1.25\,g\,cm^{-3}$) suspensions prepared by bead milling [prepared using Synperonic NP1800, nonylphenol with 13 mol of poly(propylene oxide) (PPO) and 27 mol of poly(ethylene oxide) (PEO)]. Tridemorph (density $\rho = 0.87\,g\,cm^{-3}$) emulsions prepared using the same surfactants (emulsification was carried out using a Silverson mixer).
2. Model suspoemulsions of polystyrene latex [mean volume diameter (VMD) = 1.84 μm] prepared using the surfactant-free method [3] and then stabilized with Synperonic PE P94 (an A–B–A block of PEO–PPO–PEO with 47 PPO units and 42 PEO units) and isoparaffinic oil-in water emulsion (stabilized with Synperonic PE L92 with 47 PPO units and 16 PEO units).
3. Model suspoemulsions of polystyrene containing grafted PEO chains (with molecular weight ~2000), prepared using the 'aquersamer method' [4] having a Z-average particle radius of 310 ± 10 nm and hexadecane-in-water emulsions (stabilized with Synperonic PE L92) having a Z-average particle radius of 310 ± 10 nm.

9.3
Experimental Methods

The particle size was determined using dynamic light scattering, referred to as photon correlation spectroscopy (PCS) (a Malvern PCS instrument was used) The equilibrium sediment and cream volumes were obtained using measuring cylinders at room temperature. Viscoelastic measurements were carried out using a Bohlin VOR rheometer.

9.4
Results and Discussion

9.4.1
Creaming/Sedimentation of Suspoemulsions

Assuming that a stable suspoemulsion (in the colloid sense) could be prepared, e.g. by the use of a polymeric dispersant and emulsifier, the creaming and/or sedimentation behavior of the suspoemulsion show different patterns depending on the density difference between the oil droplets and suspension particles and also the total volume fraction ϕ of the whole system.

The above behavior could be illustrated by using agrochemical suspoemulsions consisting of an oil, namely tridemorph (an aliphatic-like oil with a density $\rho = 0.87\,\text{g cm}^{-3}$) and two different suspensions, namely dichlobutrazol (with $\rho = 1.25\,\text{g cm}^{-3}$) and chlorothalonil (with $\rho = 1.85\,\text{g cm}^{-3}$).

To ensure colloid stability and absence of heteroflocculation, both the emulsion and suspension were prepared using an A–B–C block copolymer (Synperonic NPE 1800) consisting of an anchor chain (B–C) of PPO (13 mol) and nonylphenol and a stabilizing chain of PEO (23 mol).

Synperonic NPE 1800 is an excellent emulsifier for tridemorph and it is also an excellent dispersant for both dichlobutrazol and chlorothalonil. The colloid stability of both emulsion and suspension was confirmed by optical microscopy, which showed no coalescence of the emulsion or flocculation of the suspensions In addition, on mixing the emulsion and suspension there was no heteroflocculation. The suspoemulsions were prepared by simple mixing of the emulsion and suspension, keeping the total volume fraction ϕ constant while varying the ratio of suspension to emulsion. Two ϕ values were investigated, <0.2 and >0.2, and a comparison was made between systems with small density difference (dichlobutrazol–tridemorph) and large density difference (chlorothalonil–tridemorph). The results are illustrated in Figures 9.2 to 9.4.

When the density difference between the suspension particles and emulsion droplets is not large and $\phi < 0.2$, the emulsion creams and the suspension sediments separately (Figure 9.2). When the density difference between the suspension particles and emulsion droplets is small but $\phi > 0.2$, the system forms a

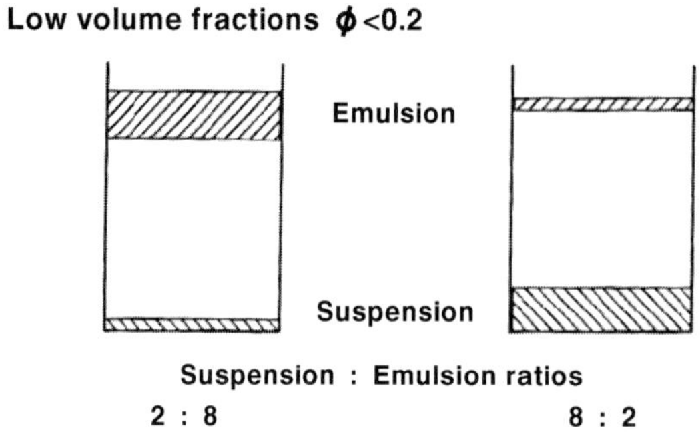

Figure 9.2 Chlobutrazol–tridemorph ($\phi < 0.2$).

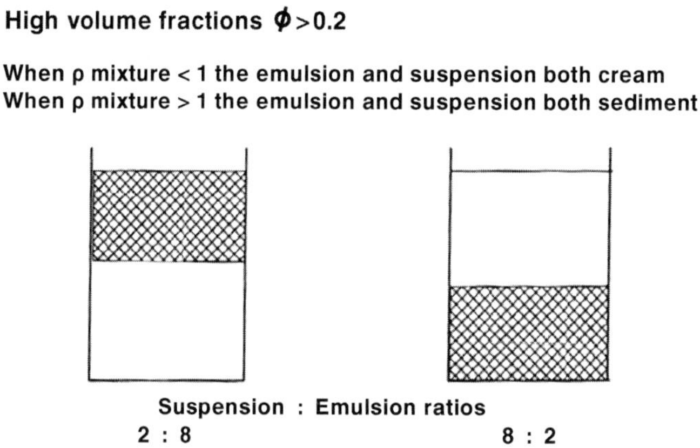

Figure 9.3 Chlobutrazol–tridemorph ($\phi > 0.2$).

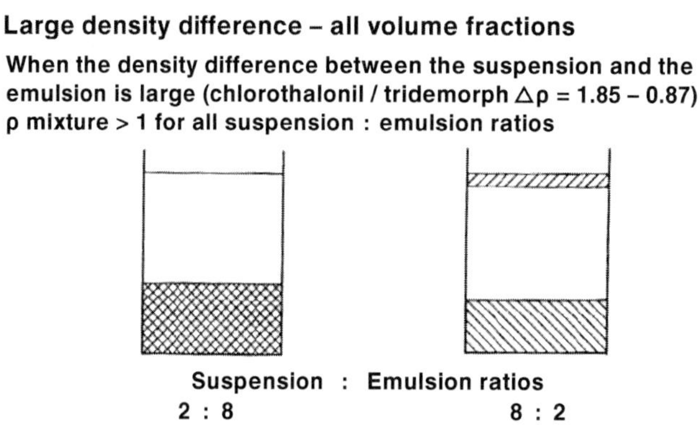

Figure 9.4 Chlorothalonil–tridemorph (all volume fractions).

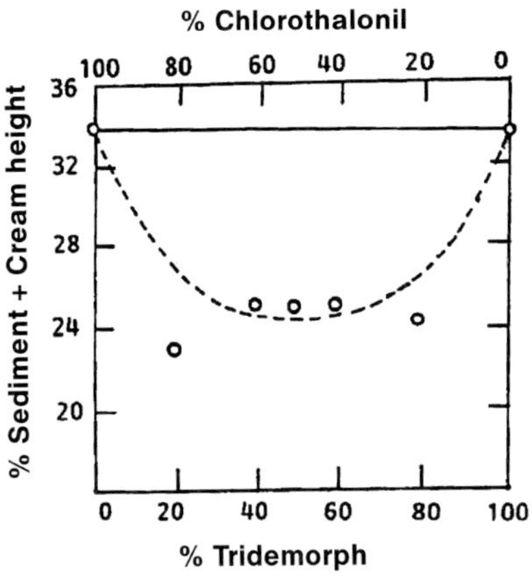

Figure 9.5 Comparison of experimental (—) and predicted (based on additivity) (- - -) sediment + cream heights for chlorothalonil–tridemorph suspoemulsions.

cream layer when the suspension: emulsion ratio is 2 : 8 and it forms a sediment layer when the ratio is 8 : 2 (Figure 9.3). When the density difference between the suspension particles and emulsion droplets is large, then the average density of the suspension particle–emulsion droplet is >1 and in this case sedimentation is observed when the ratio is 2 : 8 but some creaming occurs when the ratio is 8 : 2 (Figure 9.4).

The above creaming/sedimentation behavior indicates some interaction between the emulsion droplets and suspension particles.

A particularly useful method to illustrate the interactions in suspoemulsions is to compare the total observed sediment plus cream height with that based on simple additivity. This is illustrated in Figure 9.5 for chlorothalonil–tridemorph suspoemulsions. It can be seen that the observed sediment + cream heights are smaller than would be expected from simple additivity. It is possible that the small suspension particles become trapped between the larger oil droplets in the cream layer and the small suspension particles become entrapped between the larger suspension particles. Some deformation of the oil droplets may also occur in the sedimented layer.

Optical microscopic investigation of some other suspoemulsions showed heteroflocculation and this could be reduced or eliminated by using Atlox 4913 (an acrylic graft copolymer of poly(methyl methacrylate) backbone and PEO chains) [5], as discussed below.

9.4.2 Reduction of Interactions in Suspoemulsions

The use of strongly 'anchored' dispersants and emulsifiers is crucial for a reduction in the interaction between the particles and droplets. The interaction can also be significantly reduced by the addition of rheology modifiers such as hydroxyethylcellulose (HEC) or xanthan gum.

The above thickeners produce a 'three-dimensional' gel network by the overlap of the polymer coils of HEC or the double helices of xanthan gum. In addition to their effect in reducing creaming and sedimentation by producing a high residual viscosity (at low shear rates), these polymers will also prevent trapping of the oil droplets in the suspension and the suspension particles in the emulsion.

9.4.3
Reduction of Heteroflocculation

Heteroflocculation results from the competitive adsorption between the dispersant and emulsifier, particularly when these are not strongly anchored to the surfaces. Displacement of some or all of the dispersant by the emulsifier and vice versa may result in attraction between the particles and droplets. The repulsive barrier is weakened in both cases. If the particles are partially wetted by the oil, they may reside at the O/W interface if the oil droplets are sufficiently large.

The above processes of attraction may continue for long periods of time and ultimately the suspoemulsion becomes physically unstable. Any flocculation will result in entrapment of the liquid between the particles in the floc structure and this causes a significant increase in the viscosity of the system.

Competitive adsorption may be reduced by using the same surfactant for dispersing the solid and emulsifying the oil. This was demonstrated above when using Synperonic NPE 1800. However since this molecule shows some reversible adsorption, interaction between the particles and droplets is not completely prevented.

A better method for reducing competitive adsorption is to use a polymeric surfactant that is strongly and irreversibly adsorbed on the suspension particles and emulsion droplets, such as the graft copolymer of poly(methyl methacrylate) backbone with several PEO chains grafted on the backbone [5]. This graft copolymer (Atlox 4913) has a weight-average molecular weight of $\sim$20 000 and it adsorbs strongly and irreversibly on hydrophobic particles, e.g. polystyrene latex and most agrochemical suspensions. By using the above graft copolymer as dispersant and an A–B–A block copolymer of PEO (A) and (PPO) (B) as emulsifier, one can obtain very stable suspoemulsions. A good polymeric stabilizer is Inutec SP1 (Orafti, Belgium) that consists of an inulin (linear polyfructose with degree of polymerization $>$23) chain on which several alky chains are grafted [6]. This polymeric surfactant adsorbs on hydrophobic particles and emulsion droplets by multi-point attachment with several alky groups leaving strongly hydrated loops and tails of polyfructose that provide an effective steric barrier.

9.4.4
Reduction of Emulsion Coalescence

Coalescence of the emulsion droplets on storage accelerates the instability of the suspoemulsion. Large oil droplets can induce heteroflocculation with the suspension particles residing at the O/W interface. Emulsion coalescence can be reduced by one or more of the following methods: (1) reduction of droplet size by using a high-pressure homogenizer; (2) use of an effective emulsion stabilizer such as Inutec SP1; (3) incorporation of an oil-soluble polymeric surfactant such as Atlox 4912 (or Arlacel P135, which can be used in cosmetics), which is an A–B–A block copolymer of poly(hydroxystearic acid) (PHS) (A) and PEO (B).

9.4.5 Summary of the Criteria for Preparation of a Stable Suspoemulsion

1. Use of a strongly adsorbed ('anchored') dispersant by multi-point attachment of a block or graft copolymer.
2. Use of a polymeric stabilizer for the emulsion (also with multi-point attachment), e.g. Inutec SP1.
3. Preparation of the suspension and emulsion separately and allowing enough time for complete adsorption (equilibrium).
4. Using low shear when mixing the suspension and emulsion.
5. When dissolving an active in an oil (e.g. with many agrochemicals), one should choose an oil in which the suspension particles are insoluble and also the oil should not wet the particles.
6. Increasing dispersant and emulsifier concentrations to ensure that the life time of any bare patches produced during collision is very short.
7. Reducing emulsion droplet size by using a high-pressure homogenizer – the smaller droplets are less deformable and coalescence is prevented. In addition, accumulation of the suspension particles at the O/W interface is prevented.
8. Use of a rheology modifier such as HEC or xanthan gum that produces a viscoelastic solution that prevents creaming or sedimentation and prevents entrapment of the oil droplets between the suspension particles or the suspension particles between the emulsion droplets.
9. If possible, it is preferable to use a higher volume fraction of the oil compared with the suspension. In many cases, flocculation is more rapid at higher solid volume fractions. The emulsion oil phase volume can be increased by incorporation of an inert oil.

9.4.6 Preparation of Suspoemulsion by Emulsification of the Oil into the Suspension

When one needs to prepare a suspoemulsion with a high volume fraction ϕ of suspension and emulsion (e.g. $\phi > 0.4$), it is preferable to emulsify the oil directly into a prepared suspension. In this approach, one prepares the suspension first (e.g. by bead milling) using the polymeric dispersant. The suspension is left to equilibrate for a sufficient time (preferably overnight) to ensure complete adsorption of the polymer. The polymeric emulsifier is then added and the oil is emulsified into the SC using, for example, a Silverson or Ultra Turrax. Overmixing, which may result in orthokinetic (or shear) flocculation and dialatancy, must be avoided and the whole system should be cooled as much as possible during emulsification.

9.4.7
Prevention of Crystallization

This is by far the most serious instability problem with suspoemulsions (particularly with many agrochemicals). It arises from the partial solubility of the suspension particles in the oil droplets. The process is accelerated at higher temperatures and also on temperature cycling.

As mentioned above, the smaller particles will have higher solubility than the larger particles, because the higher the curvature the higher is the solubility as described by the Kelvin equation [7]:

$$S(r) = S(\infty) \exp\left(\frac{2\gamma V_{\mathrm{m}}}{rRT}\right) \tag{1}$$

where $S(r)$ is the solubility of a particle with radius r, $S(\infty)$ is the solubility of a particle with infinite radius (the bulk solubility), γ is the S/L interfacial tension, R is the gas constant and T is the absolute temperature.

On storage, the smaller particles will dissolve in the oil and they recrystallize on the larger particles which may be at the vicinity of the O/W interface. Some crystal habit modification may be produced and large plates or needles are formed which can reach several microns in size.

Several procedures may be applied to inhibit recrystallization:

1. Diluting the oil phase with another inert oil in which the particles are insoluble.
2. Use of a strongly adsorbed polymeric surfactant such as Atlox 4913 or Inutec SP1 that prevent entry of the suspension particles into the oil droplets.
3. Addition of electrolytes to the continuous phase. This has the effect of enhancing the polymeric surfactant adsorption, thus preventing particle entry into the oil droplets.
4. Use of crystal growth inhibitors, e.g. flat dye molecules which are insoluble in the oil and are strongly adsorbed on the particle surface. This prevents particle entry into the oil droplets.
5. Use of analogues of the solid active ingredient (having the same basic structure) that are insoluble in the oil and become incorporated on the surface of the solid particles.
6. Use of thickeners such as HEC and xanthan gum. This will increase the low shear rate viscosity of the medium and hence slow the diffusion of the small particles, thus preventing their entry into the oil droplets. These thickeners can produce gels in the continuous phase that is viscoelastic and this can prevent particle diffusion.

9.4.8
Model Suspoemulsion of Polystyrene Latex and Isoparaffinic Oil Stabilized with Synperonic PE(PEO–PPO–PEO A–B–A Block Copolymer)

The interaction was investigated using viscoelastic measurements [8, 9]. As an illustration, Figure 9.6 shows typical plots of G^* (complex modulus), G' (storage or elastic modulus), G'' (loss or viscous modulus) and η' (dynamic viscosity) as a function of frequency (Hz) for 90% emulsion and 10% latex both stabilized with Synperonic PE block copolymer. The results in Figure 9.6 were obtained at low strain amplitudes (i.e. in the linear viscoelastic region). Both G^* and G' show a rapid increase with increase in frequency above 0.1 Hz. However, G'', which is higher than G' at frequencies below 1 Hz, starts to decrease with increase in frequency above 1 Hz. The dynamic viscosity shows a decrease with increase in frequency, as expected (shear thinning system).

A well-defined cross-over point ($G' = G''$) can be identified from which the relaxation time of the system can be calculated:

$$\tau = \frac{1}{\omega^*} \tag{2}$$

where ω^* is the characteristic frequency (rad s^{-1}) at the cross-over point. Note that $\omega = 2\pi\nu$, where ν is the frequency in Hz.

Similar results were obtained for other suspoemulsions containing various emulsion:latex ratios. The same trend was also obtained for the emulsion and latex dispersions alone. Addition of the latex to the emulsion causes a shift in τ values to higher frequencies, indicating stronger interaction between the latex particles and emulsion droplets.

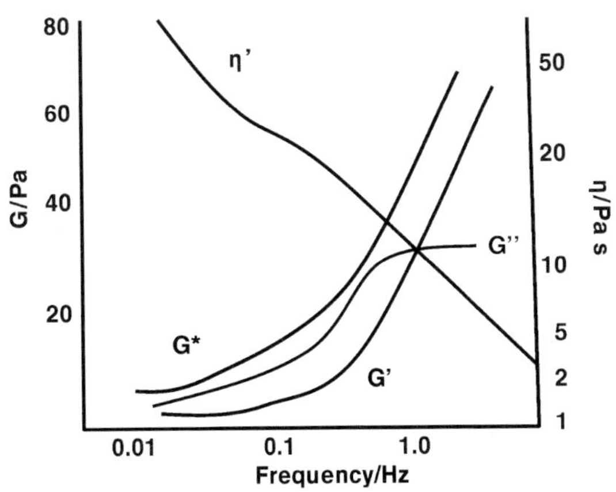

Figure 9.6 Variation of G^*, G', G'' and η' with frequency for a suspoemulsion of 90% isoparaffinic oil-in-water emulsion and 10% polystyrene latex.

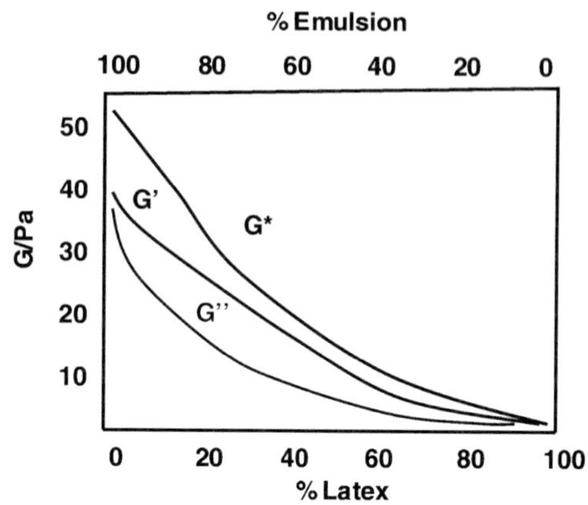

Figure 9.7 Variation of G*, G' and G'' with percentage of emulsion and latex; $\phi = 0$.

Figure 9.7 shows the variation of G^*, G' and G'' at $\phi = 0.57$ and $\nu = 1$ Hz with percentage of emulsion and latex in the suspoemulsion. The emulsion has much higher moduli than the latex at the same volume fraction. Although the emulsion has a VMD (0.98 μm) that is close to the latex (1.18 μm), the former is much more polydisperse than the latter. The much smaller emulsion droplets present may account for the higher moduli of the emulsion compared with the latex. As the proportion of the latex in the suspoemulsion is increases, the moduli decrease. Replacement of emulsion with latex would mean replacing a proportion of the smaller emulsion droplets with larger latex particles and this results in reduction of the moduli values. It should be mentioned, however, that the mixture of emulsion and latex becomes relatively more elastic than viscous, indicating stronger interaction between the emulsion droplets and the latex particles.

9.4.9
Model Systems of Polystyrene Latex with Grafted PEO Chains and Hexadecane Emulsions

It is clear from the above discussion that the interaction between suspension particles and emulsion droplets depends on the nature of the stabilizer used for the particles and droplets. For that reason, we investigated model systems where the latex particles contain grafted PEO chains (with no possible desorption) and the emulsion was based on hexadecane stabilized with Pluronic PE L92 (containing 20% PEO). The particle and droplet radii were similar (315 and 280 nm respectively), in order to avoid complications arising from the change in particle size distribution on mixing the suspension and emulsion.

Steady-state shear stress–shear rate curves were used to obtain the relative viscosity (η_r)–volume fraction (ϕ) relationship for the latex and emulsion. The results are shown in Figure 9.8, which also contains the theoretically predicted

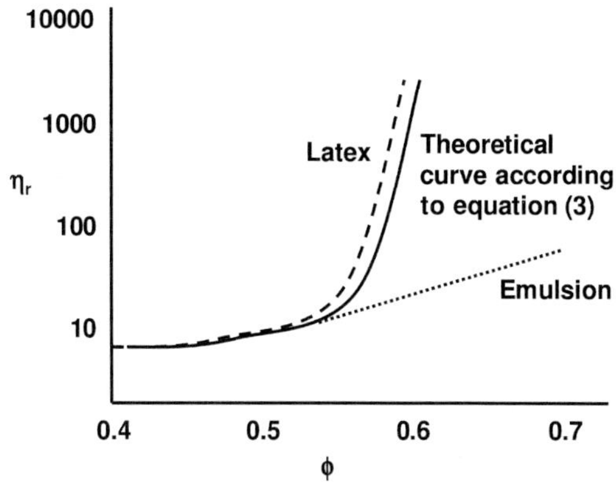

Figure 9.8 η_r–ϕ curves for latexes and emulsions.

curve based on the Dougherty–Krieger equation [10]:

$$\eta_r = \left[1 - \left(\frac{\phi}{\phi_p}\right)\right]^{-[\eta]\phi_p} \tag{3}$$

where ϕ_p is the maximum packing fraction and $[\eta]$ is the intrinsic viscosity that is equal to 2.5 for hard-spheres. Two values for ϕ_p were used, namely 0.60 and 0.61 for the emulsion and latex, respectively. Reasonable agreement between the experimental $\eta_r - \phi$ curve and the theoretical curve based on equation 3 for the latex dispersions was obtained, indicating that this system behaves as near hard spheres. However, the results for the emulsion deviates from the theoretical curves since the emulsion droplets are deformable.

Figure 9.9 shows log–log plots of the elastic modulus G' (measured at low strain amplitudes to ensure the linear viscoelastic region and frequency of 1 Hz) versus volume fraction of the latex, emulsion and various mixtures of emulsion–latex systems. All results are similar for volume fractions below ~0.62, indicating that the interactions between the emulsion droplets and latex particles are of the same nature and both particles and droplets behave as near hard spheres. However, at $\phi > 0.62$, the behaviors of the latexes and emulsions differ significantly, as indicated by the much reduced slope of the log G'–log ϕ curve for the emulsions compared with the latexes. Above this volume fraction, the interaction between the emulsion droplets is fairly high and the system can reduce this interaction by deformation of the emulsion droplets. This situation is not possible with the latexes, where the particles are rigid. Similar behavior is observed for the suspoemulsion when the percentage of the emulsion in the mixture is greater than 60%. This implies that the behavior of emulsion and suspoemulsions with more than 60% emulsion is close in its rheological behavior to concentrated emulsions and this has implications in the formulation of suspoemulsions.

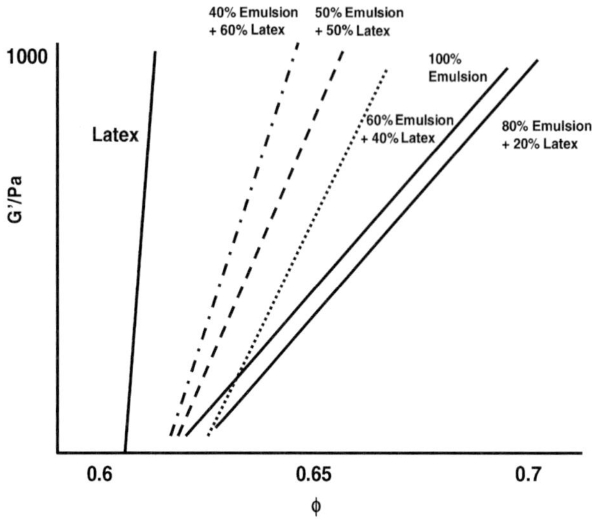

Figure 9.9 Log–log plots of G' versus ϕ for latexes, emulsions and their mixtures.

The G'–ϕ curves can be analyzed using the model suggested by Princen and Kiss [11], who assumed a system of cylinders arranged in a hexagonal array. When such an arrangement is strained, the total interface is increased; this creates a restoring force that is proportional to the interfacial tension γ. Using this model, Princen and Kiss [11] derived an expression relating the shear modulus G_0 (which can be replaced by the high frequency modulus G') to the volume fraction ϕ:

$$G' = a\left(\frac{\gamma}{R_{32}}\right)\phi^{\frac{1}{3}}(\phi - b) \tag{4}$$

where a and b are constants that are equal to 1.7 and 0.71, respectively. The origin of the constant b is the value of the maximum packing of undistorted cylinders in the array. R_{32} is the average volume-to-surface ratio of the radius R (the Sauter radius).

For the latex-emulsion systems studied, R_{32} for the emulsion is constant and, due to the excess surfactant in the emulsion, γ is expected to be little dependent on the composition. The values of the parameters to fit the data in Figure 9.9 are given in Table 9.1.

The parameter b corresponds to the volume fraction at the onset of elasticity.

The mixture of latex and emulsion may be regarded as two elastic elements in series with the appropriate volume fractions:

$$\frac{1}{G_m} = \frac{f_e}{G_e} + \frac{(1 - f_e)}{G_L} \tag{5}$$

Table 9.1 Parameters of the fit of equation 4 for emulsions, latexes and their mixtures.

Emulsion (%)	$a\gamma/R_{32}$	b
100	9896 ± 690	0.63 ± 0.01
80	9900 ± 1700	0.63 ± 0.01
60	17500 ± 2000	0.63 ± 0.03
50	20700 ± 2200	0.63 ± 0.01
40	23700 ± 2200	0.62 ± 0.02
20	~100 000	0.61 ± 0.01
0	~1 000 000	0.62 ± 0.02

where G_m, G_e and G_L are the elastic moduli of the mixture, emulsion and latex, respectively, and f_e is the weight fraction of the emulsion in the mixture.

A plot of the linear lines of Figure 9.9 (which is equal to $a\gamma/R_{32}$) is shown in Figure 9.10 together with the predicted line based on equation 5. The agreement between the values of the slopes and those predicted using equation 5 is good.

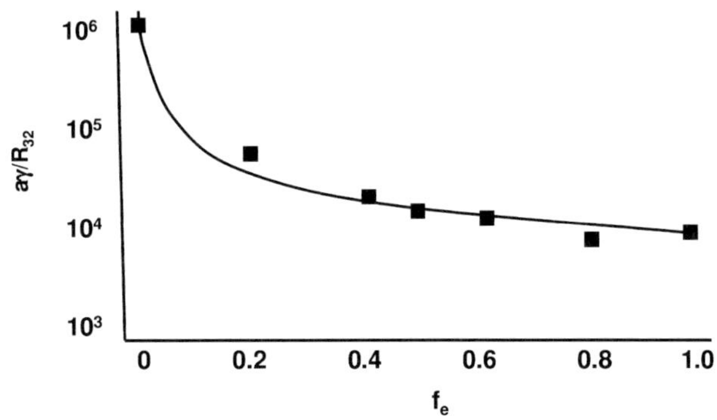

Figure 9.10 $a\gamma/R_{32}$ versus f_e. Line drawn using equation 5.

9.5 Conclusions

Mixtures of suspensions and emulsions (suspoemulsions) may undergo interaction (homo- or heteroflocculation) that depends on the nature of the dispersant–emulsifier system, the particle size distribution and the density difference between the particles and droplets. In cases where heteroflocculation occurs, strongly anchored dispersing agent for the suspension (mostly graft copolymers) are necessary to prevent their displacement from the particles. An efficient emulsifier is also required to prevent any coalescence of the droplets. Crystallization of the particles may occur on storage, particularly when the solid particles have some appreciable solubility in the oil droplets. To prevent this, it is necessary to incorporate an oil with much lower solubility for the solid. In addition, strongly anchored polymeric surfactants to the solid particles are necessary. Crystal growth inhibitors such as some dyes may also be incorporated.

To illustrate the interaction between particles and oil droplets in a suspoemulsion, model systems of latex suspensions and oil-in-water emulsions based on paraffinic oils were investigated using rheological techniques. The results illustrated the possible deformation of the oil droplets at high volume fractions. This was reflected in the viscoelastic properties of the latex–emulsion mixtures. The results could be analyzed using a model based on a system of cylinders arranged in a hexagonal array.

References

1 Th.F. Tadros, *Langmuir*, 6, 28 (1990).

2 Th.F. Tadros, W. Liang, B. Costello and P.F. Luckham, *Colloids Surf.*, *79*, 105 (1993).

3 J.W. Goodwin, J. Hearn, C.C. Ho and R.H. Ottewill, *Colloid Polym. Sci.*, *252*, 464 (1974).

4 C. Bromley, *Colloids Surf.*, *17*, 1 (1986).

5 D. Heath, R.D. Knott, D.A. Knowles and Th.F. Tadros, *ACS Symp. Ser.*, *254*, 2 (1984).

6 Th.F. Tadros, A. Vandamme, B. Levecke, K. Booten and C.V. Stevens, *Adv. Colloid Interface Sci.*, *108–109*, 207 (2004).

7 W. Thompson(Lord Kelvin), *Philos. Mag.*, *42*, 448 (1871).

8 R. Pons, P. Rossi and Th.F. Tadros, *J. Phys. Chem.*, *99*, 12624 (1995).

9 Th.F. Tadros, J. Cutler, R. Pons and P. Rossi, Investigations of the interactions between suspensions and emulsions (suspoemulsions), in *Modern Aspects of Colloidal Dispersions*, ed. R.H. Ottewill and A.R. Rennie, Kluwer Academic, Dordrecht (1998).

10 I.M. Krieger, *Adv. Colloid Interface Sci.*, *3*, 111 (1972).

11 H.M. Princen and A.D. Kiss, *J. Colloid Interface Sci.*, *112*, 427 (1986).

10
Microemulsions

10.1
Introduction

Microemulsions are a special class of 'dispersions' (transparent or translucent) which actually have little in common with emulsions. They are better described as 'swollen micelles'. The term microemulsion was first introduced by Hoar and Schulman [1, 2], who discovered that by titration of a milky emulsion (stabilized by a soap such as potassium oleate) with a medium-chain alcohol such as pentanol or hexanol, a transparent or translucent system was produced. A schematic representation of the titration method adopted by Hoar and Schulman is given below:

O/W emulsion stabilized by soap	→	Add co-surfactant e.g. $C_5H_{11}OH$, $C_6H_{13}OH$	→	Transparent or translucent system

The final transparent or translucent system is a W/O microemulsion.

A convenient way to describe microemulsions is to compare them with micelles. The latter, which are thermodynamically stable, may consist of spherical units with a radius that is usually less than 5 nm. Two types of micelles may be considered: normal micelles with the hydrocarbon tails forming the core and the polar head groups in contact with the aqueous medium, and reverse micelles (formed in nonpolar media) with a water core containing the polar head groups and the hydrocarbon tails now in contact with the oil. The normal micelles can solubilize oil in the hydrocarbon core forming an O/W microemulsion, whereas the reverse micelles can solubilize water forming a W/O microemulsion. A schematic representation of these systems is shown in Figure 10.1.

A rough guide to the dimensions of micelles, micellar solutions and macroemulsions is as follows: micelles, $R < 5$ nm (they scatter little light and are transparent); macroemulsions, $R > 50$ nm (opaque and milky); micellar solutions or microemulsions, $R = 5$–50 nm (transparent, 5–10 nm; translucent, 10–50 nm).

The classification of microemulsions based on size is not adequate. Whether a system is transparent or translucent depends not only on the size but also on the difference in refractive index between the oil and the water phases. A

Colloids and Interface Science Series, Vol. 5
Colloids in Agrochemicals. Tharwat F. Tadros

ISBN: 978-3-527-31465-2

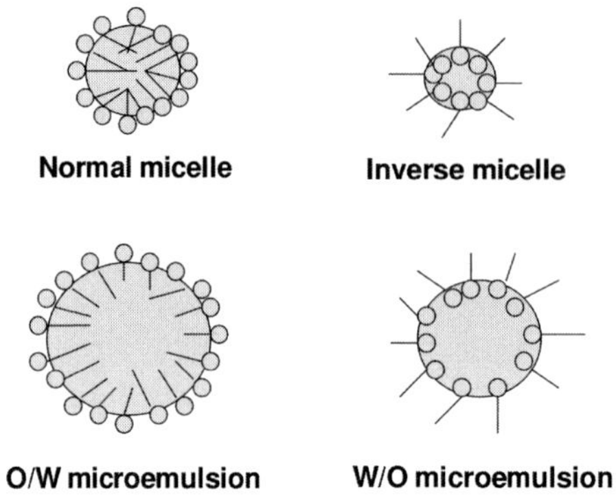

Figure 10.1 Schematic representation of microemulsions.

microemulsion with small size (in the region of 10 nm) may appear translucent if the difference in refractive index between the oil and the water is large (note that the intensity of light scattered depends on the size and an optical constant that is given by the difference in refractive index between oil and water). Relatively large-sized microemulsion droplets (in the region of 50 nm) may appear transparent if the refractive index difference is very small. The best definition of microemulsions is based on the application of thermodynamics, as discussed below.

10.2 Thermodynamic Definition of Microemulsions

A thermodynamic definition of microemulsions can be obtained from consideration of the energy and entropy terms for the formation of microemulsions, represented schematically in Figure 10.2, which shows the process of formation of microemulsion from a bulk oil phase (for an O/W microemulsion) or bulk water phase (for a W/O microemulsion).

The increase in surface area on going from state I to state II is ΔA $(=A_2 - A_1)$ and the surface energy increase is equal to $\Delta A\gamma_{12}$. The increase in entropy on

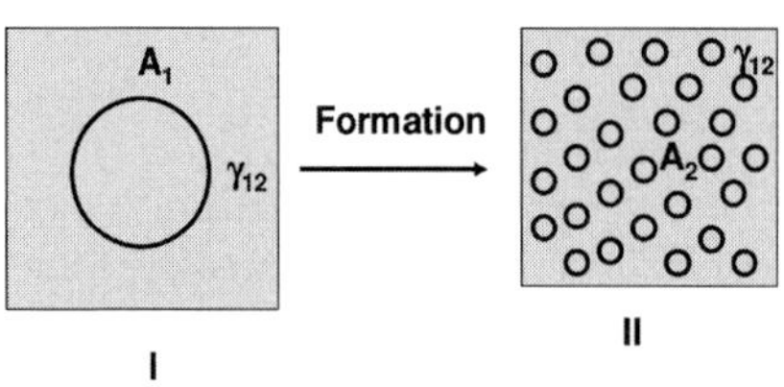

Figure 10.2 Schematic representation of microemulsion formation. A_1 is the surface area of the bulk oil phase and A_2 is the total surface area of all the microemulsion droplets. Γ_{12} is the O/W interfacial tension.

going from state I to state II is $T\Delta S^{\text{conf}}$ (note that state II has higher entropy since a large number of droplets can arrange themselves in several ways, whereas state I with one oil drop has much lower entropy).

According to the second law of thermodynamics, the free energy of formation of microemulsions, ΔG_{m}, is given by the following expression:

$$\Delta G_{\text{m}} = \Delta A\gamma_{12} - T\Delta S_{\text{conf}} \tag{1}$$

With macroemulsions, $\Delta A\gamma_{12} \gg -T\Delta S^{\text{conf}}$ and $\Delta G_{\text{m}} > 0$. The system is non-spontaneous (it requires energy for formation of the emulsion drops) and it is thermodynamically unstable. With microemulsions, $\Delta A\gamma_{12} < -T\Delta S^{\text{conf}}$ (this is due to the ultralow interfacial tension accompanied with microemulsion formation) and $\Delta G_{\text{m}} < 0$. The system is produced spontaneously and it is thermodynamically stable.

The above analysis shows the contrast between emulsions and microemulsions: with emulsions, an increase in the mechanical energy and an increase in surfactant concentration usually result in the formation of smaller droplets which become kinetically more stable. With microemulsions, neither mechanical energy nor increase in surfactant concentration can result in its formation. The latter is based on a specific combination of surfactants and specific interaction with the oil and the water phases and the system is produced at optimum composition.

Hence microemulsions have nothing in common with macroemulsions and in many cases it is better to describe the system as 'swollen micelles'. The best definition of microemulsions is as follows [3]: 'system of water + oil + amphiphile that is a single optically isotropic and thermodynamically stable liquid solution'. An amphiphile refers to any molecule that consists of a hydrophobic and a hydrophilic portion, e.g. surfactants and alcohols.

The driving force for microemulsion formation is the low interfacial energy which is overcompensated by the negative entropy of dispersion term. The low (ultralow) interfacial tension is produced in most cases by combination of two molecules, referred to as the surfactant and co-surfactant (e.g. medium-chain alcohol).

10.3 Application in Agrochemicals

As mentioned in Chapter 6, many agrochemicals are formulated as emulsion concentrates (EWs). The agrochemical, which may be an oil, is emulsified into water and a water-based concentrated emulsion (EW) is produced. In cases where the agrochemical is very viscous or semisolid, a small amount of an oil (which may be aliphatic in nature) may be added before the emulsification process. Unfortunately, EWs suffer from a number of problems, such as the difficulty of emulsification and their long-term physical stability. As discussed in Chapter 6, emulsions are thermodynamically unstable systems since their formation involves a high interfacial energy that is not overcome by the relatively small entropy of dispersion (see below). With time, the emulsion tends to reduce its interfacial area by several breakdown processes such as flocculation and coalescence. In addition, most emulsion systems undergo sedimentation or creaming on standing and they require the use of thickeners to modify their rheology. A further complication may arise from Ostwald ripening where the smaller droplets dissolve and become deposited on the larger droplets. Clearly, to produce a physically stable emulsion of an agrochemical requires a great deal of process control and also prevention of the various breakdown processes.

A very attractive alternative for the formulation of agrochemicals is to use microemulsion systems. As mentioned above, these are single optically isotropic and thermodynamically stable dispersions consisting of oil, water and amphiphile (one or more surfactants). As will be seen later, the origin of the thermodynamic stability arises from the low interfacial energy of the system, which is outweighed by the entropy of dispersion. These systems offer a number of advantages over O/W emulsions, for the following reasons. Once the composition of the microemulsion has been identified, the system is prepared by simply mixing all the components without the need for any appreciable shear. Due to their thermodynamic stability, these formulations undergo no separation or breakdown on storage (within a certain temperature range depending on the system). The low viscosity of the microemulsion systems ensures their ease of pourability and dispersion on dilution and they leave little residue in the container. Another main attraction of microemulsions is their possible enhancement of biological efficacy of many agrochemicals. This, as will be seen later, is due to the solubilization of the pesticide by the microemulsion droplets.

This chapter summarizes the basic principles involved in the preparation of microemulsions and the origin of their thermodynamic stability. A section is devoted to emulsifier selection for both O/W and W/O microemulsions. The physical methods that may be applied for the characterization of microemulsions is briefly described. Finally, a section is devoted to the possible enhancement of biological efficacy using microemulsions. The role of microemulsions in enhancing wetting, spreading and penetration is discussed. Solubilization is also another factor that may enhance the penetration and uptake of an insoluble agrochemical.

10.4
Basic Principles of Microemulsion Formation and Thermodynamic Stability

10.4.1
Mixed Film Theories [4]

The film (which may consist of surfactant and co-surfactant molecules) is considered as a liquid 'two-dimensional' third phase in equilibrium with both oil and water. Such a monolayer could be a duplex film, i.e. giving different proper-

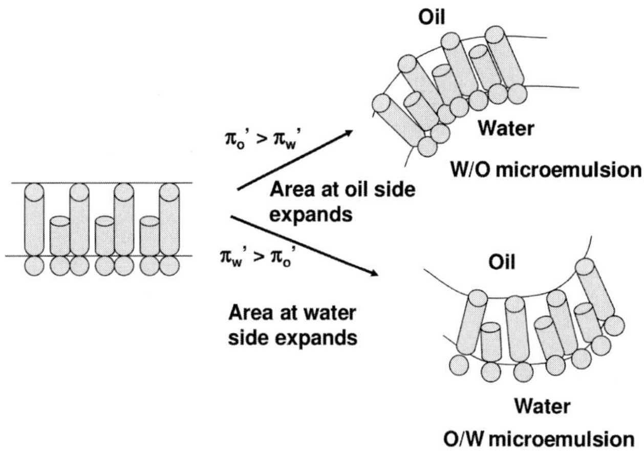

Figure 10.3 Schematic representation of film bending.

ties on the water side and the oil side. The initial 'flat' duplex film (see Figure 10.3) has different tensions at the oil and water sides. This is due to the different packing of the hydrophobic and hydrophilic groups (these groups have different sizes and cross-sectional areas).

It is convenient to define a two-dimensional surface pressure π:

$$\pi = \gamma_0 - \gamma \qquad (2)$$

where γ_0 is the interfacial tension of the clean interface and γ is the interfacial tension with adsorbed surfactant. One can define two values for π at the oil and water phases, π_O and π_W, which for a flat film are not equal, i.e. $\pi'_O \neq \pi'_W$. As a result of the difference in tensions, the film will bend until $\pi_O = \pi_W$. If $\pi'_O > \pi'_W$, the area at the oil side has to expand (resulting in reduction of π'_O) until $\pi_O = \pi_W$. In this case, a W/O microemulsion is produced. If $\pi'_W > \pi'_O$, the area at the water side expands until $\pi_W = \pi_O$. In this case, an O/W microemulsion is produced.

According to the duplex film theory, the interfacial tension γ_T is given by the following expression [5]:

$$\gamma_T = \gamma_O/W - \pi \qquad (3)$$

where $(\gamma_{O/W})_a$ is the interfacial tension that is reduced by the presence of the alcohol. The value of $(\gamma_{O/W})_a$ is significantly lower than $\gamma_{O/W}$ in the absence of the alcohol (for example, for hydrocarbon/water $\gamma_{O/W}$ is reduced from 50 to 15–20 mN m^{-1} on the addition of a significant amount of a medium-chain alcohol such as pentanol or hexanol).

Contributions to π are considered to be due to crowding of the surfactant and co-surfactant molecules and penetration of the oil phase into the hydrocarbon chains of the interface.

According to equation 3, if $\pi > (\gamma_{O/W})_a$, γ_T becomes negative and this leads to expansion of the interface until γ_T reaches a small positive value. Since $(\gamma_{O/W})_a$ is of the order of 15–20 mN m^{-1}, surface pressures of this order are required for γ_T to approach a value of zero.

The above duplex film theory can explain the nature of the microemulsion: The surface pressures at the oil and water sides of the interface depend on the interactions of the hydrophobic and hydrophilic portions of the surfactant molecule at both sides of the interface. If the hydrophobic groups are bulky in nature relative to the hydrophilic groups, then for a flat film such hydrophobic groups tend to crowd, forming a higher surface pressure at the oil side of the interface; this results in bending and expansion at the oil side, forming a W/O microemulsion. An example of a surfactant with bulky hydrophobic groups is Aerosol OT (dioctyl sulfosuccinate). If the hydrophilic groups are bulky, such as is the case with ethoxylated surfactants containing more than five ethylene oxide units, crowding occurs at the water side of the interface. This produces an O/W microemulsion.

10.4.2 Solubilization Theories

These concepts were introduced by Shinoda and Friberg [6], who considered microemulsions to be swollen micelles that are directly related to the phase diagram of their components.

Consider the phase diagram of a three-component system of water, ionic surfactant and medium chain alcohol as shown in Figure 10.4. At the water corner and at low alcohol concentration, normal micelles (L_1) are formed since in this case there are more surfactant than alcohol molecules. At the alcohol (co-surfactant) corner, inverse micelles (L_2) are formed, since in this region there are more alcohol than surfactant molecules. These L_1 and L_2 are not in equilibrium but are separated

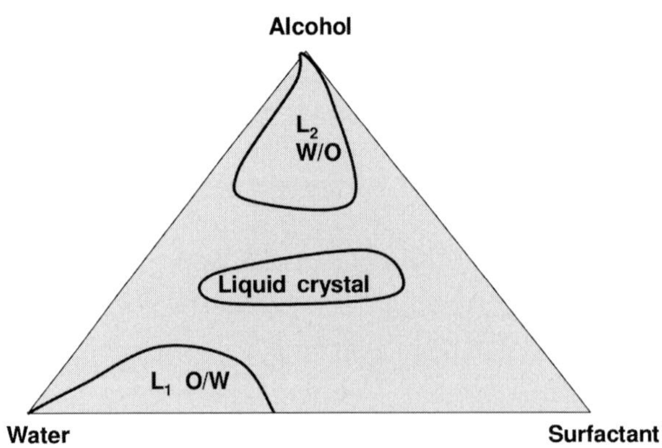

Figure 10.4 Schematic representation of three-component phase diagram.

by a liquid crystalline region (lamellar structure with equal number of surfactant and alcohol molecules). The L_1 region may be considered as an O/W microemulsion, whereas the L_2 region may be considered as a W/O microemulsion.

Addition of a small amount of oil miscible with the co-surfactant, but not with the surfactant and water, changes the phase diagram only slightly. The oil may be simply solubilized in the hydrocarbon core of the micelles. Addition of more oil leads to fundamental changes to the phase diagram, as illustrated in Figure 10.5, where 50 : 50 W : O is used. To simplify the phase diagram, the $^{50}W/^{50}O$ are presented at one corner of the phase diagram.

Near the co-surfactant (co) corner, the changes are small compared with the three-phase diagram (Figure 10.4). The O/W microemulsion near the water-surfactant (sa) axis is not in equilibrium with the lamellar phase, but with a non-colloidal oil + co-surfactant phase.

If co is added to such a two-phase equilibrium at fairly high surfactant concentration, all oil is taken up and a one-phase microemulsion appears. Addition of co at low sa concentration may lead to separation of an excess aqueous phase before all oil is taken up in the microemulsion. A three-phase system is formed, containing a microemulsion that cannot be clearly identified as W/O or W/O and that presumably is similar to the lamellar phase swollen with oil or to a more irregular intertwining of aqueous and oily regions (bicontinuous or middle phase microemulsion).

The interfacial tensions between the three phases are very low ($0.1–10^{-4}$ m Nm^{-1}). Further addition of co to the three phase system makes the oil phase disappear and leaves a W/O microemulsion in equilibrium with a dilute aqueous sa solution. In the large one-phase region, continuous transitions from O/W to middle phase to W/O microemulsions are found.

Solubilization can also be illustrated by considering the phase diagrams of nonionic surfactants containing poly(ethylene oxide) (PEO) head groups. Such surfactants do not generally need a co-surfactant for microemulsion formation.

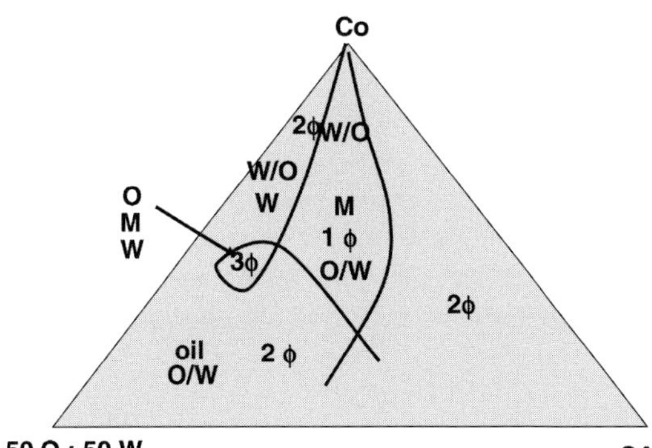

Figure 10.5 Schematic representation of the pseudoternary phase diagram of oil–water–surfactant–co-surfactant.

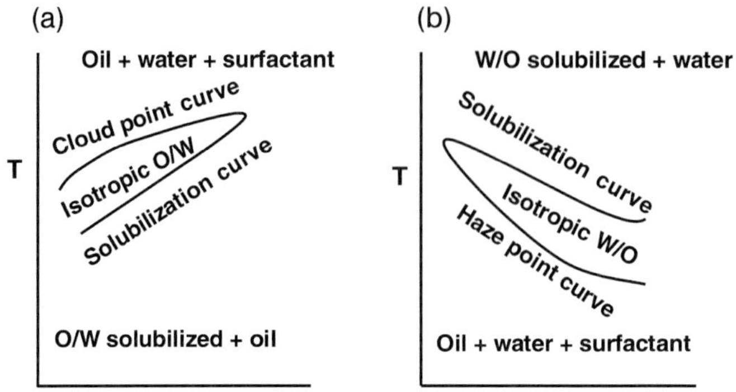

Figure 10.6 Schematic representation of solubilization: (a) oil solubilized in a nonionic surfactant solution; (b) water solubilized in an oil solution of a nonionic surfactant.

A schematic representation of oil and water solubilization by nonionic surfactants is given in Figure 10.6. At low temperatures, the ethoxylated surfactant is soluble in water and at a given concentration is capable of solubilizing a given amount of oil. The oil solubilization increases rapidly with increase in temperature near the cloud point of the surfactant. This is illustrated in Figure 10.6, which shows the solubilization and cloud point curves of the surfactant. Between these two curves, an isotropic region of O/W solubilized system exists.

At any given temperature, any increase in the oil weight fraction above the solubilization limit results in oil separation (oil solubilized + oil). At any given surfactant concentration, any increase in temperature above the cloud point results in separation into oil, water and surfactant.

If one starts from the oil phase with dissolved surfactant and adds water, solubilization of the latter takes place and solubilization increases with reduction of temperature near the haze point. Between the solubilization and haze point curves, an isotropic region of W/O solubilized system exists. At any given temperature, any increase in water weight fraction above the solubilization limit results in water separation (W/O solubilized + water). At any given surfactant concentration, any decrease in temperature below the haze point results in separation to water, oil and surfactant.

With nonionic surfactants, both types of microemulsions can be formed depending on the conditions. With such systems, temperature is the most crucial factor since the solubility of surfactant in water or oil depends on temperature. Microemulsions prepared using nonionic surfactants have a limited temperature range.

10.4.3
Thermodynamic Theory of Microemulsion Formation and Stability

The spontaneous formation of the microemulsion with decrease in free energy can only be expected if the interfacial tension is so low that the remaining free energy of the interface is overcompensated for by the entropy of dispersion of the droplets in the medium [7, 8].

The above concept forms the basis of the thermodynamic theory proposed by Ruckenstein and Chi [7] and Overbeek [8].

10.4.4
Reason for Combining Two Surfactants

Single surfactants lower the interfacial tension γ, but in most cases the critical micelle concentration (cmc) is reached before γ is close to zero. Addition of a second surfactant of a completely different nature (i.e. predominantly oil soluble, such as an alcohol) then lowers γ further and very small, even transiently negative, values may be reached [9]. This is illustrated in Figure 10.7, which shows the effect of addition of the co-surfactant on the γ–log C_{sa} curve.

It can be seen that addition of co-surfactant shifts the whole curve to low γ values and the cmc is shifted to lower values. The reason for the lowering of γ when using two surfactant molecules can be understood from consideration of the Gibbs adsorption equation for multicomponent systems [9].

For a multicomponent system i, each with an adsorption Γ_i (mol m^{-2}, referred to as the surface excess), the reduction in γ, i.e. dγ, is given by the following expression:

$$d\gamma = -\sum \Gamma_i \, d\mu_i = -\sum \Gamma_i RT d \ln C_i \tag{4}$$

where μ_i is the chemical potential of component i, R is the gas constant, T is the absolute temperature and C_i is the concentration (mol dm^{-3}) of each surfactant component. For two components, sa (surfactant) and co (co-surfactant), equation 4 becomes

$$d\gamma = -\Gamma_{sa} RT d \ln C_{sa} - \Gamma_{co} RT d \ln C_{co} \tag{5}$$

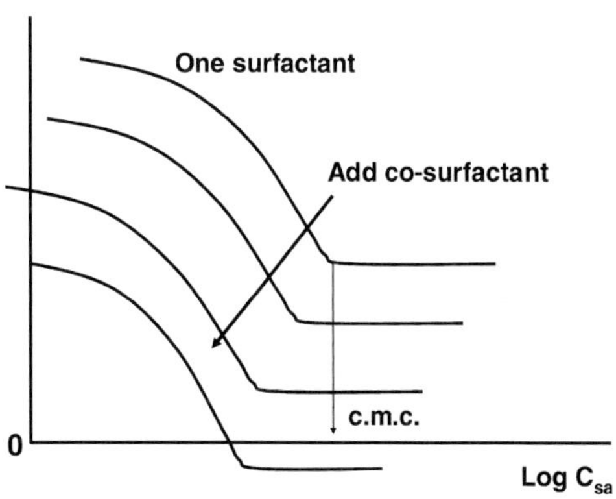

Figure 10.7 γ–log C_{sa} curves for surfactant + co-surfactant.

Integration of equation 5 gives

$$\gamma = \gamma_0 - \int_0^{C_{sa}} \Gamma_{sa} RT \mathrm{d} \ln C_{sa} - \int_0^{C_{co}} \Gamma_{co} RT \mathrm{d} \ln C_{co} \quad (6)$$

which clearly shows that γ_0 is lowered by two terms, from both surfactant and co-surfactant.

The two surfactant molecules should adsorb simultaneously and they should not interact with each other, otherwise they lower their respective activities. Hence the surfactant and co-surfactant molecules should vary in nature, one predominantly water soluble (such as an anionic surfactant) and the other predominantly oil soluble (such as a medium-chain alcohol).

In some cases a single surfactant may be sufficient for lowering γ far enough for microemulsion formation to become possible, e.g. Aerosol OT (sodium diethylhexyl sulfosuccinate) and many nonionic surfactants.

10.4.5
Factors Determining W/O Versus O/W Microemulsions

The duplex film theory predicts that the nature of the microemulsion formed depends on the relative packing of the hydrophobic and hydrophilic portions of the surfactant molecule, which determine the bending of the interface. For example, a surfactant molecule such as Aerosol OT favors the formation of a W/O microemulsion, without the need for a co-surfactant. As a result of the presence of a stumpy head group and large volume-to-length ratio (V/l) of the nonpolar group, the interface tends to bend with the head groups facing onwards, thus forming a W/O microemulsion.

The molecule has $V/l > 0.7$ which is considered necessary for the formation of a W/O microemulsion. For ionic surfactants such as SDS for which $V/l < 0.7$, microemulsion formation needs the presence of a co-surfactant (the latter has the effect of increasing V without changing l).

The importance of geometric packing was considered in detail by Mitchell and Ninham [10], who introduced the concept of the packing ratio P:

$$P - \frac{V}{l_c a_0} \quad (7)$$

where a_0 is the head group area and l_c is the maximum chain length.

P gives a measure of the hydrophilic–lipophilic balance. For values of $P < 1$ (usually $P \approx 1/3$), normal or convex aggregates are produced (normal micelles). For values of $P > 1$, inverse micelles are produced. P is influenced by many factors: hydrophilicity of the head group, ionic strength and pH of the medium and temperature. P also explains the nature of the microemulsion produced using nonionic surfactants of the ethoxylate type: P increases with increase in temperature (as a result of the dehydration of the PEO chain). A critical temperature (PIT) is

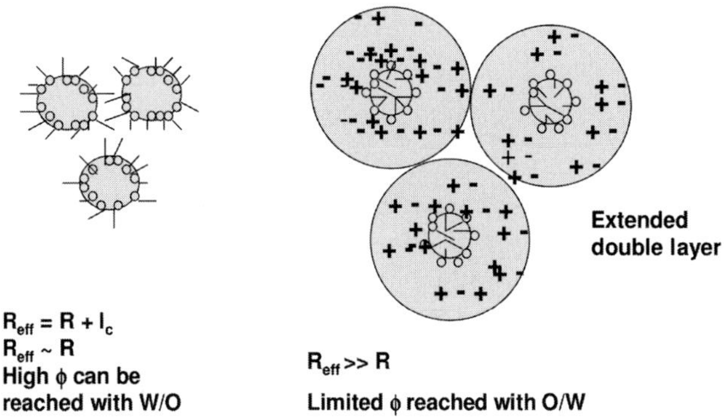

Figure 10.8 Schematic representation of W/O and O/W microemulsion droplets.

reached at which P reaches 1 and above this temperature inversion occurs to a W/O system.

The influence of the surfactant structure on the nature of the microemulsion can also be predicted from the thermodynamic theory. The most stable microemulsion would be that in which the phase with the smaller volume fraction forms the droplets (the osmotic pressure increases with increase in ϕ). For a W/O microemulsion prepared using an ionic surfactant such as Aerosol OT, the effective volume (hard-sphere volume) is only slightly higher that the water core volume, since the hydrocarbon tails may penetrate to a certain extent when two droplets come together. For an O/W microemulsion, the double layers may expand to a considerable extent, depending on the electrolyte concentration (the double layer thickness is of the order of 100 nm in 10^{-5} mol dm^{-3} 1 : 1 electrolyte and 10 nm in 10^{-3} mol dm^{-3} electrolyte). Hence the effective volume of O/W microemulsion droplets can be significantly higher than the core oil droplet volume and this explains the difficulty of preparation of O/W microemulsions at high ϕ values when using ionic surfactants.

A schematic representation of the effective volume for W/O and O/W microemulsions is shown in Figure 10.8.

10.5 Characterization of Microemulsions Using Scattering Techniques

Scattering techniques provide the most obvious methods for obtaining information on the size, shape and structure of microemulsions. The scattering of radiation, e.g. light, neutrons, X-rays, by particles has been successfully applied for the investigation of many systems such as polymer solutions, micelles and colloidal particles.

In all of the above methods, measurements can be made at sufficiently low concentration to avoid complications arising from particle–particle interactions. The results obtained are extrapolated to infinite dilution to obtain the desirable property such as the molecular weight and radius of gyration of a polymer coil, the size and shape of micelles, etc.

Unfortunately, the above dilution method cannot be applied for microemulsions, which depend on a specific composition of oil, water and surfactants. The microemulsions cannot be diluted by the continuous phase since this dilution results in breakdown of the microemulsion. Thus, when applying the scattering techniques to microemulsions, measurements have to be made at finite concentrations and the results obtained have to be analyzed using theoretical treatments to take into account the droplet–droplet interaction.

Below, two scattering methods are discussed: time-averaged (static) light scattering and dynamic (quasi-elastic) light scattering, referred to as photon correlation spectroscopy (PCS).

10.5.1
Time-averaged (Static) Light Scattering

The intensity of scattered light $I(Q)$ is measured as a function of scattering vector Q [11]:

$$Q = \left(\frac{4\pi n}{\lambda}\right)\sin\left(\frac{\theta}{2}\right) \quad (8)$$

where n is the refractive index of the medium, λ is the wavelength of light and θ is the angle at which the scattered light is measured.

For a fairly dilute system, $I(Q)$ is proportional to the number of particles N, the square of the individual scattering units V_p and some property of the system (material constant) such as its refractive index:

$$I(Q) = [(\text{material constant})(\text{instrument constant})]N{V_p}^2 \quad (9)$$

The instrument constant depends on the geometry of the apparatus (the light pathlength and the scattering cell constant).

For more concentrated systems, $I(Q)$ also depends on the interference effects arising from particle-particle interaction:

$$I(Q) = [(\text{material constant})(\text{instrument constant})]N{V_p}^2 P(Q)S(Q) \quad (10)$$

where $P(Q)$ is the particle form factor which allows the scattering from a single particle of known size and shape to be predicted as a function of Q. For a spherical particle of radius R,

$$P(Q) = \left[\frac{3\sin QR - QR\cos QR}{(QR)^3}\right]^2 \quad (11)$$

$S(Q)$ is the so-called 'structure factor', which takes into account the particle–particle interaction. $S(Q)$ is related to the radial distribution function $g(r)$ (which gives the number of particles in shells surrounding a central particle) [12]:

$$S(Q) = 1 - \frac{4\pi N}{Q} \int_0^\infty [g(r) - 1] r \sin QR \, dr \tag{12}$$

For a hard-sphere dispersion with radius R_{HS} (which is equal to $R + t$, where t is the thickness of the adsorbed layer):

$$S(Q) = \frac{1}{1 - NC(2QR_{HS})} \tag{13}$$

where C is a constant.

One usually measures $I(Q)$ at various scattering angles θ and then plots the intensity at some chosen angle (usually 90°), i_{90}, as a function of the volume fraction ϕ of the dispersion. Alternatively, the results may be expressed in terms of the Rayleigh ratio R_{90}:

$$R_{90} = \left(\frac{i_{90}}{I_0}\right) {r_s}^2 \tag{14}$$

where I_0 is the intensity of the incident beam and r_s is the distance from the detector.

$$R_{90} = K_0 MCP(90)S(90) \tag{15}$$

where K_0 is an optical constant (related to the refractive index difference between the particles and the medium) and M is the molecular mass of scattering units with weight fraction C.

For small particles (as is the case with microemulsions), $P(90) \approx 1$ and

$$M = \frac{4}{3}\pi {R_c}^3 N_A \tag{16}$$

where N_A is Avogadro's number.

$$C = \phi_c \rho_c \tag{17}$$

where ϕ_c is the volume fraction of the particle core and ρ_c is the particle density.

Equation 15 can be written in the simple form

$$R_{90} = K_1 \phi_c {R_c}^3 S(90) \tag{18}$$

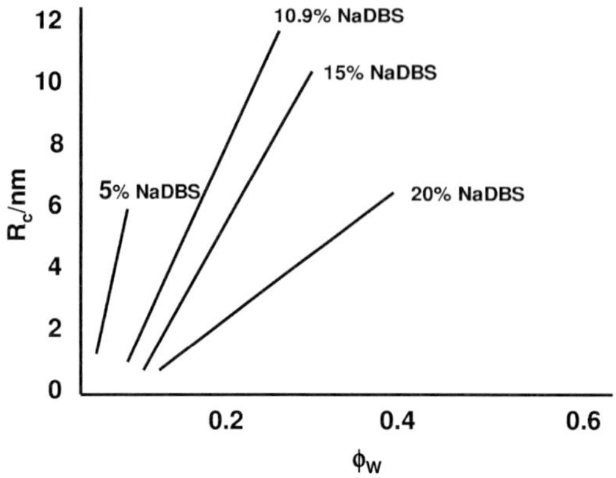

Figure 10.9 Variation of R_c with ϕ_W for W/O microemulsions.

where $K_1 = K_0\ (4/3)N_A\rho_c^2$. Equation 18 shows that to calculate R_c from R_{90} one needs to know $S(90)$. The latter can be calculated using equations 12 and 13.

The above calculations were obtained using a W/O microemulsion of water–xylene–sodium dodecyl benzenesulfonate (NaDBS)–hexanol [11]). The microemulsion region was established using the quaternary phase diagram. W/O microemulsions were produced at various water volume fractions using increasing amounts of NaDBS: 5, 10.9, 15 and 20%.

The results for the variation of R_{90} with the volume fraction of the water core droplets at various NaDBS concentrations showed that, with the exception of the 5% NaDBS results, all the others showed an initial increase in R_{90} with increase in ϕ, reaching a maximum at a given ϕ, after which R_{90} decreased with further increase in ϕ.

The above results were used to calculate R as a function of ϕ using the hard-sphere model discussed above (equation 13). This is shown in Figure 10.9. It can be seen that with increase in ϕ, at constant surfactant concentration, R increases (the ratio of surfactant to water decreases with increase in ϕ). At any volume fraction of water, an increase in surfactant concentration results in a decrease in the microemulsion droplet size (the ratio of surfactant to water increases).

10.5.1.1 **Calculation of Droplet Size from Interfacial Area**

If one assumes that all surfactant and co-surfactant molecules are adsorbed at the interface, it is possible to calculate the total interfacial area of the microemulsion from a knowledge of the area occupied by surfactant and co-surfactant molecules:

> total interfacial area = (total number of surfactant molecules × area per surfactant molecule A_s) + (total number of co-surfactant molecules × area per co-surfactant molecule A_{co})

The total interfacial area A per kilogram of microemulsion is given by the expression

$$A = \frac{n_s N_A A_s + n_{co} N_A A_{co}}{\phi} \tag{19}$$

where n_s and n_{co} are the number of moles of surfactant and co-surfactant. A is related to the droplet radius R (assuming all the droplets are of the same size) by

$$A - \frac{3}{R\rho} \tag{20}$$

Using reasonable values for A_s and A_{co} (30 Å^2 for NaDBS and 20 Å^2 for hexanol), R was calculated and the results were compared with those obtained using light scattering. Two conditions were considered: (a) all hexanol molecules were adsorbed; (b) part of the hexanol was adsorbed to give a molar ratio of hexanol to NaDBS of 2:1. Good agreement was obtained between the two results.

10.5.2
Dynamic Light Scattering (Photon Correlation Spectroscopy, PCS)

In this technique, one measures the intensity fluctuation of scattered light by the droplets as they undergo Brownian motion [13]. When a light beam passes through a colloidal dispersion, an oscillating dipole movement is induced in the particles, thereby radiating the light. Due to the random position of the particles, the intensity of scattered light, at any instant, appears as random diffraction ('speckle' pattern)

As the particles undergo Brownian motion, the random configuration of the pattern will fluctuate, such that the time taken for an intensity maximum to become a minimum (the coherence time) corresponds approximately to the time required for a particle to move one wavelength λ. Using a photomultiplier with active area of about the diffraction maximum (i.e. one coherent area), this intensity fluctuation can be measured. The analogue output is digitized (using a digital correlator) that measures the photocount (or intensity) correlation function of scattered light.

The photocount correlation function $G^{(2)}(\tau)$ is given by

$$g^{(2)} = B[1 + \gamma^2 g^{(1)}(\tau)]^2 \tag{21}$$

where τ is the correlation delay time. The correlator compares $g^{(2)}(\tau)$ for many values of τ. B is the background value to which $g^{(2)}(\tau)$ decays at long delay times. $g^{(1)}(\tau)$ is the normalized correlation function of the scattered electric field and γ is a constant ($\sim$1).

For monodispersed non-interacting particles,

$$g^{(1)}(\tau) = \exp(-\Gamma\gamma) \tag{22}$$

where Γ is the decay rate or inverse coherence time, which is related to the translational diffusion coefficient D:

$$\Gamma = DK^2 \tag{23}$$

where K is the scattering vector:

$$K = \left(\frac{4\pi n}{\lambda_0}\right)\sin\left(\frac{\theta}{2}\right) \tag{24}$$

The particle radius R can be calculated from D using the Stokes–Einstein equation:

$$D = \frac{kT}{6\pi\eta_0 R} \tag{25}$$

where η_0 is the viscosity of the medium.

The above analysis only applies for very dilute dispersions. With microemulsions that are concentrated dispersions, corrections are needed to take into account the interdroplet interaction. This is reflected in plots of $\ln g^{(1)}(\tau)$ versus τ, which become nonlinear, implying that the observed correlation functions are not single exponentials.

As with time-averaged light scattering, one needs to introduce a structure factor in calculating the average diffusion coefficient. For comparative purposes, one calculates the collective diffusion coefficient D, which can be related to its value at infinite dilution, D_0, by [14],

$$D = D_0(1 + \alpha\phi) \tag{26}$$

where α is a constant that is equal to 1.5 for hard spheres with repulsive interaction.

10.6
Characterization of Microemulsions Using Conductivity

Conductivity measurements may provide valuable information on the structural behavior of microemulsions. In the early applications of conductivity measurements, the technique was used to determine the nature of the continuous phase. O/W microemulsions should give fairly high conductivity (which is determined by that of the continuous aqueous phase), whereas W/O microemulsions should

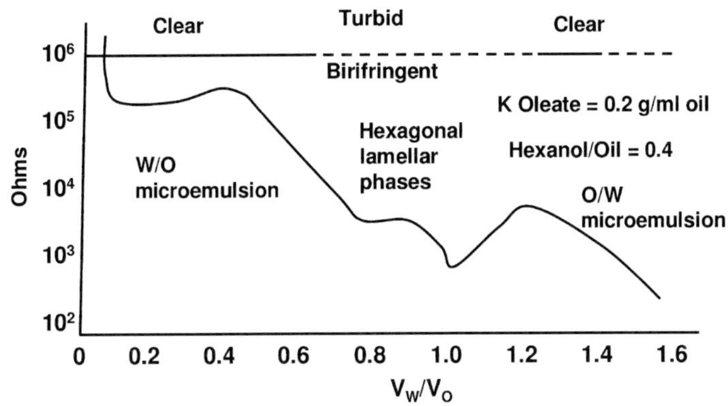

Figure 10.10 Electrical resistance versus V_w/V_0.

give fairly low conductivity (which is determined by that of the continuous oil phase).

As an illustration, Figure 10.10 shows the change in electrical resistance (reciprocal of conductivity) with the ratio of water to oil (V_w/V_0) for a microemulsion system prepared using the inversion method [15]. Figure 10.10 indicates the change in optical clarity and birifringence with the ratio of water to oil.

At low V_w/V_0, a clear W/O microemulsion is produced with a high resistance (oil continuous). As V_w/V_0 increases, the resistance decreases and, in the turbid region, hexanol and lamellar micelles are produced. Above a critical ratio, inversion occurs and the resistance decreases, producing an O/W microemulsion.

Conductivity measurements were also used to study the structure of the microemulsion, which is influenced by the nature of the co-surfactant. This is illustrated in Figure 10.11 for systems based on various co-surfactants with different alkyl chain lengths. The co-surfactant chain length was gradually increased from C_2 (ethanol) to C_7 (heptanol).

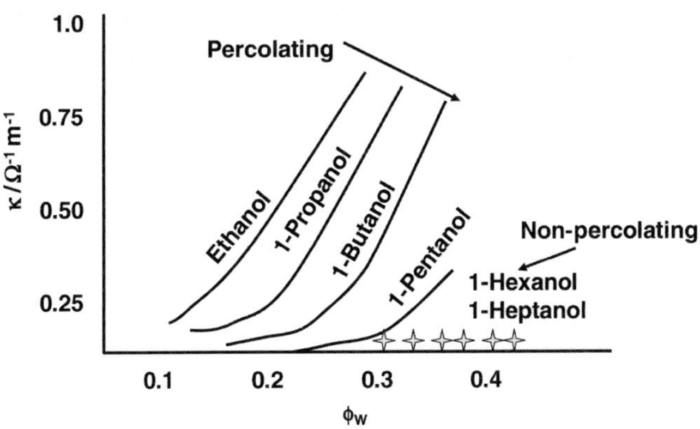

Figure 10.11 Variation of conductivity with water volume fraction for various co-surfactants.

With the short-chain alcohols ($C < 5$), the conductivity shows a rapid increase above a critical ϕ value. With longer chain alcohols, namely hexanol and heptanol, the conductivity remains very low up to a high water volume fraction [16, 17].

With the short-chain alcohols, the system shows percolation above a critical water volume fraction. Under these conditions, the microemulsion is 'bicontinuous'. With the longer chain alcohols, the system is non-percolating and one can define definite water cores. This is sometimes referred to as a 'true' microemulsion.

10.7 NMR Measurements

Lindman and co-workers [18–20] demonstrated that the organization and structure of microemulsions can be elucidated from self-diffusion measurements of all the components (using pulse gradient or spin-echo NMR techniques). Within a micelle, the molecular motion of the hydrocarbon tails (translational, reorientation and chain flexibility) is almost as rapid as in a liquid hydrocarbon. In a reverse micelle, water molecules and counterions are also highly mobile.

For many surfactant–water systems, there is a distinct spatial separation between hydrophobic and hydrophilic domains. The passage of species between different regions is an improbable event and occurs very slowly. Therefore, self-diffusion, if studied over macroscopic distances, should reveal whether the process is rapid or slow depending on the geometric properties of the inner structure. For example, a phase that is water continuous and oil discontinuous should exhibit rapid diffusion of hydrophilic components, whereas the hydrophobic components should diffuse slowly. An oil continuous but water discontinuous system should exhibit rapid diffusion of the hydrophobic components. One would expect that a bicontinuous structure should give rapid diffusion of all components.

Using the above principle, Lindman and co-workers [18–20] measured the self-diffusion coefficients of all components consisting of various components, with particular emphasis on the role of the co-surfactant. For microemulsions consisting of water, hydrocarbon, an anionic surfactant and a short-chain alcohol (C_4 and C_5), the self-diffusion coefficient of water, hydrocarbon and co-surfactant was fairly high, of the order of $10^{-9}\,m^2\,s^{-1}$, i.e. two orders of magnitude higher than the value expected for a discontinuous medium ($10^{-11}\,m^2\,s^{-1}$). This high diffusion coefficient was attributed to three main effects: bicontinuous solutions, easily deformable and flexible interface and absence of any large aggregates.

With microemulsions based on long-chain alcohols (e.g. decanol), the self-diffusion coefficient for water was low, indicating the presence of definite (closed) water droplets surrounded by surfactant anions in the hydrocarbon medium. Hence NMR measurements could clearly distinguish between the two types of microemulsion systems.

10.8
Selection of Surfactants for Formulation of Microemulsions

The formulation of microemulsions is still an art, since understanding the interactions, at a molecular level, at the oil and water sides of the interface is far from being achieved. However, some rules may be applied for the selection of emulsifiers for formulating O/W and W/O microemulsions. These rules are based on the same principles as applied for the selection of emulsifiers for macroemulsions described in Chapter 6. Three main methods may be applied for such selection, namely the hydrophilic–lipophilic balance (HLB), the phase inversion temperature (PIT) and the cohesive energy ratio (CER) concepts. As mentioned before, the HLB concept is based on the relative percentage of hydrophilic to lipophilic (hydrophobic) groups in the surfactant molecule. Surfactants with a low HLB number (3–6) normally form W/O emulsions, whereas those with a high HLB number (8–18) form O/W emulsions. Given an oil to be microemulsified, the formulator should first determine its required HLB number. Several procedures may be applied for determining the HLB number depending on the type of surfactant that needs to be used. These procedures were described in Chapter 5. Once the HLB number of the oil is known, one must try to find the chemical type of emulsifier which best matches the oil. Hydrophobic portions of surfactants which are similar to the chemical structure of the oil should be looked at first.

The PIT system provides information on the type of oil, phase volume relationships and concentration of the emulsifier. The PIT system is established on the proposition that the HLB number of a surfactant changes with temperature and that the inversion of the emulsion type occurs when the hydrophile and lipophile tendencies of the emulsifier just balance. At this temperature, no emulsion is produced. From a microemulsion viewpoint, the PIT has an outstanding feature since it can throw some light on the chemical type of the emulsifier needed to match a given oil. Indeed, the required HLB values for various oils estimated from the PIT system compare very favorably with those prepared using the HLB system described above. This shows a direct correlation between the HLB number and the PIT of the emulsion.

As discussed in Chapter 6, the CER concept provides a more quantitative method for the selection of emulsifiers. The same procedure can also be applied for microemulsions.

10.9
Role of Microemulsions in Enhancement of Biological Efficacy

The role of microemulsions in enhancement of biological efficiency can be described in terms of the interactions at various interfaces and their effect on transfer and performance of the agrochemical. This will be described in detail in

Chapter 13 and only a summary is given here. The application of an agrochemical as a spray involves a number of interfaces, where interaction with the formulation plays a vital role. The first interface during application is that between the spray solution and the atmosphere (air), which governs the droplet spectrum, rate of evaporation, drift, etc. In this respect, the rate of adsorption of the surfactant at the air/liquid interface is of vital importance. Since microemulsions contain high concentrations of surfactant and mostly more than one surfactant molecule is used for their formulation, then on diluting a microemulsion on application, the surfactant concentration in the spray solution will be sufficiently high to cause efficient lowering of the surface tension γ. As discussed above, two surfactant molecules are more efficient in lowering γ than either of the two components alone. Hence the net effect will be the production of small spray droplets, which as will be seen later, adhere better to the leaf surface. In addition, the presence of surfactants in sufficient amounts will ensure that the rate of adsorption (which is the situation under dynamic conditions) is fast enough to ensure coverage of the freshly formed spray by surfactant molecules.

The second interaction is between the spray droplets and the leaf surface, where the droplets impinging on the surface undergo a number of processes that determine their adhesion and retention and further spreading on the target surface. The most important parameters that determine these processes are the volume of the droplets and their velocity, the difference between the surface energy of the droplets in flight, E_0, and their surface energy after impact, E_s. As mentioned above, microemulsions which are effective in lowering the surface tension of the spray solution ensure the formation of small droplets which do not usually undergo reflection if they are able to reach the leaf surface. Clearly, the droplets should not to be too small, otherwise drift may occur. One usually aims at a droplet size range in the region of 100–400 μm. As will be discussed in Chapter 13, the adhesion of droplets is governed by the relative magnitude of the kinetic energy of the droplet in flight and its surface energy as it lands on the leaf surface. Since the kinetic energy is proportional to the third power of the radius (at constant droplet velocity), whereas the surface energy is proportional to the second power, one would expect that sufficiently small droplets will always adhere. For a droplet to adhere, the difference in surface energy between free and attached drop ($E_0 - E_s$) should exceed the kinetic energy of the drop, otherwise bouncing will occur. Since E_s depends on the contact angle, θ, of the drop on the leaf surface, it is clear that low values of θ are required to ensure adhesion, particularly with large drops that have high velocity. Microemulsions when diluted in the spray solution usually give low contact angles of spray drops on leaf surfaces as a result of lowering the surface tension and their interaction with the leaf surface.

Another factor which can affect the biological efficacy of foliar spray application of agrochemicals is the extent to which a liquid wets and covers the foliage surface. This, in turn, governs the final distribution of the agrochemical over the areas to be protected. Several indices may be used to describe the wetting of a surface by the spray liquid, of which the spread factor and spreading coefficient are probably the most useful. The spread factor is simply the ratio of the diameter of the area wetted on the leaf, D, to the diameter of the drop, d. This ratio is determined by the contact angle of the drop on the leaf surface. The lower the value of θ, the higher is the spread factor. As mentioned above, microemulsions usually give a low contact angle for the drops produced from the spray. The spreading coefficient is determined by the surface tension of the spray solution and also the value of θ. Again, with microemulsions diluted in a spray both γ and θ are sufficiently reduced and this results in a positive spreading coefficient. This ensures rapid spreading of the spray liquid on the leaf surface.

Another important factor for the control of biological efficacy is the formation of 'deposits' after evaporation of the spray droplets, which ensure the tenacity of the particles or droplets of the agrochemical. This will prevent removal of the agrochemical from the leaf surface by the falling rain. Many microemulsion systems form liquid crystalline structures after evaporation, which have high viscosity (hexagonal or lamellar liquid crystalline phases). These structures will incorporate the agrochemical particles or droplets and ensure their 'stickiness' to the leaf surface.

One of the most important roles of microemulsions in enhancing biological efficacy is their effect on penetration of the agrochemical through the leaf. Two effects may be considered which are complementary. The first effect is due to enhanced penetration of the chemical as a result of the low surface tension. For penetration to occur through fine pores, a very low surface tension is required to overcome the capillary (surface) forces. These forces produce a high pressure gradient that is proportional to the surface tension of the liquid. The lower the surface tension, the lower is the pressure gradient and the higher the rate of penetration. The second effect is due to solubilization of the agrochemical within the microemulsion droplet. Solubilization results in an increase in the concentration gradient, thus enhancing the flux due to diffusion. This can be understood from a consideration of Fick's first law:

$$J_D = D\left(\frac{\partial C}{\partial x}\right) \tag{27}$$

where J_D is the flux of the solute (amount of solute crossing a unit cross-section in unit time), D is the diffusion coefficient and $\partial C/\partial x$ is the concentration gradient. The presence of the chemical in a swollen micellar system will lower the diffusion coefficient. However, the presence of the solubilizing agent (the microemulsion droplet) increases the concentration gradient in direct proportionality to the increase in solubility. This is because Fick's law involves the absolute gradient of concentration which is necessarily small provided that the solubility is small, but not its relative rate. If the saturation is noted by S, Fick's law may be written as

$$J_D = D \times 100S\left(\frac{\partial \%S}{\partial x}\right) \tag{28}$$

where $\partial\%S/\partial x$ is the gradient in relative value of S. Equation 28 shows that for the same gradient of relative saturation, the flux caused by diffusion is directly proportional to saturation. Hence solubilization will in general increase transport by diffusion, since it can increase the saturation value by many orders of magnitude (which outweighs any reduction in D). In addition, the solubilization enhances the rate of dissolution of insoluble compounds and this will have the effect of increasing the availability of the molecules for diffusion through membranes.

References

1 T.P. Hoar and J.H. Schulman, *Nature*, *152*, 102 (1943).

2 L.M. Prince, *Microemulsion Theory and Practice*, Academic Press, New York (1977).

3 I. Danielsson and B. Lindman, *Colloids Surf.*, *3*, 391 (1983).

4 J.H. Schulman, W. Stoeckenius and L.M. Prince, *J. Phys. Chem.*, *63*, 1677 (1959).

5 L.M. Prince, *Adv. Cosmet. Chem.*, *27*, 193 (1970).

6 K. Shinoda and S. Friberg, *Adv. Colloid Interface Sci.*, *4*, 281 (1975).

7 E. Ruckenstein and J.C. Chi, *J. Chem. Soc., Faraday Trans. 2*, *71*, 1690 (1975).

8 J. Th. G. Overbeek, *Faraday Discuss. Chem. Soc.*, *65*, 7 (1978).

9 J. T. G. Overbeek, P.L. de Bruyn and F. Verhoeckx, in *Surfactants*, ed. Th.F. Tadros, Academic Press, London, pp. 111–132 (1984).

10 D.J. Mitchell and B.W. Ninham, *J. Chem. Soc., Faraday Trans. 2*, *77*, 601 (1981).

11 R.C. Baker, A.T. Florence, R.H. Ottewill and Th.F. Tadros, *J. Colloid Interface Sci.*, *100*, 332 (1984).

12 N.W. Ashcroft and J. Lekner, *Phys. Rev.*, *45*, 33 (1966).

13 P.N. Pusey, in *Industrial Polymers: Characterization by Molecular Weights*, ed. J.H.S. Green and R. Dietz, Transcripta Books, London (1973).

14 A.N. Cazabat and D. Langevin, *J. Chem. Phys.*, *74*, 3148 (1981).

15 L.M. Prince, *Microemulsions*, Academic Press, New York (1977).

16 B. Lagourette, J. Peyerlasse, C. Boned and M. Clausse, *Nature*, *281*, 60 (1969).

17 M. Clausse, J. Peyerlasse, C. Boned, J. Heil, L. Nicolas-Margantine and A. Zrabda, in *Solution Properties of Surfactants*, Vol. 3, ed. K.L. Mittal and B. Lindman, Plenum Press, New York, p. 1583 (1984).

18 B. Lindman and H. Winnerstrom, *Top. Curr. Chem.*, *87*, 1 (1980).

19 H. Winnerstrom and B. Lindman, *Phys. Rep.*, *52*, 1 (1970).

20 B. Lindman, P. Stilbs and M.E. Moseley, *J. Colloid Interface Sci.*, *83*, 569 (1981).

11
Multiple Emulsions

11.1
Introduction

Multiple emulsions are complex systems of 'emulsions of emulsions'. Both water-in-oil-in-water (W/O/W) and oil-in-water-in-oil (O/W/O) emulsions have potential applications in agrochemicals. The most common are W/O/W, which may be considered as a W/W emulsion where the internal water droplets are separated by an 'oily layer' (membrane). The internal droplets could also consist of a polar solvent such glycol or glycerol, which may contain a dissolved or dispersed agrochemical active ingredient (a.i.).

Multiple emulsions are ideal systems for application in agrochemicals. One may have three a.i.s in one formulation and one can incorporate adjuvants in three compartments. Multiple emulsions can be usefully applied for controlled release by controlling the rate of the breakdown process of the multiple emulsion on application. Initially, one prepares a stable multiple emulsion (with a shelf-life of 2 years), which on dilution breaks down in a controlled manner, releasing the a.i. also in a controlled manner (slow or sustained release). As will be discussed later, the formulated W/O/W multiple emulsion is osmotically balanced but on dilution the system breaks down as a result of the lack of this balance.

11.1.1
Preparation of Multiple Emulsions

These multiple emulsions are usually prepared in a two-stage process. For example, a W/O/W multiple emulsion is formulated by first preparing a W/O emulsion using a surfactant with a low HLB number (5–6) using a high-speed mixer (e.g. an Ultra Turrax or Silverson). The resulting W/O emulsion is further emulsified in an aqueous solution containing a surfactant with a high HLB number (9–12) using a low-speed stirrer (e.g. a paddle stirrer). A schematic representation of the preparation of multiple emulsions is given in Figure 11.1.

The yield of the multiple emulsion can be determined using dialysis for W/O/W multiple emulsions. A water-soluble marker is used and its concentration in the

Colloids and Interface Science Series, Vol. 5
Colloids in Agrochemicals. Tharwat F. Tadros

ISBN: 978-3-527-31465-2

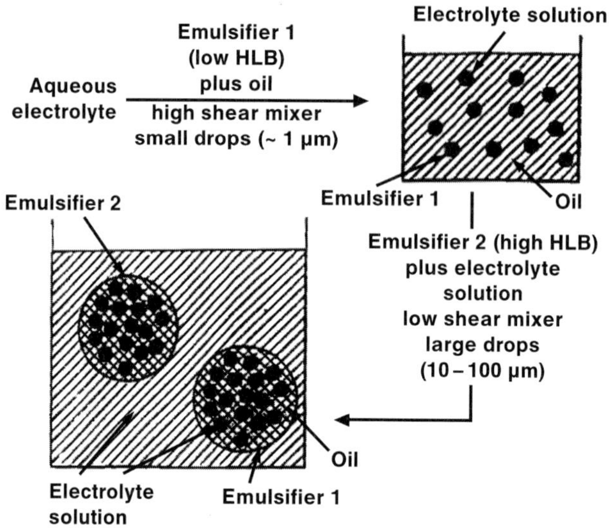

Figure 11.1 Scheme for preparation of W/O/W multiple emulsion.

outside phase is determined:

$$\text{multiple emulsion}(\%) = \frac{C_\mathrm{i}}{C_\mathrm{i} + C_\mathrm{e}} \times 100 \tag{1}$$

where C_i is the amount of marker in the internal phase and C_e is that in the external phase. It has been suggested that if a yield of more than 90% is required, the lipophilic (low HLB) surfactant used to prepare the primary emulsion must be ~10 times higher in concentration than the hydrophilic (high HLB) surfactant.

11.1.2
Types of Multiple Emulsions

Florence and Whitehill [1] classified multiple emulsions into three main categories, A, B and C, as illustrated in Figure 11.2. Type A contains one large internal droplet similar to that described by Matsumoto *et al.* [2]. This type was produced when polyoxyethylene (4) lauryl ether (Brij 30) was used as emulsifier at 2%. Type B contains several small internal droplets. These were prepared using 2% polyoxyethylene (16.5) nonyl phenyl ether (Triton X-165). Type C drops entrapped a very large number of small internal droplets. These were prepared using a 3:1 Span 80–Tween 80 mixture. It should be mentioned that type A multiple emulsions are not encountered much in practice. Type C is difficult to prepare since a large number of small water internal droplets (which are produced in the primary emulsification process) result in a large increase in vis-

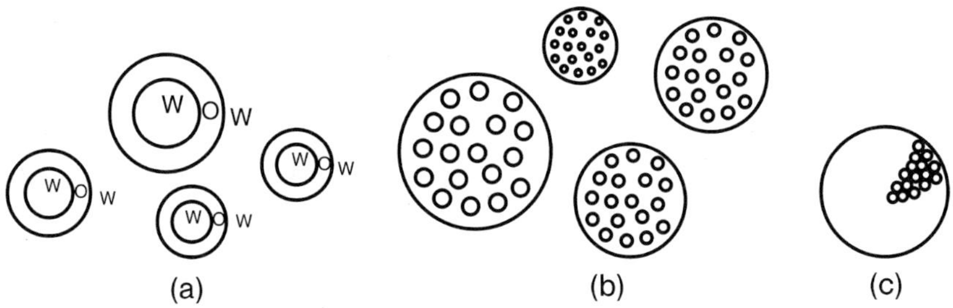

Figure 11.2 Schematic representation of three structures of W/O/W multiple emulsions. (a) One large internal droplet (Brij 30); (b) several small internal droplets; (c) large number of very small droplets (3 : 1 Span 80–Tween 80).

cosity. Hence the most common multiple emulsions used in practice are those represented by type B.

11.1.3
Breakdown Processes of Multiple Emulsions

Florence and Whitehill [1] identified several types of breakdown processes. The external oil drops may coalesce with other oil drops (which may or may not

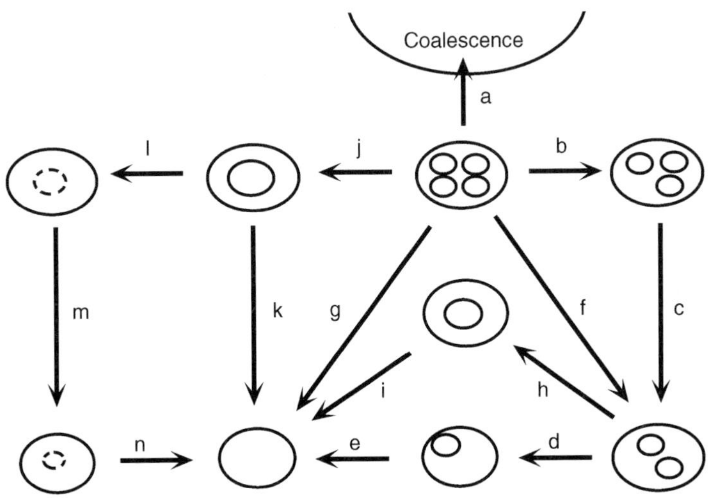

Figure 11.3 Schematic representation of the possible breakdown pathways in W/O/W multiple emulsions: (a) coalescence; (b–e) expulsion of one or more internal aqueous droplets; (f) expulsion; (g) less frequent expulsion; (h, i) coalescence of water droplets before expulsion; (j, k) diffusion of water through the oil phase; (l–n) shrinking of internal droplets.

contain internal aqueous droplets). Alternatively, the internal aqueous droplets may be individually expelled or more than one expelled, or they may be less frequently expelled in one step. The internal droplets may coalesce before being expelled or water may pass by diffusion through the oil phase, resulting in shrinkage of the internal droplets. A schematic diagram of the breakdown processes in multiple emulsions [1] is given in Figure 11.3.

All the processes are influenced by the nature of the two emulsifiers used for preparation of the multiple emulsion. Most papers on multiple emulsions are based on conventional nonionic surfactants. Unfortunately, most of these surfactant systems produce multiple emulsions with a limited shelf-life, particularly if the system is subjected to large temperature variations. During the past few years, we have formulated multiple emulsions using polymeric surfactants for both primary and multiple emulsion preparation. These polymeric surfactants proved to be superior to the conventional nonionic surfactants in maintaining the physical stability of the multiple emulsion and they could now be successfully applied for the formulation of agrochemical multiple emulsions. The results obtained using these polymeric surfactants offer several potential applications in agrochemical formulations. The key in the latter cases is to use polymeric surfactants that are approved by the US Environmental Protection Agency (EPA).

11.1.4
Factors Affecting Stability of Multiple Emulsions and Criteria for Their Stabilization

It is clear that the stability of the resulting multiple emulsion depends on a number of factors: (a) the nature of the emulsifiers used for preparation of the primary and multiple emulsion; (b) the osmotic balance between the aqueous droplets in the multiple emulsion drops and those in the external aqueous phase; (c) the volume fractions of the disperse water droplets in the multiple emulsion drops and the final volume fraction of the multiple emulsions; (d) the temperature range to which the multiple emulsion is subjected; (e) the process used to prepare the system; and (f) the rheology of the whole system, which can be modified by the addition of thickeners to the external aqueous phase.

The main criteria for the preparation of a stable multiple emulsion are as follows: (a) two emulsifiers, one with low (emulsifier I) and one with high (emulsifier II) HLB number; (b) emulsifier I should provide a very effective barrier against coalescence of the water droplets in the multiple emulsion drop, and emulsifier II should also provide an effective barrier against flocculation and/or coalescence of the multiple emulsion drops; (c) the amount of emulsifiers used in the preparation of primary and multiple emulsions is critical: excess emulsifier I in the oil phase may result in further emulsification of the aqueous phase into the multiple emulsion with the ultimate production of a W/O emulsion, and excess emulsifier II in the aqueous phase may result in solubilization of the low HLB number surfactant with the ultimate formation of an O/W emulsion; (d) optimum osmotic balance of the internal and external aqueous phases; if the osmotic pressure of the internal aqueous droplets is higher than that of the external aqueous phase, water will flow to the internal droplets, resulting in 'swelling' of the multiple emulsion drops with the ultimate production of a W/O emulsion; in contrast, if the osmotic pressure in the outside external phase is higher, water will diffuse in the opposite direction and the multiple emulsion will revert to an O/W emulsion.

This chapter deals with the use of polymeric surfactants for the preparation of W/O/W multiple emulsions for agrochemical applications. It starts with a section giving a brief description of polymeric surfactants. This is followed by a brief section on the fundamental principles of stabilization of emulsions using polymeric surfactants. The last section gives some examples of formulations using these polymeric surfactants.

It should be mentioned that the above principles also apply to the preparation of stable O/W and W/O emulsions and also other related applications.

11.2
General Description of Polymeric Surfactants

The most convenient polymeric surfactants are those of the block (A–B or A–B–A) and graft (BA_n) types, which exhibit considerable surface activity. For an O/W emulsifier the B chain is chosen to have high affinity to the oil phase (or to be soluble in it), whereas the A chain is chosen to be highly soluble in the aqueous medium and strongly solvated by the water molecules. This configuration is the most suitable for effective stabilization, as will be discussed below. The high affinity of the B chain to the oil phase ensures strong 'anchoring' of the chain to the surface, and this prevents any displacement of the molecule on the approach of two oil droplets, thus preventing any flocculation and/or coalescence. Examples of the A–B–A block copolymers for stabilization of O/W emulsions are the Pluronics (trade name of BASF) and Synperonic PE (trade name of ICI), represented schematically in Figure 11.4, which also shows the adsorption

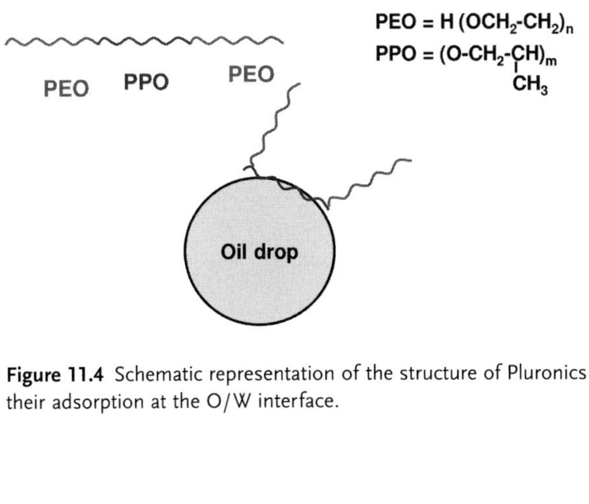

Figure 11.4 Schematic representation of the structure of Pluronics and their adsorption at the O/W interface.

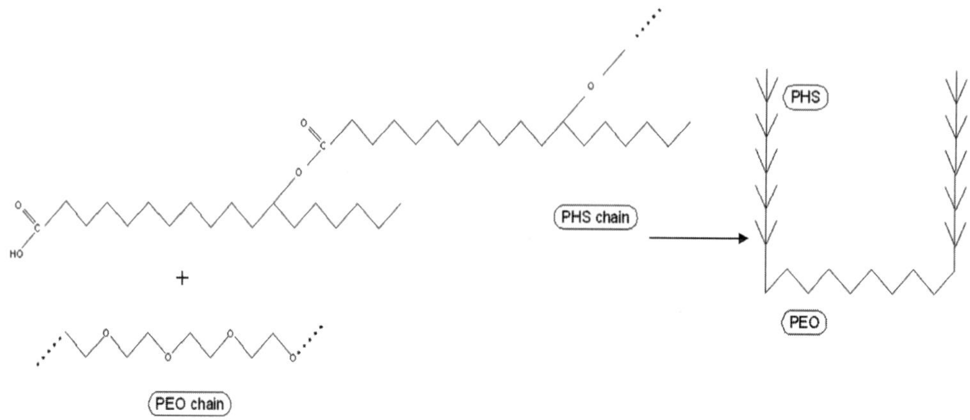

Figure 11.5 Structure of PHS–PEO–PHS block copolymer.

and conformation of the polymer. These triblock copolymers consist of two poly-A blocks of poly(ethylene oxide) (PEO) and one poly-B block of poly(propylene oxide) (PPO). Several chain lengths of PEO and PPO are available. The PPO chain, which is hydrophobic, adsorbs at the oil surface either due to its high affinity to the oil molecules or as a result of 'rejection anchoring' (since the PPO chain is insoluble in water). The two PEO chains reside in the aqueous medium and they become strongly solvated by the water molecules (as a result of hydrogen bonding with water molecules). These PEO chains provide strong repulsion on close approach of two emulsion droplets, as will be discussed later.

For W/O emulsions, the B chain should have high affinity to the water droplets (or be soluble in water), whereas the A chains should be soluble in the oil phase

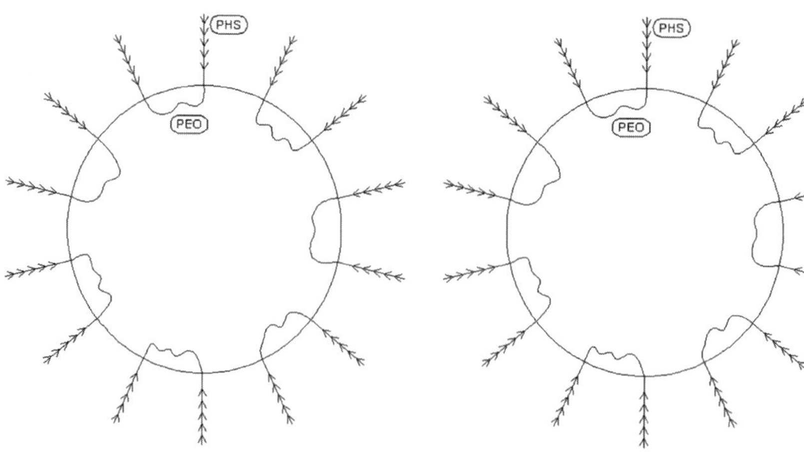

Figure 11.6 Conformation of PHS–PEO–PHS block copolymer at the W/O interface.

and strongly solvated by the oil molecules. A good example of such a triblock is poly(hydroxystearic acid)–poly(ethylene oxide)–poly(hydroxystearic acid) (PHS–PEO–PHS), which was marketed by ICI under the trade name Hypermer B246 or Arlacel P135. A schematic representation of the structure of this block copolymer is shown in Figure 11.5. Its adsorption and conformation at the W/O interface are shown in Figure 11.6. The PEO chain is soluble in the water droplets and it provides a strong 'anchor' to the interface, whereas the PHS chains are highly soluble in most hydrocarbon solvents and also some of the polar ones. These PHS chains provide strong repulsion on approach of the water droplets. The PHS–PEO–PHS molecules also lower the interfacial tension of the W/O interface to very low values (approaching zero), hence emulsification of water in oil is very efficient, allowing one to prepare highly concentrated W/O emulsions that have low viscosity.

11.3 Interaction Between Oil or Water Droplets Containing an Adsorbed Polymeric Surfactant – Steric Stabilization

Let us consider the case of two water droplets containing an adsorbed polymeric surfactant such as Arlacel P135 (which is the molecule that is used for the preparation of the primary W/O emulsion). As discussed above, this molecule has a very high surface activity at the W/O interface, because the PEO chain (which has a molecular weight of ~1500) resides in the water droplets, leaving the two PHS chains (each with a molecular weight of ~1000) in the oil phase. Surface pressure (π)–area per molecule (A) isotherms at the water/air (W/A) and water/oil (W/A) interfaces obtained using a semiautomatic Langmuir trough [3] are shown in Figure 11.7. The W/O isotherm is more expanded than the W/O isotherm, although the two curves tend to approach each other as the surface

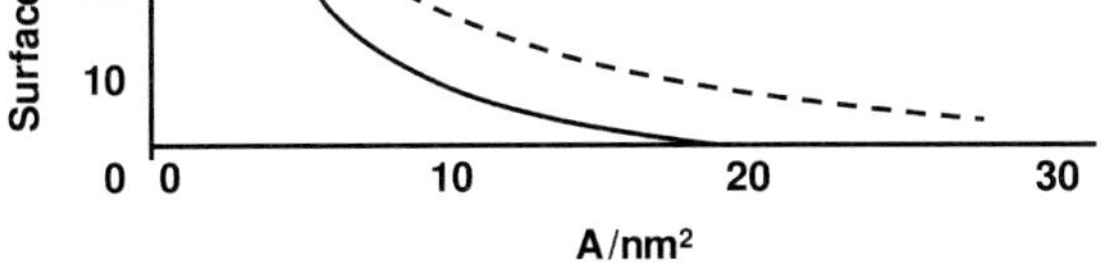

Figure 11.7 π–A curves (25 °C) for PHS–PEO–PHS block copolymer at the air/water and oil/water interfaces.

pressure is increased. The W/O monolayer would be expected to be more expanded since the oil will reduce the van der Waals attraction between the hydrocarbon chains. Collapse at the W/A interface was well defined, occurring at a surface pressure of 38.5 mN m^{-1}. This possibly involves the formation of bilayers of PHS–PEO–PHS at the surface, particularly in view of the lamellar liquid crystalline structure of the bulk material.

For the W/O case, the maximum surface pressure approached 51 mN m^{-1}, implying an interfacial pressure approaching zero. These interfacial tension results show that the PHS–PEO–PHS block copolymer is an excellent W/O emulsifier. Indeed W/O emulsions with a water volume fraction ϕ of 0.9 can be prepared using this block copolymer.

The PHS chains are highly solvated by the oil molecules and they extend at the W/O interface giving a layer thickness of the order of 7–8 nm. This was confirmed using film thickness measurements [4]. A thin film consisting of oil plus surfactant was formed between two aqueous droplets, when they were brought into contact, and the thickness of the film was measured using light reflectance. A film thickness of 14.2 nm was measured, indicating a layer thickness of the PHS chains of the order of 7 nm. The layer thickness could also be measured using rheological measurements. Shear stress (τ)–shear rate (γ) curves were obtained for W/O emulsions at various volume fractions of water. The results showed a change from Newtonian to non-Newtonian flow behavior as the volume fraction of the water in the emulsion increased. The non-Newtonian flow reflects the droplet-droplet interaction as these approach to closer distances. The data were analyzed using the Bingham model [5]:

$$\tau = \tau_\beta + \eta_{\text{pl}}\gamma \tag{2}$$

Figure 11.8 shows plots of η_{pl} versus ϕ. The results show that η_{pl} increases gradually with increase in ϕ, but above $\phi = 0.6$, there is a rapid increase in both parameters with further increase in the volume fraction. Such behavior is typical of concentrated dispersions [6, 7], which show a rapid increase in the rheological parameters when the separation distance between the particles becomes comparable to the range of repulsive interaction.

Assuming that the W/O emulsion behaves as a near 'hard-sphere' dispersion, it is possible to apply the Dougherty–Krieger equation [8, 9] to obtain the effective volume fraction, ϕ_{eff}. The assumption that the W/O emulsion behaves as a near hard-sphere dispersion is reasonable since the water droplets are stabilized with a block copolymer with relatively short PHS chains (of the order of 10 nm and less). Any lateral displacement of the polymer will be opposed by the high Gibbs elasticity of the adsorbed polymer layer and the droplets will maintain their spherical shape up to high volume fractions.

For hard-sphere dispersions, the relative viscosity, η_r, is related to the effective volume fraction by the following expression:

$$\eta_r = \left[1 - \left(\frac{\phi_{\text{eff}}}{\phi_p}\right)\right]^{-[\eta]\phi_p} \tag{3}$$

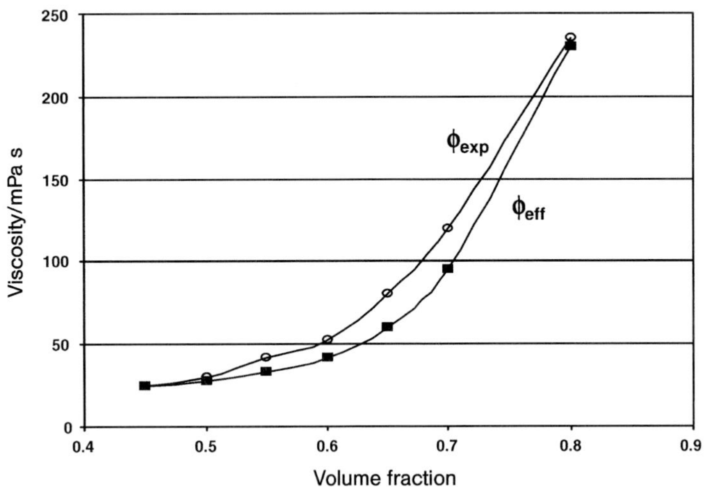

Figure 11.8 Viscosity–volume fraction curves for W/O emulsions stabilized with PHS–PEO–PHS block copolymer.

where ϕ is replaced by ϕ_{eff}, which includes the contribution from the adsorbed layer; $[\eta]$ is the intrinsic viscosity, which for hard spheres is equal to 2.5, and ϕ_p is the maximum packing fraction. It was shown that a plot of $1/\eta_r{}^{1/2}$ versus ϕ is linear with an intercept that is equal to ϕ_p: For the present W/O emulsion, such a plot gave a ϕ_p value of 0.84. This value is higher than the theoretical maximum packing fraction for monodisperse spheres (0.74 for hexagonal packing). However, this high value is not unreasonable considering the polydispersity of the W/O emulsion. The high ϕ_p value shows without doubt that the PHS–PEO–PHS block copolymer is very suitable for the preparation of high volume fraction W/O emulsions.

Using ϕ_p and the measured η_r, ϕ_{eff} was calculated at each ϕ value using equation 3 and the results are plotted in Figure 11.8. From ϕ_{eff}, the adsorbed layer thickness, δ, was calculated using the following equation:

$$\phi_{eff} = \phi\left[1 + \left(\frac{\delta}{R}\right)\right]^3 \qquad (4)$$

where R is the droplet size (which could be determined using dynamic light scattering).

A plot of δ versus ϕ showed a linear decrease in δ with increase in ϕ. The value of δ at $\phi = 0.4$ is $\sim$10 nm, which is a measure of the fully extended PHS chains. At such a low ϕ value, there will be no interpenetration of the PHS chains since the distance between the droplets is relatively large. The reduction in δ with increase in ϕ is due to the possible interpenetration and/or compression of the PHS chains on close approach of the droplets. This is also possible in the thin liquid film studies, which showed a layer thickness in the region of 7–8 nm.

When two water droplets each containing adsorbed copolymers of PHS–PEO–PHS approach to a separation distance h that is smaller than twice the fully extended PHS chains, i.e. h becomes smaller than about 20 nm, repulsion occurs as a result of two main effects [10]. The first repulsive force arises from the unfavorable mixing of the PHS chains when these are in a good solvent. The PHS chains are soluble in most hydrocarbon solvents and are strongly solvated by their molecules over a wide range of temperatures. The unfavorable mixing of polymer chains in good solvent conditions was considered by Flory and Krigbaum [11], whose theory could be applied to the present case of mixing two PHS chains in a hydrocarbon solvent.

Before overlap, the chains have a volume fraction ϕ_2 and the solvent has a chemical potential μ_1^{α}. In the overlap region, the volume fraction of the chains is ϕ_2', which is higher than ϕ_2, and the solvent has a chemical potential μ_1^{β}, which is lower than μ_1^{α}. This is equivalent to an increase in the osmotic pressure in the overlap region. As a result, solvent diffuses from the bulk to the overlap region and the two water droplets become separated, i.e. this results in strong repulsion. The latter is referred to mixing or osmotic repulsion.

Using the Flory–Krigbaum theory [11], one can calculate the free energy of mixing, G_{mix}, due to this unfavorable overlap:

$$\frac{G_{mix}}{kT} = \frac{4\pi}{3V_1}\phi_2{}^2 N_A\left(\frac{1}{2}-\chi\right)\left(\delta-\frac{h}{2}\right)^2\left(3R+2\delta+\frac{h}{2}\right) \tag{5}$$

where k is Boltzmann's constant, T is the absolute temperature, V_1 is the molar volume of the solvent and N_A is Avogadro's number.

It is clear from equation 5 that when the Flory–Huggins interaction parameter, χ, is less than 0.5, i.e. the chains are in good solvent conditions, G_{mix} is positive and the interaction is repulsive and it increases very rapidly with decreasing h, when the latter is lower than 2δ. This explains the strong repulsion obtained between water droplets surrounded by PHS chains. The latter are highly soluble in the hydrocarbon medium and any attempt to overlap the chains results in very strong repulsion as a result of the above mentioned unfavorable mixing.

Equation 5 also shows that when $\chi > 0.5$, i.e. when the solvency of the medium for the chains becomes poor, G_{mix} is negative and the interaction becomes attractive. Therefore, one has to be sure that the solvent used for preparation of the W/O emulsion is a good solvent for the PHS chains, otherwise flocculation of the water droplets (which could be followed by their coalescence) may occur. Fortunately, the PHS chains are soluble in most hydrocarbon solvents used in agrochemical formulations. The condition $\chi = 0.5$ is referred to as a θ-solvent and this denotes the onset of change of repulsion to attraction. Therefore, to ensure steric stabilization by the above mechanism, one has to ensure that the chains are kept in better than a θ-solvent.

The second repulsive force arises from the loss of configuration entropy when the chains overlap to some extent. When the two surfaces of the water droplets are separated at an infinite distance, each chain will have a number of configurations, Ω_∞, that are determined by the volume of the hemisphere swept by the PHS chain. When the two surfaces approach to a distance h that is significantly smaller than the radius of the hemisphere swept by the PHS chain, the volume available to the chain becomes smaller and this results in a reduction in the configurational entropy to a value Ω (which is smaller than Ω_∞). This results in strong repulsion and the effect is referred to as entropic, volume restriction or elastic repulsion and is given by the following expression:

$$G_{el} = 2\nu \ln \frac{\Omega}{\Omega_\infty} \tag{6}$$

where ν is the number of polymer chains per unit area of the surface. It should be mentioned that G_{el} is always repulsive in any solvent and it becomes very high on considerable overlap of the PHS chains. This can be illustrated from rheological measurements, as will be discussed later.

Plots of G_{mix}, G_{el} and G_A (the van der Waals energy) versus h are illustrated in Figure 11.9, which shows that G_{mix} is positive (when $\chi < 0.5$) and it increases very rapidly with decrease in h as soon as the latter becomes less than 2δ ($\sim$20 nm for the PHS chains). G_{el} (which is always positive) increases with decrease in h, becoming very large at short distances (when h becomes smaller than δ; about 10 nm for PHS).

Combination of G_{mix}, G_{el} and G_A results in the total G_T–h curve shown in Figure 11.9. This curve shows a shallow minimum, G_{min} (weak attraction), at $h \approx 2\delta$, i.e. at $h \approx 20$ nm for the present W/O emulsion based on PHS–PEO–PHS block copolymer. When $h < 2\delta$, G_T increases very rapidly with further decrease in h. The depth of the minimum, G_{min}, depends on the adsorbed layer thickness. In the present W/O emulsion based on a PHS layer thickness of about 10 nm, G_{min} is very small (fraction of kT). This shows that with the present sterically stabilized W/O emulsion, there is only very weak attraction at relatively long distance of separation between the water droplets, which is overcome by the Brownian diffusion (which is of the order of $1kT$). Hence one can say that the net interaction is repulsive and this ensures the long-term physical stability of the W/O emulsion (which approaches thermodynamic stability).

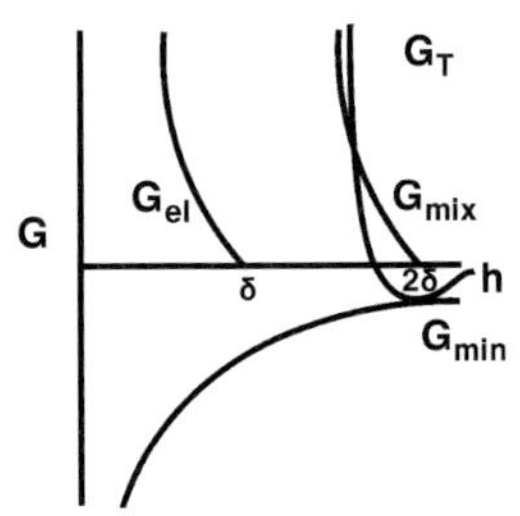

Figure 11.9 Schematic representation of the variation of G_{mix}, G_{el}, G_A and G_T with h.

Another important use of the PHS–PEO–PHS block copolymer is the formation of a viscoelastic film around the water droplets [12]. This results from the dense packing of the molecule at the W/O interface, which results in an appreciable interfacial viscosity. This viscoelastic film prevents transport of water from the internal water droplets in the multiple emulsion drop to the external aqueous medium and this ensures the long-term physical stability of the multiple emulsion when using polymeric surfactants. The viscoelastic film can also reduce the transport of any active ingredient in the internal water droplets to the external phase. This is desirable in many cases when protection of the ingredient in the internal aqueous droplets is required and release is provided on application of the multiple emulsion.

When the W/O emulsion is emulsified in an aqueous solution containing another polymeric surfactant with high HLB number, the multiple emulsion drops become surrounded with another polymer surfactant layer. This is illustrated, for example, when using Pluronic PEF127 as the secondary polymer emulsifier. This triblock copolymer consists of two PEO chains of ~100 EO units each and a PPO chain of ~55 units. The PPO chain adsorbs relatively strongly at the O/W interface, leaving the two PEO chains in the aqueous continuous phase. The PEO layer thickness is probably larger than 10 nm and hence at separation distances h between the multiple emulsion drops smaller than ~20 nm strong repulsion between the multiple emulsion drops becomes very strong, thus preventing any flocculation and/or coalescence.

It should be emphasized that polymeric surfactants prevent coalescence of the water droplets in the multiple emulsion drops and also coalescence of the latter drops themselves. This is due to the interfacial rheology of the polymeric surfactant films. As a result of the strong lateral repulsion between the stabilizing chains at the interface (PHS chains at the W/O interface and PEO chains at the O/W interface), these films resist deformation under shear and hence they produce a viscoelastic film. On approach of two droplets, this film prevents deformation of the interface and hence coalescence is prevented.

From the above discussion, one can summarize the role of polymeric surfactants in stabilization of W/O and W/O/W multiple emulsion. First, the polymeric surfactant ensures complete coverage of the droplets and, by virtue of its strong adsorption, displacement of the film on close approach is prevented. This is essential for eliminating coalescence of the emulsion droplets. Second, since the stabilizing chains (PHS for the W/O emulsion and PEO for the W/O/W multiple emulsion) are in good solvent conditions, the mixing interaction is positive, leading to strong repulsion between the drops on close approach. This, together with the elastic interaction, provides a system that is repulsive at short separation distances. This prevents any flocculation and/or coalescence between the drops. The polymeric surfactant chains should be sufficiently long to prevent any weak flocculation, which may result if the depth of the minimum becomes large, i.e. when δ becomes small (say less than 5 nm).

As mentioned above, the strong repulsive force between sterically stabilized emulsion droplets can be investigated using rheological measurements, in particular dynamic (oscillatory) measurements. In these measurements, the emulsion is placed in the gap between two concentric cylinders or a cone and plate geometry. A sinusoidal strain with small amplitude γ_0 is applied on one of the platens (say the cup of the concentric cylinder or the plate of the cone and plate geometry). The stress on the other platen is measured simultaneously during the oscillation. The response in stress of a viscoelastic material subjected to a sinusoidally varying strain is monitored as a function of strain amplitude and frequency. The stress amplitude, τ_0, is also a sinusoidally varying function in time, but in a viscoelastic material it is shifted out of phase with the strain. The phase angle shift between stress and strain, δ, is given by

$$\delta = \Delta t \omega \tag{7}$$

where ω is the frequency in rad s^{-1} ($\omega = 2\pi v$, where v is the frequency in hertz).

From the amplitudes of stress and strain and the phase angle shift, one can obtain the following viscoelastic parameters:

$$|G^*| = \frac{\tau_0}{\gamma_0} \tag{8}$$

$$G' = |G^*| \cos \delta \tag{9}$$

$$G'' = |G^*| \sin \delta \tag{10}$$

where G^* is the complex modulus, G' is the elastic component of the complex modulus (which is a measure of the energy stored by the system in a cycle) and G'' is the viscous component of the complex modulus (which is a measure of the energy dissipated as viscous flow in a cycle).

In viscoelastic measurements, one measures the viscoelastic parameters as a function of strain amplitude (at a fixed frequency) in order to obtain the linear viscoelastic region. The strain amplitude is gradually increased from the smallest possible value at which a measurement can be made and the rheological parameters are monitored as a function of strain amplitude γ_0: Initially, the rheological parameters remain virtually constant and independent of the strain amplitude. However, above a critical value of strain amplitude (γ_{cr}), the rheological parameters show a change with further increase in the amplitude above γ_{cr}: The linear viscoelastic region is the range of strain amplitudes below γ_{cr}: Once this region has been established, measurements are made as a function of frequency, keeping γ_0 below γ_{cr}: By fixing the frequency region while changing the volume fraction of the emulsion, one can obtain information on the interdroplet interaction.

As an illustration, Figure 11.10 shows the variation of G' and G'' (measured in the linear viscoelastic region and at a frequency of 1 Hz) versus the water volume fraction ϕ. The results show a transition from predominantly viscous to predominantly elastic response as ϕ exceeds 0.67. This is a direct manifestation of the

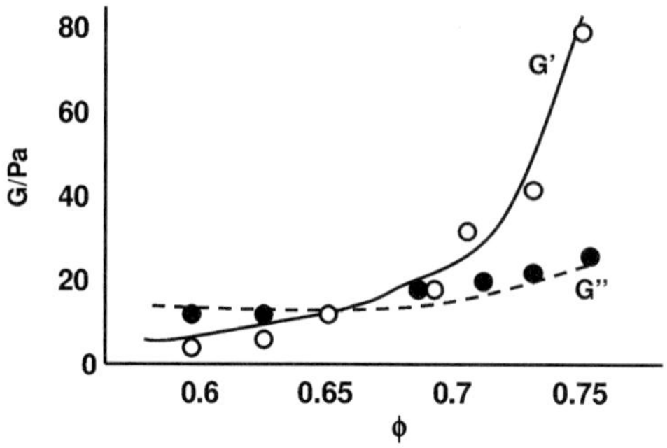

Figure 11.10 Variation of G' and G'' with ϕ for W/O emulsions stabilized with a PHS–PEO–PHS block copolymer.

strong elastic interaction that occurs at and above this critical ϕ. At this volume fraction, the interdroplet distance is comparable to twice the thickness of PHS chains, resulting in their interpenetration and/or compression. As ϕ exceeds 0.7, the storage modulus increases very sharply with further increase in ϕ and this is a reflection of the very strong repulsion between the water droplets.

11.4 Examples of Multiple Emulsions Using Polymeric Surfactants

Several examples of W/O emulsions and W/O/W multiple emulsions based on the PHS–PEO–PHS block copolymer have been produced. As an illustration, a typical formulation of W/O/W multiple emulsion is described below, using two different thickeners, namely Keltrol (xanthan gum from Kelco) and Carbopol 980 (a cross-linked polyacrylate gel produced by Goodrich). These thickeners were added to reduce creaming of the multiple emulsion. A two-step process was used in both cases.

The primary W/O emulsion was prepared using Arlacel P135. A 4 g amount of Arlacel P135 was dissolved in 30 g of a hydrocarbon oil. For quick dissolution, the mixture was heated to 75 °C. The aqueous phase consisted of 65.3 g of water, 0.7 g of $MgSO_4{:}7H_2O$ and a preservative. This aqueous solution was also heated to 75 °C. The aqueous phase was added to the oil phase slowly while stirring vigorously using a high-speed mixer. The W/O emulsion was homogenized for 1 min and allowed to cool to 40–45 °C followed by further homogenization for a further 1 min and stirring was continued till the temperature reached ambient.

The primary W/O emulsion was emulsified in an aqueous solution containing the polymeric surfactant Pluronic PEF127. A 2 g amount of the polymeric surfactant was dissolved in 16.2 g of water containing a preservative with stirring at 5 °C, then 0.4 g of $MgSO_4{:}7H_2O$ was added to the aqueous polymeric surfactant solution. Then 60 g of the primary W/O emulsion were slowly added to the aqueous Pluronic PFE127 solution while stirring slowly at 700 rpm (using a paddle stirrer). An aqueous Keltrol solution was prepared by slowly adding 0.7 g of Keltrol powder to 20.7 g of water, while stirring. The resulting thickener solution was further stirred for 30–40 min until a homogeneous gel was produced. The thickener solution was slowly added to the multiple emulsion while stirring at low speed (400 rpm) and the whole system was homogenized for 1 min followed by gentle stirring at 300 rpm until the thickener completely dispersed in the multiple emulsion (stirring for about 30 min was sufficient). The final system was investigated using optical microscopy to ensure that the multiple emulsion was produced. The formulation was left to stand for several months and the droplets of the multiple emulsion were investigated using optical microscopy. The rheology of the multiple emulsion was also measured at various intervals to ensure that the consistency of the product remained the same on prolonged storage.

The above multiple emulsion was prepared under the same conditions except using Carbopol 980 as a thickener (gel). In this case, no $MgSO_4$ was added, since the Carbopol gel was affected by electrolytes. The aqueous Pluronic PEF127 polymeric surfactant solution was prepared by dissolving 2 g of the polymer in 23 g of water. A 15 g amount of 2% master gel of Carbopol was added to the Pluronic PEF127 solution while stirring until the Carbopol was completely dispersed. A 60 g amount of the primary W/O emulsion was slowly added to the aqueous solution of Pluronic PEF127–Carbopol solution, while stirring thoroughly at 700 rpm. Triethanolamine was added slowly, while stirring gently until the pH of the system reached 6.0–6.5.

References

1 A.T. Florence and D. Whitehill, *J. Colloid Interface Sci.*, *79*, 243 (1981).
2 S. Matsumoto, Y. Kita and D. Yonezawa, *J. Colloid Interface Sci.*, *57*, 353 (1976).
3 Th.F. Tadros, *Int. J. Cosmet. Sci.*, *14*, 93 (1992).
4 M. Aston, PhD Thesis, Reading University (1987).
5 R.W. Whorlow, *Rheological Techniques*, Ellis Horwood, Chichester (1980).
6 Th.F. Tadros, *Adv. Colloid Interface Sci.*, *12*, 121 (1980).
7 J. Mewis and A.J.B. Spaul, *Adv. Colloid Interface Sci.*, *6*, 173 (1976).
8 I.M. Krieger and M. Dougherty, *Trans. Soc. Rheol.*, *3*, 137 (1959).
9 I.M. Krieger, *Adv. Colloid Interface Sci.*, *3*, 111 (1972).
10 D.H. Napper, *Polymeric Stabilization of Dispersions*, Academic Press, London (1983).
11 P.J. Flory, *Principles of Polymer Chemistry*, Cornell University Press, Ithaca, NY (1953).
12 Th.F. Tadros, *Colloids Surf.*, *91*, 39 (1994).

12
Controlled-release Formulations

12.1
Introduction

Controlled-release formulation of agrochemicals offer a number of advantages, of which the following are worth mentioning: (1) improvement of residual activity; (2) reduction of application dosage; (3) stabilization of the core active ingredient (a.i.) against environmental degradation; (4) reduction of mammalian toxicity by reducing worker exposure; (5) reduction of phytotoxicity; (6) reduction of fish toxicity; (7) reduction of environmental pollution.

One of the main advantages of using controlled-release formulations, in particular microcapsules, is the reduction of physical incompatibility when mixtures are used in the spray tank. They also can reduce biological antagonism when mixtures are applied in the field.

Several types of controlled-release systems can be identified: (1) microcapsules with particles in the size range 1–100 μm that consist of a distinct capsule wall (mostly a polymer) surrounding the agrochemical core; (2) microparticles (size range 1–100 μm) consisting of a matrix in which the agrochemical is uniformly dissolved or dispersed; (3) granules with matrix particles of 0.2–2.0 mm with the agrochemical uniformly dissolved or dispersed within the matrix.

This chapter gives a brief account of the different types of slow-release systems. For more detail, the reader can refer to the text edited by Scher [1].

12.2
Microencapsulation

Microencapsulation of agrochemicals is mainly carried out by interfacial condensation, *in situ* polymerization and coacervation. Interfacial condensation [2] is perhaps the most widely used method for encapsulation in industry. The active ingredient (a.i.), which may be oil soluble, oil dispersible or an oil itself, is first emulsified in water using a convenient surfactant or polymer. A hydrophobic monomer A is placed in the oil phase (oil droplets of the emulsion) and a hydrophilic monomer B is placed in the aqueous phase. The two monomers interact at

Colloids and Interface Science Series, Vol. 5
Colloids in Agrochemicals. Tharwat F. Tadros

ISBN: 978-3-527-31465-2

the interface between the oil and the aqueous phase, forming a capsule wall around the oil droplet. Two main types of systems may be identified. For example, if the material to be encapsulated is oil soluble, oil dispersible or an oil itself, an oil-in-water (O/W) emulsion is first prepared. In this case, the hydrophobic monomer is dissolved in the oil phase which forms the dispersed phase. The role of surfactant in this process is crucial since an oil–water emulsifier [with high hydrophilic–lipophilic-balance (HLB)] is required.

Alternatively, a polymeric surfactant such as partially hydrolyzed poly(vinyl acetate) [referred to as poly(vinyl alcohol) (PVA)] or a poly(ethylene oxide)–poly(propylene oxide)–poly(ethylene oxide) (PEO–PPO–PEO) (Pluronic) block copolymer can be used. The emulsifier controls the droplet size distribution and hence the size of capsules formed. On the other hand, if the material to be encapsulated is war soluble, a water-in-oil (W/O) emulsion is prepared using a surfactant with low HLB number or an A–B–A block copolymer of poly(hydroxystearic acid)–poly(ethylene oxide)–poly(hydroxystearic acid) (PHS–PEO–PHS). In this case, the hydrophilic monomer is dissolved in the aqueous internal phase droplets.

In interfacial polymerization, the monomers A and B are polyfunctional monomers capable of causing polycondensation or polyaddition reaction at the interface [2]. Examples of oil-soluble monomers are polybasic acid chloride, bis-haloformate and polyisocyanates, whereas water-soluble monomers can be polyamines or polyols. Thus, a capsule wall of polyamide, polyurethane or polyurea may be formed. Some trifunctional monomers are present to allow cross-linking reactions. If water is the second reactant with polyisocyanates in the organic phase, polyurea walls are formed. The latter modification has been termed *in situ* interfacial polymerization [3].

One of the most useful microencapsulation processes involves reactions that produce urea–formaldehyde (UF) resins. Urea and other ingredients such as amines, maleic anhydride copolymers or phenols are added to the aqueous phase that contains oily droplets of the a.i. that is to be encapsulated. Formaldehyde or formaldehyde oligomers are added and the reaction conditions are adjusted to form UF condensates, sometimes referred to as aminoplasts, which should preferentially wet the disperse phase [1]. The reaction is continued to completion over several hours. Products with fairly high activity can be obtained. A modification of this technique is the use of etherified UF resins. The UF prepolymers are dissolved in the organic phase, along with the a.i., through the use of protective colloids (such as PVA), and the reaction is initiated through heating and an acid catalyst. This promotes the formation of the shell in the organic phase adjacent to the interface between the bulk oil phase droplets and the aqueous phase solution [1].

It should be mentioned that the role of surfactants in the encapsulation process is very important. Apart from their direct role in the preparation of microcapsule dispersions, surfactants can be used to control the release of the a.i. from the microcapsule dispersion. It has been shown that the efficacy of an edifenphos suspension can be improved by addition of a surfactant either to the aqueous medium or to the core. This was attributed to the possible solubilization of the a.i. by the surfactant micelles, thus increasing the release rate.

12.3 Mechanism of Release of Active Ingredient from Microcapsules

There are generally two mechanisms for release of the a.i. from a capsule: (1) diffusion of the a.i. through the microcapsule wall and (2) destruction of the microcapsule wall by either physical means, e.g. mechanical power, or by chemical means, e.g. hydrolysis, biodegradation or thermal degradation. The release behavior is controlled by several factors, such particle size, wall thickness, type of wall material, wall structure (porosity, degree of polymerization, cross-link density, additives, etc.), type of core material (chemical structure, physical state, presence or absence of solvents) and amount or concentration of the core material. The release behavior is determined by interaction of these factors and optimization is essential for achieving the desirable release rate.

In order to achieve better performance of the microcapsule for biological efficacy, time-dependent or site-specific release is desirable. It is essential in this case to develop various functional microcapsules that are specific to the target organism (e.g. for insecticides). Temperature-, pH-, light- and enzyme-responsive microcapsules are desirable.

The simplest release kinetics of microcapsules are diffusion controlled, as predicted by Fick's first law. The amount of a.i. that diffuses through the wall of a microcapsule, dm/dt (mol s^{-1}) is proportional to the diffusion coefficient D, the surface area A and the concentration gradient dc/dx (where dc is the difference in concentration between the inside and outside wall and dx is the thickness of the capsule wall):

$$\frac{dm}{dt} = -DA\frac{dc}{dx} \tag{1}$$

Equation 1 clearly shows that the release rate increases with increase in A (i.e. when using small capsules) and decrease in dx (thinner capsule wall). To decrease the rate of diffusion, one has to use larger capsules and a thicker capsule wall.

12.4 Encapsulation by Phase Separation from Aqueous Solution

There are four types of encapsulation utilizing the system of phase separation from aqueous solution [2]: (1) complex coacervation or phase separation resulting from two oppositely charged colloids neutralizing one another; (2) simple coacervation where a non-electrolyte such an alcohol causes the formation of a separate polymer-rich phase; (3) salt coacervation, where a polymer separates as a result of salting-out process; and (4) precipitation and insolubilization of a polymer by changing the pH of the aqueous solution system.

An example of complex coacervation is the interaction between gelatin and gum arabic [4]. In this case, a dispersion of oil in a dilute solution of gelatin–gum arabic mixture is prepared. Gelatin usually has an isoelectric point (IEP) at about pH = 4.8, whereas gum arabic which contains only carboxylic groups is usually negative over a wide range of pH values. Thus, by lowering the pH to a value below the IEP of gelatin, say to pH 4.0, the gelatin acquires a positive charge and become coacervated with gum arabic, forming capsules around the oil droplets. Various other anionic polyelectrolytes may be used, such as sodium alginate, agar and poly(vinylbenzenesulfonic acid). In general, effective materials include polymers, surface-active agents and organic compounds which have acid groups in the molecule [2].

Encapsulation by phase separation can also be applied in nonaqueous media. This is particularly suitable for encapsulation of water-soluble materials. A W/O emulsion of the a.i. is dissolved in the water droplets using an oil-soluble polymer. Phase separation of the oil-soluble polymer may be induced by addition of another polymer, a non-solvent, or by changing the temperature.

Salt coacervation is best exemplified by the formation of calcium alginate capsules. In this process, a drop of a solution, an emulsion or suspension containing the a.i. and sodium alginate is dropped into a solution of calcium chloride. When the drop touches the calcium chloride solution, a membrane of calcium alginate forms instantaneously, maintaining the drop shape in this aqueous–aqueous system. Calcium ions diffuse in, gelling the entire drop. This drop is then placed in a solution of a polycation, which displaces the calcium ions from the outer surface, forming a permanent membrane. This capsule is then placed in sodium citrate, which slowly solubilizes the calcium through the formation of a soluble citrate complex, ungelling the internal portion of the drop. By controlling the molecular weight of the reactants and the times of reaction, the thickness and size selectivity of the permanent wall can be controlled over a wide range.

12.5 Microencapsulation of Solid Particles

This is by far the most challenging process of encapsulation since one has to coat the particles individually without any aggregation. These particles cover the size range 0.1–5 μm with an average of 1–2 μm. Clearly, when encapsulating these particles one has to make sure that the smallest size fraction is retained without any aggregation. This is vital for biological efficacy since the smaller particles are more effective for disease control (due to their higher solubility compared with the larger particles). Beetsman [5] suggested an injection treatment method of coating for encapsulation of solid particles. This method utilizes air at sonic velocity to atomize the coating material and accelerate the particles in such a manner that they become coated on all surfaces. The liquid coating material may be melted wax or resins, solutions of polymers or coating materials or suspensions of film-forming solids such as polymer latexes. Coating is accomplished by metering the solid particles in the shear zone concurrently with metering the liquid coating material into the air stream. The latter is accelerated to the speed of sound through a restriction zone to give a shear zone of sufficient intensity to affect coating. The mixing action within the shear zone coats the solid particles individually with the coating material. On-line particle size measurement of the encapsulated solid particles showed that the particle size range of the solid particles remains virtually unchanged by this injection coating process, indicating that individual particles of all sizes are discretely coated.

Another method that can be applied to encapsulate solid particles is a modification of the coacervation process described above. In this method, a technique of solvent evaporation is used to precipitate the polymers as intact coatings. The solid particles are suspended in a solvent solution of the polymer and emulsified into a liquid. The emulsion is then heated to evaporate the solvent, causing the polymer to insolubilize as a coating around the suspended particles. Alternatively, a non-solvent for the polymer is added to the suspension of particles in polymer solution, causing the solvent to phase separate and the polymers to insolubilize to coatings.

12.6
Controlled Release of Agrochemicals from Matrix-based Microparticles

Matrix-based microparticles are of three main types [6], as follows. (1) Matrix powders where the a.i. is dispersed throughout the matrix and the mixture is ground (if necessary) to form a powder that can be applied as a wettable powder. Surface-active agents are incorporated to aid wetting and dispersion of the microparticles. The matrices used include polymers such as lignin, starch, proteins, high molecular weight natural polymers such as waxes, cyclodextrin, synthetic polymers such as urea–formaldehyde resins or acrylic acid polymers. Inorganic materials such as glass, silica or diatomaceous earth can also be used. These inorganic materials can also act as carriers. (2) Carriers plus matrix where the particles are based on a porous powder that is used as a carrier. Two types can be distinguished, co-loaded (where the a.i.–matrix mixture is loaded into the carrier) and post-coated (where the a.i. is loaded to the carrier and the matrix is then loaded separately). (3) Matrix emulsions where the microparticles are made by emulsifying a hot solution of the a.i. plus matrix, typically in water. On cooling the emulsion, droplets solidify producing an aqueous suspension of the microparticles.

Generally, one component of the formulation, the 'matrix', will be responsible for the controlled release of the formulation. It is convenient to consider the controlled release as being due to interaction among the a.i., the matrix and the environment. Matrix systems where the a.i. is uniformly dispersed through a matrix material are the basis of commercial formulations [7]. Three models may be used for describing the behavior of such systems. The first two mechanisms apply when the a.i. is uniformly dispersed throughout the matrix and is essentially impermeable to water or the external environment. Leaching of the a.i. occurs at the edge of the particle, setting up a concentration gradient within the particle that provides the driving force for diffusion of the a.i. to the edge of the particle and into the external environment. In such a system, the rate of release is governed by the solubility of the a.i. in the matrix, the diffusion coefficient for the transport of the a.i. through the matrix and the geometry of the particle. Matrix particles usually contain pores and cracks, thus increasing the effective surface area between the particle and the external environment and hence the release rate. The second mechanism applies to rigid, often glassy matrices where diffusion of the a.i. within the matrix of the active is negligible. Leaching is controlled by surface exposure of the a.i through biological or chemical degradation of the matrix. The third mechanism applies to systems where the matrix material is permeable to the external environment, e.g. water. This corresponds to a system where the a.i. is dispersed in a latex. In this case, water permeates the matrix through a combination of capillary and osmotic effects. The a.i. dissolves and diffuses to the edge of the particle into the surrounding medium. The process is diffusion controlled and is governed by the solubility and diffusion coefficient of the a.i. in water.

12.7
Mechanism of Controlled Release from Microparticles

The release of a.i. from conventional formulations generally follows an exponential decay, i.e. the release rate is proportional to the concentration of the a.i. remaining in the formulation. This decay follows first-order kinetics, which means that the initial concentration in the environment is initially very high (often resulting in an undesirable toxic effect) and decreases rapidly to a low (ineffective) level. In contrast, a controlled-release formulation generally exhibits lower initial concentrations and a longer time before the concentration decreases to an ineffective level. This is illustrated schematically in Figure 12.1, which clearly shows that when using a conventional formulation several treatments are required for biological control. During these treatments, the a.i. concentration may reach an undesirably high level that is above the toxic limit. In contrast, a controlled-release formulation maintains an effective concentration that is sufficient for bioefficacy without reaching the toxic limit. Thus, with a conventional treatment a higher dose of a.i. is required to maintain the bioefficacy. This dose is significantly reduced when using a controlled-release formulation. The high a.i. concentrations reached with conventional formulations can also have adverse effects on toxicity to humans, birds, fish, etc.

Most controlled-release systems rely on diffusion of a.i. through a rate-controlling membrane or polymer matrix. Transport through a polymer membrane or

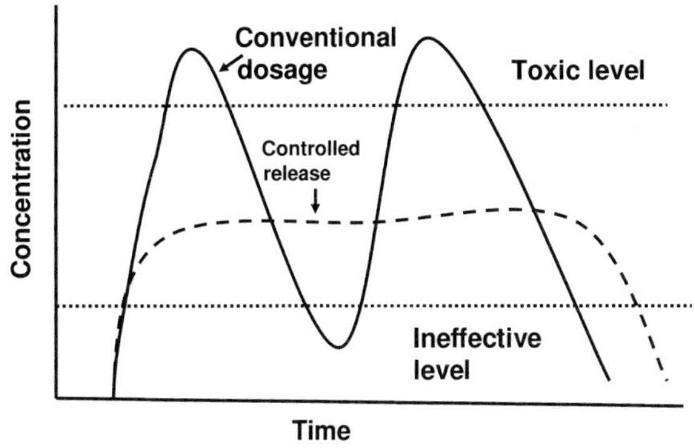

Figure 12.1 Release of active ingredient from conventional and controlled-release formulations.

matrix occurs by a solution–diffusion process, where the a.i. first dissolves in the polymer and then diffuses across the polymer to the external surface, where the concentration is lower. As discussed before, the process follows Fick's first law of diffusion (equation 1).

The rate of a.i. release from controlled release systems can follow a variety of patterns, ranging from first-order (exponential) decay (as with conventional systems) to zero-order kinetics in which the release rate is constant over most of the lifetime of the device. In the latter case, the release rate decreases proportionally to the square root of time. A comparison of the release kinetics is shown in Figure 12.2. As can be clearly seen, the zero-order kinetics (membrane-coated reservoir) result in a lower peak concentration and more extended release compared with the case of first-order kinetics.

As mentioned before, the most common types of controlled-release microparticles are membrane-coated reservoirs and polymeric matrices. A reservoir system consists of a core of pure or saturated a.i. surrounded by a rate-controlling

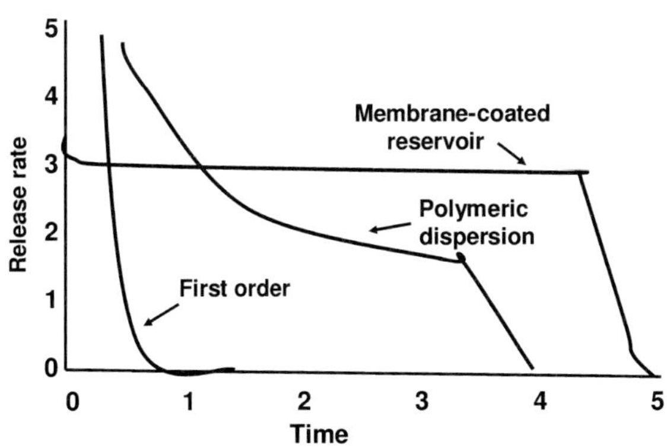

Figure 12.2 Comparison of release kinetics observed from controlled-release formulations.

membrane (polymer shells). The a.i. is released from the reservoir system by diffusion through the rate-controlling membrane at a constant rate (zero-order kinetics) and this follows Fick's law. For a spherical capsule with outer and inner radii r_2 and r_1, respectively, and hence a membrane thickness of $r_2 - r_1$, the release rate at time t, $\mathrm{d}M_t/\mathrm{d}t$, is given by the following equation:

$$\frac{\mathrm{d}M_t}{\mathrm{d}t} = 4\pi D \frac{\mathrm{d}c}{\mathrm{d}x} \frac{r_2 r_1}{r_2 - r_1} \qquad (2)$$

where D is the diffusion coefficient of the a.i. and $\mathrm{d}c/\mathrm{d}x$ is the concentration gradient within the membrane.

The release rate remains constant as long as D and $\mathrm{d}c/\mathrm{d}x$ remain constant. However, the latter values are temperature dependent and hence the release rate will increase with increase in temperature. In many cases, the release rate doubles for every 10 °C increase in temperature. In addition, $\mathrm{d}c/\mathrm{d}x$ remains constant as long as the activity of the a.i. remains constant within the reservoir. If the activity of the a.i. decreases as a result of release of the chemical, the release rate will decrease.

Another important factor that affects the release rate of membrane-coated reservoir-type microparticles is the polydispersity of the system. The release rate from each microparticle may be constant and hence initially the release rate may also be constant. However, with time the release rate may change as a result of polydispersity. The quantity of the a.i. in the microparticle is a function of its volume, i.e. the cube of the radius, but the release rate is a function of the radius. Hence the duration, which is approximately equal to the mass of the a.i. divided by the release rate, is a function of the square of the radius. Therefore, the smaller microparticles become depleted before the larger ones and this results in a decrease in the overall release rate from a collection of microparticles with different sizes.

It should also be mentioned that in order to maintain a constant release rate, the membrane must remain intact. In general, large microparticles and those with high loading of a.i. are more susceptible to rupture, resulting in rapid release.

With matrix-type microparticles where the a.i. is dispersed or dissolved in a polymeric matrix, the release of the a.i. occurs by diffusion through the matrix to the surface and hence the process follows Fick's law (equation 1). However, the release kinetics can depend on the quantity of the a.i. and whether this is dispersed or dissolved in the matrix.

12.8 Controlled Release from Granules

Many agrochemicals are formulated as water-dispersible granules (WGs) that disperse quickly and completely when added to water. The main advantage of

WGs is that they avoid the use of solvents. thus reducing the risk during manufacture and to farm workers during application. In addition, they can be applied for slow release, as will be discussed below.

Several processes can be applied to produce WGs of insoluble a.i.: (1) those in which the starting materials are essentially dry and are subsequently making them wet and then redrying and (2) those in which the starting materials are wet and are granulated and dried. A typical composition of a WG is one or two a.i.s, dispersing agent, suspending agent, wetting agent, binder (such as lignosulfonate or a gum) and a filler (mineral filler or water-soluble salt).

As mentioned above, granulation is carried out using a dry or wet route process. Several dry route processes are possible, such as pan granulation, fluid-bed granulation, Schugi granulator, extrusion and peg or pin granulator [8]. The wet route process can be carried out by spray drying or spray granulation [8].

Approaches to achieve controlled release in granules fall into two main categories: (1) the matrix (monolith) with the a.i. dispersed throughout the structure and (2) the reservoir in which a polymeric coating entraps the a.i. with or without a support. [9]. Particle size and uniformity are very important, especially in applications where the duration of release is critical. Three types of granule dimensions can be distinguished, namely fine granules of diameter 0.3–2.5 mm, microgranules of 0.1–0.6 mm and macrogranules of 2–6 mm. A formulation containing a range of particle sizes (from dusts to macrogranules) will have an extended period of effectiveness. A controlled-release system based on a monolithic polymer granule made from extruding the a.i. with a release rate-modifying inert material ('porisogen') in a thermoplastic matrix can play an important role in pest management for periods (following a single treatment of a non-persistent agrochemical) of up to 2–3 years.

Although the above approach based on synthetic polymers is the most successful of the controlled-release granules, natural polymers have shown great success in matrix formulations for a.i. delivery. Examples of natural polymers are cross-linked starch, polysaccharides, cross-linked alginates and cellulose derivatives. To provide effective delay of release, alginate gels cross-linked with calcium require the incorporation of absorbents such as silica, alumina, clays or charcoal. Further control of the release rate could be achieved by combining kaolin clay with linseed oil in the granule. Other gel-forming polymers include carboxymethylcellulose stabilized with gelatin and cross-linked with cupric or aluminum ions. Coating of granules with rate-controlling polymer film can also be applied. Controlled delivery of agrochemicals has also been achieved with superabsorbent acrylamide and acrylate polymers.

The biodegradability of the formulating material is an important aspect of controlled release for environmental applications. Several synthetic and natural polymers used for formulating granules are biodegradable. The delivery of bioactives from controlled-release granules can be enhanced by inclusion of biosurfactants.

Several lignin-based granules have been introduced for controlled release of several a.i.s. Lignin is a polyphenolic material that occurs in the cell wall of most terrestrial plants, where it is strongly associated with carbohydrates. It is a polymer produced by random dehydrogenation of a number of phenolic precursors linked to the polysaccharide component of the plant cell. This produces a complex structure without any regular repeating monomer. Lignin is separated from lignocellulosic plants by physical or chemical means.

Several agrochemicals are formulated as granules using lignin, in particular for oil application. The a.i. is characterized by some physicochemical properties such as moderate sorption on soil components, low volatility, moderate to high melting point, crystallinity and low to moderate water solubility. Such properties make them compatible with alkali lignins for preparing matrices by melting the components together. This produces a glassy matrix upon cooling.

The compatibility of a lignin and an agrochemical can be assessed by examining a film of the melt mixture under a microscope for the presence of unsolvated lignin particles. Where solvation occurs, the melting point of the agrochemical is depressed and this can be determined using differential scanning calorimetry (DSC). The density of the glassy adhesive matrix is usually lower than that of the lignin and often less than that of the a.i. This can be explained by the presence of voids or pores that cannot be observed by microscopy.

The effect of water on the matrix formulation varies according to the compatibility of the agrochemical with the lignin and the ratio of a.i. to lignin. With a highly compatible a.i. such as diuron, the surface of the matrix changes from dark brown to dull light brown on exposure to water. On further exposure some swelling occurs and the outer region is very porous. Diffusion of diuron is enhanced compared with that in the unswollen glassy interior. The swelling and water uptake depend to a large extent on the type of lignin used.

The mechanism of release from lignin matrix granules intended for use in soil and aqueous media is studied by immersing the granule in water under static, stirred or flowing conditions. Granules prepared from various lignin types always show release rates that decrease with time. This is illustrated in Figure 12.3, which also shows the dependence on lignin type.

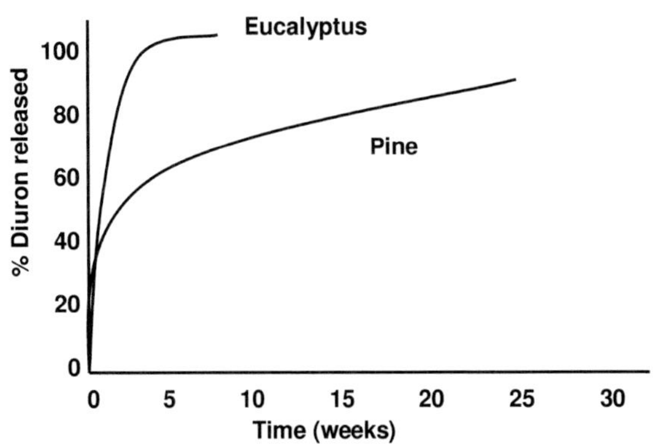

Figure 12.3 Release of diuron (50%) from granules based on two different lignin types.

The release kinetics fitted the generalized model [9]

$$\frac{M_t}{M_z} = kt^n + c \tag{3}$$

where M_t/M_z is the proportion of a.i. released at time t, the constant k incorporates the polymer properties, the exponent n characterizes the transport mechanism and c is a constant.

The results in Figure 12.3 could be fitted to equation 3, giving an exponent n ranging from 0.35 to 0.53, indicating that the release is mainly diffusion controlled.

References

1 H.B. Scher (ed.), *Controlled-release Delivery Systems for Pesticides*, Marcel Dekker, New York (1999).

2 A. Kondo, *Microcapsule Processing and Technology*, Marcel Dekker, New York (1979).

3 P.W. Morgan and S.L. Kvolek, *J. Polym. Sci.*, **2**, 90 (1947).

4 H.C. Bunderberg de Jong, Complex colloid systems, in *Colloid Science*, Vol. II, ed. H.R. Ktuyt, Elsevier, Amsterdam (1949).

5 G.B. Beetsman, in *Controlled-release Delivery Systems for Pesticides*, ed. H.B. Scher, Marcel Dekker, New York (1999).

6 D.J. Park, W.R. Jackson, I.R. McKinnon and M. Marshall, in *Controlled-release Delivery Systems for Pesticides*, ed. H.B. Scher, Marcel Dekker, New York (1999).

7 M. Bahadir and G. Pfister, Controlled release formulations of pesticides, in *Controlled Release, Biochemical Effects of Pesticides and Inhibition of Plant Pathogenic Fungi*, ed. W.S. Bowers, W. Ebing and D. Martin, Springer, Berlin, pp. 1–64 (1990).

8 A.R. Woodford, Dispersible granules, in *Pesticide Formulations*, Chapter 9, ed. W. Van Valkenberg, B. Sugavanan and S.K. Khetan, UNIDO, Vienna and New Age International, New Delhi (1998).

9 R.M. Wilkins (ed.), *Controlled Delivery of Crop Protection Agents*, Taylor and Francis, London (1990).

13
Basic Principles of Selection of Adjuvants in Agrochemical Formulations

13.1
Introduction

The discovery and development of effective agrochemicals that can be used with maximum efficiency and minimum risk to the user require the optimization of their transfer to the target during application. In this way, the agrochemical can be used in an effective way, thus minimizing any waste during application. Optimization of the transfer of the agrochemical to the target requires careful analysis of the steps involved during application. Most agrochemicals are applied as liquid sprays, particularly for foliar application. The spray volumes applied range from high values of the order of 1000 L ha^{-1} (where the agrochemical concentrate is diluted with water) to ultralow volumes of the order of 1 L ha^{-1} (when the agrochemical formulation is applied without dilution). Various spray application techniques are used, of which spraying using hydraulic nozzles is probably the most common. In this case, the agrochemical is applied as spray droplets with a wide range of droplet sizes (usually in the range 100–400 μm in diameter). On application, parameters such as droplet size range, their impaction and adhesion, sliding and retention, wetting and spreading are of prime importance in ensuring maximum capture by the target surface and adequate coverage of the target surface. These factors are discussed in some detail below. In addition to these 'surface chemical' factors, i.e. the interaction with various interfaces, other parameters that affect biological efficacy are deposit formation, penetration and interaction with the site of action. As will be seen later, deposit formation, i.e. the residue left after evaporation of the spray droplets, has a direct effect on the efficacy of the pesticide, since such residues act as 'reservoirs' of the agrochemical and hence they control the efficacy of the chemical after application. The penetration of the agrochemical and its interaction with the site of action are very important for systemic compounds. Enhancement of penetration is sometimes crucial to avoid removal of the agrochemical by environmental conditions such as rain and or/wind. All these factors are influenced by surfactants and polymers and this is discussed in detail below.

In addition, some adjuvants that are used in combination with the formulation consist of mixtures of oils and/or surfactants. The role of these adjuvants in the

Colloids and Interface Science Series, Vol. 5
Colloids in Agrochemicals. Tharwat F. Tadros

ISBN: 978-3-527-31465-2

enhancement of biological efficacy is far from being understood and in most cases they are arrived at by a trial and error procedure. A great deal of research is required in this area, which would involve understanding the surface chemical processes, both static and dynamic, e.g. static and dynamic surface tension and contact angles, and also their effect on penetration and uptake of the chemical. In recent years, some progress has been made in the techniques that could be applied to such complex problems and these should hopefully lead to a better understanding of the role of adjuvants. The role of these complex mixtures of oils and/or surfactants in controlling agrochemical efficiency is important from a number of points of view. First, there is a greater demand to reduce the application rate of chemicals and to make better use of the present agrochemicals, for example in greater selectivity. In addition, environmental pressure on the hazards produced for the operator and the long-term effects of such residues and wastage demands a better understanding of the role of the adjuvants in application of agrochemicals. This should lead to optimization of the efficacy of the chemical and a reduction in hazards for the operators, the crops and the environment.

There are generally two main approaches for the selection of adjuvants: (1) an interfacial (surface) physicochemical approach which is designed to increase the dose of the agrochemical received by the target plant or insect, i.e. enhancement of spray deposition, wetting, spreading, adhesion and retention; and (2) uptake activation that is enhanced by addition of surfactant which is the result of specific interactions between the surfactant, the agrochemical and the target species. These interactions may not be related to the intrinsic surface-active properties of the surfactant/adjuvant.

The above two approaches must be considered when selecting an adjuvant for a given agrochemical and the type of formulation that is being used. The most important adjuvants are surface-active agents and polymers. In some cases these are used in combination with crop oils (e.g. methyl oleate). Several complex recipes may be used and in many cases the exact composition of an adjuvant is not exactly known.

Adjuvants are applied in two different ways: (1) incorporated in the formulation, mostly the case with flowables (SCs and EWs); and (2) used in tank mixtures during application. Such adjuvants can be complex mixtures of several surfactants, oils, polymers, etc.

The choice of an adjuvant depends on (1) the nature of the agrochemical, water soluble or insoluble (lipophilic), where its solubility and log P values are important; (2) the mode of action of the agrochemical, i.e. systemic or non-systemic, selective or non-selective; and (3) the type of formulation that is used, i.e. flowable, EC, grain, granule, capsule, etc.

As mentioned above, most important adjuvants are surface-active agents of the anionic, nonionic or zwitterionic type. In some cases, polymers are added as stickers or anti-drift agents. As mentioned in Chapter 4, the surfactant molecules accumulate at various interfaces as a result of their dual nature. Basically, a surfactant molecule consists of a hydrophobic chain (usually a hydrogenated or fluorinated alkyl or alkylaryl chain with 8–18 carbon atoms) and a hydrophilic group or chain [ionic or polar nonionic such as poly(ethylene oxide)]. At the air/water interface (as for spray droplets) and the solid/liquid interface (such as the leaf surface), the hydrophobic group points towards the hydrophobic surface (air or leaf), leaving the hydrophilic group in bulk solution. This results in lowering of the air/liquid surface tension, γ_{LV}, and the solid/liquid interfacial tension, γ_{SL}: As the surfactant concentration is gradually increased, both γ_{LV} and γ_{SL} decrease until the critical micelle concentration (cmc) is reached, after which both values remain virtually constant. This situation represents the conditions under equilibrium where the rate of adsorption and desorption are the same. The situation under dynamic conditions, such as during spraying, may be more complicated since the rate of adsorption is not equal to the rate of formation of droplets. Above the cmc, micelles are produced, which at low C values are essentially spherical (with an aggregation number in the region of 50–100 monomers). Depending on the conditions (e.g. temperature, salt concentration, structure of the surfactant molecules), other shapes may be produced, e.g. rod-shaped and lamellar micelles. Since micelles play a vital role when considering adjuvants, it is essential to understand their properties in some detail. As mentioned in Chapter 3, micelle formation is a dynamic process, i.e. a dynamic equilibrium is set up where surface-active agent molecules are constantly leaving the micelles while others enter the micelles (the same applies to the counterions). The dynamic process of micellization is described by two relaxation processes: (1) a short relaxation time τ_1 (of the order of 10^{-8}–10^{-3} s), which is the lifetime for a surfactant molecule in a micelle; and (2) a longer relaxation time τ_2 (of the order of 10^{-3}–1 s), which is a measure of the micellization–dissolution process; τ_1 and τ_2 depend on the surfactant structure and its chain length and these relaxation times determine some of the important factors in selecting adjuvants, such as the dynamic surface tension (discussed below).

The cmc of nonionic surfactants is usually two orders of magnitude lower than that of the corresponding anionic surfactants of the same alkyl chain length. This explains why nonionics are generally preferred when selecting adjuvants. For a given series of nonionics, with the same alkyl chain length, the cmc decreases with decrease in the number of ethylene oxide (EO) units in the chain. Under equilibrium conditions, the γ–log C curves shift to lower values as the EO chain length decreases. However, under dynamic conditions, the situation may be reversed, i.e. the dynamic surface tension could become lower for the surfactant with the longer EO chain. This trend is understandable if one considers the dynamics of micelle formation. The surfactant with the longer EO chain has a higher cmc and it forms smaller micelles than the surfactant containing a shorter EO chain. This means that the lifetime of a micelle with a longer EO chain is shorter than of a micelle that with a longer EO chain. This explains why the dynamic surface tension of a solution of a surfactant containing a longer EO chain can be lower than that of a solution of an analogous surfactant (at the same concentration) with a shorter EO chain.

For a series of anionic surfactants with the same ionic head group, the lifetime of a micelle decreases with decrease in the alkyl chain length of the hydrophobic component. Branching of the alkyl chain could also play an important role in the lifetime of a micelle. It is important, therefore, to carry out dynamic surface tension measurements when selecting a surfactant as an adjuvant as this may play an important role in spray retention.

However, the above measurements should not be taken in isolation as other factors may also play an important role, e.g. solubilization, which may require larger micelles. The selection of a surfactant as an adjuvant requires knowledge of the factors involved, which will be discussed in some detail below.

At high surfactant concentrations (usually above 10%), several liquid crystalline phases are produced, as discussed in Chapter 3. Three main types of liquid crystals may be distinguished: (1) hexagonal (middle) phase that consists of cylindrical anisotropic units with high viscosity; (2) cubic, body-centered isotropic phase with a viscosity that is higher than that of the hexagonal phase; and (3) lamellar (neat) phase consisting of sheet-like units that are anisotropic, but with a viscosity that is lower than that of the hexagonal phase.

The above phases may form during evaporation of a spray drop. In some cases, a middle phase is produced first, which on further evaporation may produce a cubic phase that in view of its very high viscosity may entrap the agrochemical. This could be advantageous for some of the systemic fungicides which require 'deposits' that act as reservoirs for the chemical. The viscous cubic phases may also enhance the tenacity of the agrochemical particles (particularly with SCs) and hence enhance rain fastness. In some other applications, a lamellar phase is preferred as this provides some mobility (due to its lower viscosity).

The various phases produced by a surfactant can be related to its structure. An important parameter that can be used to predict the phase behavior of surfactants is the critical packing parameter (CPP), which was described above ($\mathrm{CPP} = v/la$, where v is the volume of the hydrocarbon chain with a length l and a is the cross-sectional area of the hydrophilic head group). For spherical micelles $\mathrm{CPP} \leq 1/3$, for cylindrical micelles $1 > \mathrm{CPP} > 1/2$ and for lamellar micelles $\mathrm{CPP} \approx 1$.

Study of the phase behavior of surfactants (which can be obtained using polarizing microscopy) is crucial in the selection of adjuvants. The interaction of the above units with the agrochemical is crucial in determining performance (e.g. solubilization). Similar interactions may also occur between the above structural units and the leaf surface (wax solubilization).

The application of an agrochemical, as a spray, involves a number of interfaces, where the interaction with the formulation plays a vital role. The first interface during application is that between the spray solution and the atmosphere (air), which governs the droplet size range, rate of evaporation, drift, etc. In this respect, the rate of adsorption of the surfactant and/or polymer at the air/liquid interface is of vital importance. This requires dynamic measurements of parameters, such as surface tension, which will give information on the rate of adsorption. This subject will be dealt with in the first part of this chapter. The second interface is that between the impinging droplets and the leaf surface (with insecticides the interaction with the insect surface may be important). The droplets impinging on the surface undergo a number of processes that determine their adhesion and retention and further spreading on the target surface. The rate of evaporation of the droplet and the concentration gradient of the surfactant across the droplet governs the nature of the deposit formed. These processes of impaction, adhesion, retention, wetting and spreading are discussed in subsequent parts of this chapter. The interaction with the leaf surface is described in terms of the various surface forces involved.

13.2 Interactions at the Air/Solution Interface and Their Effect on Droplet Formation

In a spraying process, a liquid is forced through an orifice (the spray nozzle) to form droplets by application of a hydrostatic pressure. Before describing what happens in a spraying process, it is beneficial to consider the processes that occur when a drop is formed at various time intervals. If the time of formation of a drop is large (greater than, say, 1 min), the volume of the drop depends on the properties of the liquid such as its surface tension and the dimensions of the orifice, but is independent of the time of its formation. However, at shorter times of formation of the drop (less than 1 min), the drop volume depends on the time of its formation. The loosening of the drop which occurs when its weight W exceeds the surface force $2\pi r\gamma$ (i.e. $W > 2\pi r\gamma$) progresses at a speed that is determined by the viscosity and surface tension of the liquid. However, during this loosening process, the hydrostatic pressure pumps more liquid into the drop and this is represented by a 'hump' in the W–t curve. The height of the 'hump' increases with increase in viscosity, perhaps because the rate of contraction diminishes as the viscosity rises.

At short t values, W becomes smaller since the liquid in the drop has considerable kinetic energy even before the drop breaks loose. The liquid coming into the drop imparts downward acceleration and this may cause separation before the drop has reached the value given by the equation

$$W = 2\pi r\gamma f\left(\frac{g\rho r^2}{2\gamma}\right) \tag{1}$$

where ρ is the density and r is the radius of the orifice. Equation 1 is the familiar equation for calculating the surface or interfacial tension from the drop weight or volume.

When the hydrostatic pressure is raised further, i.e. when at even shorter t values than those described above, no separate drops are formed at all and a continuous jet issues from the orifice. Then at even higher hydrostatic pressure, the jet breaks up into droplets, the phenomenon usually referred to as spraying. The process of break-up of jets (or liquid sheets) into droplets is the result of surface forces. The surface area and consequently the surface free energy (area × surface tension) of a sphere are smaller than those of a less symmetrical body. Hence small liquid volumes of other shapes tend to give rise to smaller spheres. For example, a liquid cylinder becomes unstable and divides into two smaller droplets as soon as the length of the liquid cylinder is greater than its circumference. This occurs on accidental contraction of the long liquid cylinder. A prolate spheroid tends to give two spherical drops when the length of the spheroid is greater than 3–9 times its width. A very long cylinder with radius r (e.g. a jet) tends to divide into drops with a volume equal to $(9/2)\pi r^3$. Since the surface area of two unequal drops is smaller than that of two equal drops with the same total volume, the formation of a polydisperse spray is more probable.

The effect of surfactants and/or polymers on the droplet size range of a spray can be described first in terms of their effect on the surface tension. Since surfactants lower the surface tension of water, one would expect that their presence in the spray solution would result in the formation of smaller droplets. This is similar to the process of emulsification described in Chapter 6. As a result of the low surface tension in the presence of surfactants, the total surface energy of the droplets produced on atomization is lower than that in the absence of surfactants. This implies that less mechanical energy is required to form the droplets when a surfactant is present. This leads to smaller droplets at the same energy input. However, the actual situation is not simple since one is dealing with a dynamic situation. In a spraying process a fresh liquid surface is continuously being formed. The surface tension of that liquid depends on the relative ratio between the time taken to form the interface and the rate of adsorption of the surfactant from bulk solution to the air/liquid interface. The rate of adsorption of a surfactant molecule depends on its diffusion coefficient and its concentration (see below). Clearly, if the rate of formation of a fresh interface is much faster than the rate of adsorption of the surfactant, the surface tension of the spray liquid will not be far from that of pure water. Alternatively, if the rate of formation of the fresh surface is much slower than that of the rate of adsorption, the surface tension of the spray liquid will be close to that of the equilibrium value of the surface tension. The actual situation is somewhere in between and the rate of formation of a fresh surface is comparable to that of the rate of surfactant adsorption. In this case, the surface tension of the spray liquid will be between that of a clean surface (pure water) and the equilibrium value of the surface tension which is reached at times longer than that required to produce the jet and the droplets. This shows the importance of the measurement of dynamic surface tension and the rate of surfactant adsorption.

The rate of surfactant adsorption may be described by application of Fick's first law. When concentration gradients are set up in the system or when the system is stirred, then the diffusion to the interface may be expressed in terms of Fick's first law:

$$\frac{d\Gamma}{dt} = \frac{D}{\delta}\frac{N_A}{100}C(1-\theta) \tag{2}$$

where Γ is the surface excess (number of moles of surfactant adsorbed per unit area), t is time, D is the diffusion coefficient of the surfactant molecule, δ is the thickness of the diffusion layer, N_A is Avogadro's number and θ is the fraction of the surface already covered by adsorbed molecules. Equation 2 shows that the rate of surfactant diffusion increases with increase in D and C. The diffusion coefficient of a surfactant molecule is inversely proportional to its molecular weight. This implies that shorter chain surfactant molecules are more effective in reducing the dynamic surface tension. However, the limiting surface tension reached by a surfactant molecule decreases with increase in its chain length and hence a compromise is usually made when selecting a surfactant molecule. Usually one chooses a surfactant with a chain length of the order of 12 carbon atoms. In addition, the greater the surfactant chain length, the lower is its cmc (see Chapter 2) and hence lower concentrations are required when using a longer chain surfactant molecule. Again, a problem with longer chain surfactants is their high Krafft temperatures (becoming soluble only at temperatures higher than ambient). Therefore, an optimum chain length is usually necessary for optimizing the spray droplet size range.

As mentioned above, the faster the rate of adsorption of surfactant molecules, the greater is the effect on reducing the droplet size. However, with liquid jets there is an important factor that may enhance surfactant adsorption. Addition of surfactants reduces the surface velocity (which is in general lower than the mean velocity of flow of the jet) below that obtained with pure water. This results from surface tension gradients, which can be explained as follows. Where the velocity profile is relaxing, the surface is expanding, i.e. it is newly formed and might even approach the composition and surface tension of pure water. Slightly further downstream, appreciable adsorption of the surfactant will have occurred, giving rise to a back spreading tendency from this part of the surface in the direction back towards the cleaner surface immediately adjacent to the nozzle. This phenomenon is thus a form of the Marangoni effect (see Chapter 6), which reduces the surface velocity near the nozzle and induces some liquid circulation that accelerates the adsorption of the surfactant molecules by as much as 10-fold. This effect casts doubt on the use of liquid jets to obtain the rate of adsorption. Indeed, under conditions of jet formation, it is likely that the surface tension approaches its equilibrium value very closely. Therefore, one should be careful in using dynamic surface tension values, as for example measured using the maximum bubble pressure method.

The influence of polymeric surfactants on the droplet size range of spray liquids is relatively more complicated since adsorbed polymers at the air/liquid interface produce other effects than simply reducing the surface tension. In addition, polymeric surfactants diffuse very slowly to the interface and it is doubtful whether they have appreciable effect on the dynamic surface tension. In most agrochemical formulations, polymers are used in combination with surfactants and this makes the situation more complicated. Depending on the ratio of polymer to surfactant in the formulation, various effects may be envisaged. If the concentration of the polymer is appreciably greater than that of the surfactant and interaction between the two components is strong, the resulting 'complex' will behave more like a polymer. On the other hand, if the surfactant concentration is appreciably higher than that of the polymer and interaction between the two molecules is still strong, one may end up with polymer–surfactant 'complexes' in addition to free surfactant molecules. The latter will behave as free molecules and the reduction in the surface tension may be sufficient even under dynamic conditions. However, the role of the polymer–surfactant 'complex' could be similar to that of the free polymer molecules. The latter produce a viscoelastic film at the air/water interface and that may modify the droplet size range and the adhesion of the droplets to the leaf surface. The situation is far from being understood and fundamental studies are required to evaluate the role of polymer in spray formation and droplet adhesion.

The above discussion is related to the case where a polymeric surfactant is used for the formulation of agrochemicals, as discussed in Chapters 6 and 7 on emulsions and suspension concentrates. However, in many agrochemical applications high molecular weight materials such as polyacrylamide, poly(ethylene oxide) and guar gum are sometimes added to the spray solution to reduce drift. It is well known that incorporation of high molecular weight polymers favors the formation of larger drops. The effect can be achieved at very low polymer concentrations when the molecular weight of the polymer is fairly high ($>10^6$). The most likely explanation of how polymers affect the droplet size range is in terms of their viscoelastic behavior in solution. High molecular weight polymers adopt spatial conformations in bulk solution, depending on their structure and molecular weight. Many flexible polymer molecules adopt a random coil configuration which is characterized by a root mean square radius of gyration, R_G: The latter depends on the molecular weight and the interaction with the solvent. If the polymer is in good solvent conditions, e.g. poly(ethylene oxide) in water, the polymer coil becomes expanded and R_G can reach high values, of the order of several tens of nanometers. At relatively low polymer concentrations, the polymer coils are separated and the viscosity of the polymer solution increases gradually with increase in its concentration. However, at a critical polymer concentration, denoted C^*, the polymer coils begin to overlap and the solutions show a rapid increase in viscosity with further increase in the concentration above C^*. This concentration C^* is defined as the onset of the semi-dilute region. C^* decreases with increase in the molecular weight of the polymer and at very high molecular weights it can be as low as 0.01%. Under this condition of polymer coil overlap, the spray jet opposes deformation and this results in the production of larger drops. This phenomenon is applied successfully to reduce drift. Some polymers also produce conformations that approach a rod-like or double helix structure. An example of this is xanthan gum, which is used with many emulsions and suspension concentrates to reduce sedimentation. If the concentration of such a polymer is appreciable in the formulation, then even after extensive dilution on spraying (usually by 100–200-fold), the concentration of the polymer in the spray solution may be sufficient to cause the production of larger drops. This effect may be beneficial if drift is a problem. However, it may be undesirable if relatively small droplets are required for adequate adhesion and coverage. Again, the ultimate effect required depends on the application methods and the mode of action of the agrochemical. Fundamental studies of the various effects are required to arrive at the optimum conditions. The effect of the various surfactants and polymers should be studied in spray application during the formulation of the agrochemical. In most cases, the formulation chemist concentrates on producing the best system that produces long-term physical stability (shelf-life). It is crucial to investigate the effect of the various formulation variables on the droplet size range, their adhesion, retention and spreading. In addition, simultaneous investigations should be made on the effect of the various surfactants on the penetration and uptake of the agrochemical.

One of the problems with many anti-drift agents is their shear degradation. At the high shear rates involved in spray nozzles (which may reach several thousand s^{-1}), the polymer chain may degrade into smaller units and this results in a considerable decrease in viscosity. This will have the effect of reducing the anti-drift effect. It is essential, therefore, to choose polymers that are stable towards the high shear rates involved in a spraying process.

13.3 Spray Impaction and Adhesion

When a drop of a liquid impinges on a solid surface, e.g. a leaf, one of several states may arise depending on the conditions. The drop may bounce or undergo fragmentation into two or more droplets, which in turn may bounce back and return to the surface with a lower kinetic energy. Alternatively, the drop may adhere to the leaf surface after passing through several stages, where it flattens, retracts, spreads and finally rests to form a hemispherical cap. In some cases, the droplet may not adhere initially but float as an individual drop for a fraction of a second or even several seconds, and can either adhere to the surface or leave it again.

The most important parameters that determine which of the above-mentioned stages is reached are the mass (volume) of the droplet, its velocity in flight, the distance between the spray nozzle and the target surface, the difference between the surface energy of the droplet in flight, E_0, and its surface energy after impact, E_s, and displacement of air between the droplet and the leaf.

Droplets in the range 20–50 μm in diameter do not usually undergo reflection if they are able to reach the leaf surface. Such droplets have a low momentum and can only reach the surface if they travel in the direction of the air stream. On the other hand, large droplets of the order of few thousand micrometers in diameter undergo fragmentation. Droplets covering the range 100–400 μm, which includes the range produced by most spray nozzles, may be reflected or retained depending on a number of parameters such as the surface tension of the spray solution, surface roughness and elasticity of the drop surface. A study by Brunskill [1] showed that with drops of 250 μm, 100% adhesion was obtained when the surface tension of the liquid was lowered (using methanol) to 39 N m^{-1}, whereas only 4% adhesion occurred when the surface tension, γ, was 57 N m^{-1}. For any given spray solution (with a given surface tension), a critical droplet diameter exists below which adhesion is high and above which adhesion is low. The critical droplet diameter increases as the surface tension of the spray solution decreases. The viscosity of the spray solution has only a small effect on the adhesion of large drops, but with small droplets adhesion increases with increase in viscosity. As expected, the percentage of adhered droplets decreases as the angle of incidence of the target surface increases.

A simple theory for bouncing and droplet adhesion was formulated by Hartley and Brunskill [2], who considered an ideal case where there are no adhesion (short-range) forces between the liquid and solid substrate and the liquid has zero viscosity. During impaction, the initially spherical droplet will flatten into an oblately spheroidal shape until the increased area has stored the kinetic energy as increased surface energy. This is often followed by an elastic recoil towards the spherical form and later beyond it with the long axis normal to the surface. During this process, energy will be transformed into upward kinetic energy and the drop may leave the surface in a state of oscillation between the spheroidal forms. This sequence was confirmed using high-speed flash illumination.

When the reflected droplet leaves in an elastically deformed condition, the coefficient of restitution must be less than unity since part of the translational energy is transformed to vibrational energy. Moreover, the distortion of droplets involves loss of energy as heat by operation of viscous forces. The effect of increasing the viscosity of the liquid is rather complex, but at a very high viscosity liquids usually have a form of elasticity operating during deformations of very short duration. Reduction of deformation as a result of increase in viscosity will affect adhesion.

As mentioned above, the adhesion of droplets is governed by the relative magnitude of the kinetic energy of the droplet in flight and its surface energy as it lands on the leaf surface. Since the kinetic energy is proportional to the third power of the radius (at constant droplet velocity) whereas the surface energy is proportional to the second power of the radius, one would expect that sufficiently small droplets will always adhere. However, this is not always the case since smaller droplets fall with lower velocities. Indeed, the kinetic energy of sufficiently small drops, in the Stokesian range, falling at their terminal velocity, is proportional to the seventh power of the radius. In the 100–400 μm range, it is nearly proportional to the fourth power of the radius.

Consider a droplet of radius r (sufficiently small for gravity to be neglected) falling on to a solid surface and spreading with an advancing contact angle θ_A and having a spherical upper surface of radius R. The surface energy of the droplet in flight, E_0, is given by

$$E_0 = 4\pi r^2 \gamma_{LA} \tag{3}$$

where γ_{LA} is the liquid/air surface tension. The surface energy of the spread drop is given by the following equation:

$$E_s = A_1\gamma_{LA} + A_2\gamma_{SL} - A_2\gamma_{SA} \tag{4}$$

where A_1 is the area of the spherical air/liquid interface, A_2 is that of the plane circle of contact with the solid surface, γ_{SL} is the solid/liquid interfacial tension and γ_{SA} is that of the solid/air interface.

From Young's equation,

$$\gamma_{SA} = \gamma_{SL} + \gamma_{LA} \cos\theta \tag{5}$$

Therefore, the surface energy of the droplet spreading on the leaf surface is given by the equation

$$E_s = \gamma_{LA}(A_1 - A_2 \cos\theta) \tag{6}$$

The volume of a free drop is $(4/3)\pi r^3$, whereas that of the spread drop is $\pi R^3[(1 - \cos\theta) \times (1/3)(\cos^3\theta)]$, so that

$$\frac{4}{3}\pi r^3 = \pi R^3 \left[(1 - \cos\theta) + \frac{1}{3}(\cos^3\theta - 1)\right] \tag{7}$$

and

$$A_1 = 2\pi R^2(1 - \cos\theta) \tag{8}$$

$$A_2 = \pi R^2 \sin^2\theta \tag{9}$$

Combining equations 6–9, one can obtain the minimum energy barrier between attached and free drops:

$$\frac{E_0 - E_s}{E_0} = 1 - 0.39[2(1 - \cos\theta) - \sin^2\theta \cos\theta]\left[1 - \cos\theta + \frac{1}{3}(\cos^3\theta - 1)\right]^{-\frac{2}{3}} \tag{10}$$

A plot of $(E_0 - E_s)/E_0$ shows that this ratio decreases rapidly from its value of unity when $\theta = 0$ to a near zero value when $\theta > 160°$. This plot can be used to calculate the critical contact angle required for adhesion of water droplets, with a surface tension $\gamma = 72\,\mathrm{mN\,m^{-1}}$ at 20 °C, of various sizes and velocities. As an illustration, consider a water droplet of 100 μm diameter falling with its terminal velocity v of ~0.25 m s^{-1}. The kinetic energy of the drop is 1.636×10^{-9} J, whereas its surface energy in flight is 2.26×10^{-9} J. The surface energy of the attached drop at which the kinetic energy is just balanced is 2.244×10^{-9} J. The contact angle at which this occurs can be obtained by calculating the fraction $(E_0 - E_s)/E_0$ and interpolation using the above-mentioned plot. This gives $(E_0 - E_s)/E_0 = 0.00723$ and $\theta \approx 160°$. Therefore, provided that droplets of this size form an angle that is less than 160°, they will stick to the leaf surface. It is therefore not surprising that droplets of this size do not need any surfactant for adhesion. For a 200 μm droplet, with a velocity of 1 m s^{-1}, the critical contact angle is 87° and this shows that in this case surfactants are required for adhesion. The higher the velocity of the drop, the smaller is the critical contact angle required for adhesion. With larger drops, this critical contact angle becomes smaller and smaller and this clearly shows the importance of surfactants for ensuring drop adhesion.

It should be mentioned, however, that the above calculations are based on 'idealized' conditions, i.e. droplets falling on a smooth surface. Deviation is expected when dealing with practical surfaces such as leaf surfaces. The latter are rough, containing leaf hairs and wax crystals that are distributed in different ways depending on the nature of the leaf and climatic conditions. Under such conditions, the adhesion of droplets may occur at critical contact angle values that are either smaller or larger than those predicted from the above calculations. The critical θ values will certainly be determined by the topography on the leaf surface. As will be seen later, the definition and measurement of the contact angle on a rough surface are not straightforward. In spite of these complications, experimental results on droplet adhesion seem to support the predictions from the above simple theory. These experimental results showed little dependence of adhesion of spray droplets on surfactant concentration. Since with most spray systems the contact angles obtained were lower than the critical value for adhesion (except for droplets that are larger the 400 μm), then in most circumstances surfactant addition had only a marginal effect on droplet adhesion. However, one should not forget that the surfactant in the spray solution determines the droplet size range. Also, addition of surfactants will certainly affect the adhesion of droplets moving at high velocities and on various plant species. The situation is further complicated by the dynamics of the process, which depend on the nature and concentration of the surfactant added. For fundamental investigations, measurements of the dynamic surface tension and contact angle are required on both model and practical surfaces. These measurements are now easy to perform due to the advances in instrumentation such as the maximum bubble pressure method for measurement of dynamic surface tension and high-speed video equipment for measurement of the dynamic contact angle. Such techniques will permit the formulation chemist and the biologist to understand the role and function of the surfactant in spray solutions.

13.4
Droplet Sliding and Spray Retention

Many agrochemical applications involve high-volume sprays, where with continuous spraying the volume of the drops continue to grow in size by impaction of more spray droplets upon them and by coalescence with neighboring drops on the surface. During this process, the amount of spray retained steadily increases, provided that the liquid drops which are impacted are also retained. However, on further spraying the drops continue to grow in size until they reach a critical value above which they begin to slide down the surface and 'drop off', the so-called 'run-off' condition. At the point of 'incipient run-off', the volume of the spray retained is a maximum. The retention at this point is governed by the movement of the liquid drops on the solid surface. Bikerman [3] stated that the percentage of droplets sticking to a plant after having touched it should depend on the tilt of the leaf, the size of the droplets and the contact angle at the plant leaf/droplet/air interface. However, such a process is complicated and governed by many other factors [4], such as droplet size range, velocity of impacting droplets, volatility and viscosity of the spray liquid and ambient conditions.

Some authors have tried to relate the resistance to movement of liquid drops on a tilted surface to the surface tension and the contact angles (advancing and receding) of liquid droplets with the solid surface. A detailed analysis was given by Furmidge [4], which is summarized below.

Consider a droplet of mass m on a plane surface that is inclined at an angle α from the horizontal (Figure 13.1). Due to gravity, the droplet will start to slide down with a slow constant velocity. Assuming the droplet will have a rectangular plan view (Figure 13.1), with width ω, and it has moved a distance dl, then the work done by the droplet in moving such a distance, W_g, is given by

$$W_g = mg \sin \alpha \tag{11}$$

The above force is opposed by the surface force resulting from wetting and dewetting of the leaf surface as the droplet slides downwards. In moving down, an area $\omega\, dl$ of the leaf is wetted by the droplet and a similar area is dewetted by the trailing edge. The work of wetting per unit area of the surface is equal to $\gamma_{LA}(\cos \theta_A + 1)$, whereas that of dewetting is given by γ_{LA} $(\cos \theta_R + 1)$, where θ_A and θ_R are the advancing and receding contact angles, respectively. Hence the surface force, W_s, is given by

$$W_s = \gamma_{LA}\omega\, dl(\cos \theta_R - \cos \theta_A) \tag{12}$$

At equilibrium, $W_g = W_s$ and

$$\frac{mg \sin \alpha}{\omega} = \gamma_{LA}(\cos \theta_R - \cos \theta_A) \tag{13}$$

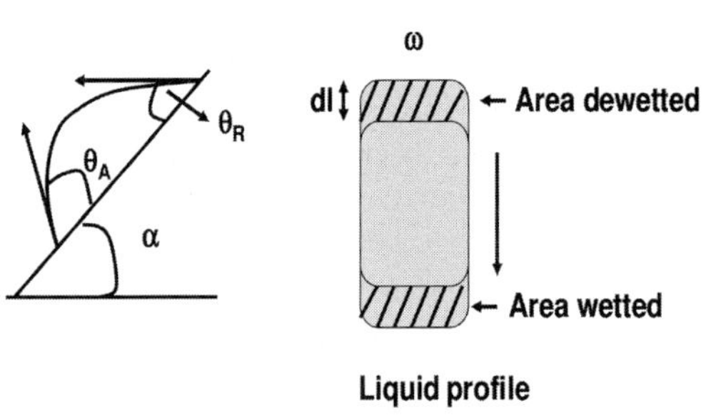

Figure 13.1 Representation of drop profile on a tilted surface.

If the impaction of the spray is uniform and the spray droplets are reasonably homogeneous in size, the total volume of spray retained in an area L^2 of surface is proportional to the time of spraying until the time when the first droplet runs off the surface. Also, the volume v of spray retained per unit area, R, at the moment of 'incipient run-off' is given by

$$R = \frac{kv}{\omega^2} \tag{14}$$

where k is a constant. Equation 14 gives the critical relationship of m/ω for the movement of liquid droplets on a solid surface. As the surface is sprayed, the adhering drops grow in size until the critical value of m/ω is reached and during this period they remain more or less circular in plan form. Since the droplets are small, the deforming effect of gravity may be ignored and they may be regarded as spherical caps whose volume v is given by the expression

$$v = \frac{\pi(1 - \cos\theta)^2(2 + \cos\theta)\omega^3}{24\sin^3\theta} = \frac{m}{\rho} \tag{15}$$

Combining equations 14 and 15, it is possible to obtain an expression for ω, the diameter of adhering droplets, in terms of the surface forces given above, i.e. γ_{LA} and θ:

$$\omega = \left[\frac{24\sin^3\theta_A\gamma_{LA}(\cos\theta_R - \cos\theta_A)}{\pi\rho(1 - \cos\theta_A)^2(2 + \cos\theta)g\sin\alpha}\right]^{\frac{1}{2}} \tag{16}$$

Combining equations 14–16, one obtains an expression of spray retention, R, in terms of γ_{LA} and θ:

$$R = k\left[\frac{\pi\gamma_{LA}(\cos\theta_R - \cos\theta_A)}{24\rho g\sin\alpha}\right]^{\frac{1}{2}}\left[\frac{(1 - \cos\theta_A)^2(2 + \cos\theta_A)}{\sin^3\theta_A}\right]^{\frac{1}{2}} \tag{17}$$

The value of k depends on the droplet size range, since it relates to the rate of build-up of critical droplets and their distribution. However, equation 17 does not take into account the flattening effect of the droplet on impact, which results in a decrease in θ and an increase in ω above the value predicted by equation 16. Hence equation 17 is only likely to be valid under conditions of small impaction velocity. In this case, retention is governed by the surface tension of the spray liquid, the difference between θ_A and θ_R (i.e. the contact angle hysteresis) and the value of θ_A:

Equation 17 can be further simplified by removing the constant terms and standardizing $\sin\alpha$ as equal to 1. A further simplification is to replace the second term in the square brackets on the right-hand side of equation 17 by θ_M, the arithmetic mean of θ_A and θ_R: In this way, a retention factor, F, may be defined by the following simple expression:

$$F = \theta_M \left[\frac{\gamma_{LA}(\cos\theta_R - \cos\theta_A)}{\rho} \right]^{\frac{1}{2}} \tag{18}$$

Equation 18 shows that F depends on γ_{LA}, the difference between θ_R and θ_A and θ_M: At any given value of θ_A and γ_{LA}, F increases rapidly with increase in $\theta_A - \theta_R$, reaches a maximum and then decreases. At any given $\theta_A - \theta_R$ and γ_{LA}, F increases rapidly with increase in θ_A (and also θ_M). With systems having the same contact angles, F increases with increase in γ_{LA} but the effect is not very large since $F \propto \gamma^{1/2}{}_{LA}$: Obviously, any variation in γ_{LA} is accompanied by a change in contact angles and hence one cannot investigate these parameters in isolation. In general, on increasing the surfactant concentration, γ_{LA}, θ_A and θ_R are reduced. The relative extent to which these three values are affected depends on the nature of the surfactant, its concentration and the surface properties of the leaf. This is a very complex problem and predictions are almost impossible.

It should also be mentioned that the above treatment does not take into account the effect of surface roughness and the presence of hairs, which play a significant role. A difference in the amount of liquid retained of up to an order of magnitude may be encountered, at constant F value, between a hairy and a smooth leaf. In addition to these large variations in surface properties between the leaves of various species, there are also variations within the same species depending on age, environmental conditions and position. However, contact angle measurements on leaf surfaces are not easy and one has to make several measurements and subject the results to statistical analysis. Therefore, at best, the measured F values can be used as a guideline to compare various surface-active agents on leaf surfaces of a particular species that are grown under standard conditions.

Several other factors affect retention, of which droplet size range, droplet velocity and wind speed are probably the most important. Usually, retention increases with decrease in droplet size, but significantly reduced at high droplet velocities and wind speeds. The impact of the velocity effect becomes more marked as the receding contact angle decreases. Wind reduces the volume of spray that can be retained, particularly when θ_A and θ_R are fairly large, because little force is required to remove the drop along the surface. As θ_A and θ_R become small, the wind effect becomes less significant and it becomes negligible when θ_A and θ_R are close to zero. The leaf structure is also important, since less spray is lost due to wind movements from leaves with a very rough surface compared with smooth leaves. Therefore, care should be taken when results are obtained on plants grown under standard conditions, such as in glasshouses. These results should not be extrapolated to field conditions, since plants grown under normal environmental conditions may have surfaces that are vastly different from those grown in glasshouses. In order to obtain a realistic picture on spray retention, measurements should be made on field-grown plants and the results obtained may be correlated with those obtained on glasshouse plants. In this case, it is possible to use glasshouse plants for the selection of surfactants, if an allowance is made for the difference between the two sets of results.

13.5
Wetting and Spreading

Another factor which can affect the biological efficacy of foliar spray application of agrochemicals is the extent to which the liquid wets, spreads and covers the foliage surface. This, in turn, governs the final distribution of the agrochemical over the area to be protected. The optimum degree of coverage in any spray application depends on the mode of action of the agrochemical and the nature of the pest to be controlled. With non-systemic agrochemicals, the cover required depends on the mobility or location of the pest. The more static the pest, the greater is the need for complete coverage on those areas of the plant liable to attack. Under those conditions, good spreading of the liquid spray with maximum coverage is required. On the other hand, with systemic agrochemicals satisfactory cover is ensured provided that the spray liquid is brought into contact with those areas of the plant through which the agrochemical is absorbed. Since, as will be seen later, high penetration requires high concentration gradients, an optimum situation may be required here, where one achieves adequate coverage of those areas where penetration occurs, without too much spreading over the total leaf surface since these usually results in 'thin' deposits. These 'thin' deposits do not give adequate 'reservoirs', which are sometimes essential to maintain a high concentration gradient, thus enhancing penetration. In addition, thick deposits which are produced from droplets with limited spreading can increase the tenacity of the agrochemical and ensure the longer term protection by the agrochemical. This situation may be required with many systemic fungicides.

Many leaf surfaces represent the most unwettable of most known surfaces. This is due to the predominantly hydrophobic nature of the leaf surface, which is usually covered with crystalline wax of straight-chain paraffinic alcohols in the range 24–35 carbon atoms. The crystals may be less than 1 μm thick and only a few microns apart, giving the surface 'microroughness', and the 'real' area of the surface can be several times the 'gross' (apparent) area. When a water drop is placed on a leaf surface, it takes the form of a spherical cap that is characterized by the contact angle θ. From the balance of tensions, one obtains the following expression:

$$\gamma_{SA} = \gamma_{SL} + \gamma_{LA} \cos \theta \tag{19}$$

Equation 19 is the familiar Young's equation, which applies to a liquid drop on a smooth surface.

Wetting is sometimes simply assessed by the value of the contact angle; the smaller the angle, the better the liquid is said to wet the solid. Complete wetting implies a contact angle of zero, whereas complete non-wetting dictates an angle of contact of 180°. However, contact angle measurements are not easy on real surfaces since great variations in the value are obtained at various locations on the surface. In addition, it is very difficult to obtain an equilibrium value, due to the heterogeneity of the surface and its roughness. Therefore, in most practical systems such as spray drops on leaf surfaces, the contact angle exhibits hysteresis, i.e. its value depends on the history of the system and varies according to whether the given liquid is tending to advance across or recede from the leaf surface. The limiting angles achieved just prior to movement of the wetting line (or just after movement ceases) are known as the advancing and receding contact angles, θ_A and θ_R, respectively.

For a given system, $\theta_A > \theta_R$ and θ can usually take any value between these two limits without discernible movement of the wetting line. Since smaller angles imply better wetting, it is clear that the contact angle always changes in such a direction as to oppose wetting line movement.

The use of contact angle measurements to assess wetting depends on equilibrium thermodynamic arguments, which unfortunately does not represent the real situation. In the practical situation of spraying, the liquid has to displace the air or another fluid attached to the leaf surface and hence measurement of dynamic contact angles, i.e. those associated with moving wetting lines, is more appropriate. Such measurements require special equipment such as video cameras and image analysis and they should enable one to obtain a more accurate assessment of wetting by the spray liquid.

As mentioned above, the contact angle often undergoes hysteresis so that θ cannot be defined unambiguously by experiment. This hysteresis is accounted for by surface roughness, surface heterogeneity and metastable configurations. Surface roughness can be taken into account by introducing a term r in Young's equation, where r is the ratio of real to apparent surface area:

$$r\gamma_{SA} = r\gamma_{SL} + \gamma_{LA} \cos \theta \tag{20}$$

Hence the contact angle on a rough surface is given by the expression

$$\cos \theta = \frac{r(\gamma_{SA} - \gamma_{SL})}{\gamma_{LA}} \tag{21}$$

In other words, the contact angle on a rough surface, θ, is related to that on a smooth surface, θ^0, by the equation

$$\cos \theta = r \cos \theta^0 \tag{22}$$

Equation (22) shows that surface roughness increases the magnitude of $\cos \theta^0$, whether its value is positive or negative. If $\theta^0 < 90°$, $\cos \theta^0$ is positive and it becomes more positive as a result of roughness, i.e. $\theta < \theta^0$, or roughness in this case enhances wetting. In contrast, if $\theta^\circ > 90°$, $\cos \theta^0$ is negative and roughness increases the negative value of $\cos \theta^0$, i.e. roughness results in $\theta > \theta^0$. This means that if $\theta^0 > 90°$, roughness makes the surface even more difficult to wet.

The influence of surface heterogeneity was analyzed by Cassie and Baxter [5], as was described in Chapter 11. The possibility of adoption of metastable configurations as a result of surface roughness was suggested by Deryaguin [6]. He considered the wetting line to move in a series of thermodynamically irreversible jumps from one metastable configuration to the next.

Assuming an idealized rough surface consisting of concentric patterns of sinusoidal corrugations, one may relate the apparent contact angle θ to that on a smooth surface θ^0 by the following simple equation:

$$\theta = \theta^0 + \alpha \tag{23}$$

where α is the slope of the solid surface at the wetting line. The value of θ is, therefore, dependent on the location of the wetting line and, hence, upon factors such as drop volume and gravitational forces. A model heterogeneous surface may be represented by a series of concentric bands having alternate characteristic contact angles θ' and θ'', such that $\theta' > \theta''$. A drop of a liquid placed on this type of a surface will spread or retract until the wetting line assumes some configuration such that $\theta' > \theta > \theta''$.

In spite of the above complications, measurement of contact angles of spray liquids on leaf surfaces is still most useful in defining the wetting and spreading of the spray. A very useful index for measurement of spreading of a liquid on a solid surface is Harkin's spreading coefficient, *S*, which is defined by the change in tension when solid/liquid and liquid/air interfaces are replaced by a solid/air interface. In other words, *S* is the work required to destroy a unit area each of the solid/liquid and liquid/air interfaces while forming a unit area of the solid/air interface, i.e.

$$S = \gamma_{SA} - (\gamma_{SL} + \gamma_{LA}) \tag{24}$$

If *S* is positive, the liquid will usually spread until it completely wets the solid. If *S* is negative, the liquid will form a non-zero contact angle. This can be clearly shown if equation 24 is combined with the Young's equation:

$$S = \gamma_{LV}(\cos\theta - 1) \tag{25}$$

Clearly, if $\theta > 0$, *S* is negative and this implies only partial wetting. In the limit $\theta = 0$, *S* is equal to zero and this represents the onset of complete wetting. A positive *S* implies rapid spreading of the liquid on the solid surface. Indeed, by measuring the contact angle only, one can define a spread factor, *SF*, which is the ratio of the diameter of the area wetted on the leaf, D, to the diameter of the drop, *D*:

$$SF = \frac{D}{d} \tag{26}$$

Provided that θ is not too small (>5°), the spread factor can be calculated from θ:

$$SF = \left[\frac{4\sin^3\theta}{(1-\cos\theta)^2(2+\cos\theta)}\right]^{\frac{1}{3}} \quad (27)$$

A plot of SF versus θ shows a rapid increase in SF when $\theta < 35°$. The most practical method of measuring the spread factor is to apply drops of known volume using a microapplicator on the leaf surface. By using a tracer material, such as a fluorescent dye, one may be able to measure the spread area directly using, for example, image analysis. This area can be converted to an equivalent sphere, allowing D to be obtained.

An alternative method of defining wetting and spreading is through measurement of the work of adhesion, W_A, which is the work required to separate a unit area of the solid/liquid interface to leave a unit area each of the liquid/air and solid/air, respectively:

$$W_a = (\gamma_{LA} + \gamma_{SA}) - \gamma_{SL} \quad (28)$$

Again using the Young's equation, one obtains the following expression for W_A:

$$W_a = \gamma_{LA}(\cos\theta + 1) \quad (29)$$

Another useful concept for assessing the wettability of surfaces is that introduced by Zisman *et al.* [7], namely the critical surface tension of wetting, γ_c: They found that for a given surface and a series of related liquids such as *n*-alkanes, siloxanes or dialkyl ethers, cos θ is a reasonably linear function of γ_{LA}: The surface tension at the point where the line cuts the $\cos\theta = 1$ axis is known as the critical surface tension of wetting, γ_c: It is the surface tension of a liquid that would just spread to give complete wetting.

Several authors have tried to relate the critical surface tension to the solid/liquid interfacial tension or at least its dispersion component, $\gamma^d{}_S$: From the above discussion, it is clear that for enhancement of wetting and spreading of liquids on leaf surfaces, one needs to lower the contact angle of the droplets. This is usually achieved by the addition of surfactants, which adsorb at various interfaces and modify the local interfacial tension. Since most leaf surfaces are nonpolar, low-energy surfaces, an increase in surfactant concentration enhances wetting. This explains why most agrochemical formulations contain high concentrations of surfactants to enhance wetting and spreading. However, as will be seen later, surfactants play other roles in deposit formation, distribution of the agrochemical on the target surface and enhancement of penetration of the chemical.

Although the role played by a surfactant is complex, these materials, sometimes referred to as wetting agents or simply adjuvants, need to be carefully selected for optimization of biological efficacy. To date, surfactants are still selected by the formulation chemist on the basis of a trial and error procedure. However, some guidelines may be applied in such selection. As discussed in

Chapter 6, the HLB system may be initially applied for choosing the most common wetting agents. The latter have HLB numbers between 7 and 9. As discussed in Chapter 2, nonionic surfactants usually have a two orders of magnitude lower critical micelle concentration (cmc) compared with their ionic counterparts with the same alkyl chain length. Since the limiting value of the surface tension is reached at concentrations above the cmc, it is clear that many nonionic surfactants are more effective as wetting agents since, after dilution of the formulation, the concentration of the nonionic surfactant in the spray solution may be higher than its cmc. However, many nonionic surfactants with HLB numbers in the range 7–9 undergo phase separation at high concentrations and/or temperatures. This may limit their incorporation in the formulation at high concentrations. In some cases, addition of a small amount of an ionic surfactant may be beneficial in reducing this phase separation and raising the cloud point of the nonionic surfactant. Thus, many agrochemical formulations contain complex mixtures of surfactants which are carefully arrived at by the formulation chemist. The composition of such mixtures is usually kept confidential.

Another important property of the surfactant that is selected for a given agrochemical is its effect on the leaf structure and the cuticle. Surfactants that cause significant damage to the leaf are described as phytotoxic and in many crops such damage must be avoided. This can sometimes limit the choice, since in some cases the best wetter may not be the best from the phytotoxicity point of view, and a compromise has to be made. This shows that selecting the surfactant can be difficult and requires careful investigation of many surface chemical properties and also its interaction with the leaf surface and the cuticle. In addition, its effects on deposit formation and penetration of the agrochemical need to be separately investigated.

13.6
Evaporation of Spray Drops and Deposit Formation

The object of spraying is often to leave a long-lasting deposit of a particulate fungicide or insecticide or a residue able to penetrate the cuticle in the case of systemic pesticides and herbicides or to be transferred locally within the crop by its own slower evaporation. The form of residue left by evaporation of the carrier liquid depends to a large extent on the rate of evaporation and most importantly on the nature and concentration of the surfactant and other ingredients in the formulations. Evaporation from a spray drop tends to occur most rapidly near the edges since these receive the necessary heat most rapidly from the air by conduction through the dry surround of the leaf. This results in a higher concentration of surfactant at the edge, causing surface tension gradients and convection (arising from the associated density difference). Surface tension gradients cause a Marangoni effect (see Chapter 5) with liquid circulation within the drop that causes the particles to be preferentially deposited at the edge. Convection within the drop leads to preferential precipitation near the edge because the particles can first become 'wedged' between the solid/liquid and liquid/air interfaces.

The type and composition of the spray deposit depends to a large extent on the type of formulation and also the concentration and type of dispersing agent (for suspensions) or emulsifier (for emulsifiable concentrates and emulsions). Additives such as wetters, humectants and stickers also affect the nature of the deposit. It should also be mentioned that during evaporation of a spray droplet containing dispersed particles or droplets, these may undergo some physical changes on drying. For example, the solid particles of a suspension may undergo recrystallization forming differently shaped particles, which will affect the final form of the deposit. Both suspension particles and emulsion droplets may also undergo flocculation, coalescence and Ostwald ripening, all of which affect the nature of the deposit. Following such changes during evaporation is not easy and requires special techniques such as microscopy and differential scanning calorimetry (DSC).

Another important factor with deposits is the tenacity of the resulting particles or droplets. Strong adhesion between the particles or droplets and the leaf surface is required to prevent removal of these particles or droplets by rain. The adhesion forces between a particle or droplet are determined by the van der Waals attraction and the area of contact between the particles and the surface. Several other factors may affect adhesion, namely electrostatic attraction and chemical and hydrogen bonding. The area of contact between the particle and the surface is determined by its size and shape. It is obvious that by reducing the particle size of a suspension, one increases the total area of contact between them and the leaf surface, compared with coarser particles of the same total mass. The shape of the particle also affects the area of contact. For example, flat or cube-like particles will have larger areas of contact than needle-shaped crystals of the same equivalent volume. Several other factors may affect adhesion, such as the water solubility of the agrochemical. In general, the lower the solubility, the greater is the rain fastness.

One of the most important factors that affect deposit formation is the phase separation that occurs during evaporation. As discussed in Chapter 2, surfactants form liquid crystalline phases when their concentration exceeds a certain value that depends on the nature of the surfactant, it hydrocarbon chain length and the nature and length of the hydrophilic portion of the molecules. During evaporation, liquid crystals of very high viscosity such as hexagonal or cubic phases may be produced first. Such highly viscous (and elastic structures) will incorporate any particles or droplets and they act as reservoirs for the chemical. As a result of solubilization of the chemical, penetration and uptake may be enhanced (see below). With further evaporation, the hexagonal and cubic phases may produce lamellar structures with lower viscosity than former phases. Such structures will affect the distribution of particles or droplets in the deposit. Hence the choice of a particular surfactant for an agrochemical formulation necessitates study of its phase diagram in order to identify the nature of the liquid crystalline phases that are produced on increasing its concentration. The effect of temperature on the liquid crystalline structures is also important. Liquid crystalline structures 'melt' above a critical temperature, producing liquid phases that contain micelles. These liquid phases have much lower viscosity and hence the particles or droplets of the agrochemical within these liquid phases become mobile. The temperature at which such melting occurs depends on the structure of the surfactant molecule and hence the choice depends on the mode of action of the agrochemical and the environmental conditions encountered, such as temperature and humidity. It should also be mentioned that the liquid crystalline structures will be affected by other additives in the formulation such as antifreeze and electrolytes. In addition, the particles or droplets of the agrochemical may affect the liquid crystalline structures produced, and this requires a detailed study of the phase diagram in the presence of the various additives and also in the presence of the agrochemical. Various methods may be applied for such investigations, such as polarizing microscopy, DSC and rheology.

13.7 Solubilization and its Effect on Transport

Solubilization is the incorporation of an 'insoluble substance' (usually referred to as the substrate) into surfactant micelles (the solubilizer). Solubilization may also be referred to as the formation of a thermodynamically stable, isotropic solution of a substance, normally insoluble or slightly soluble in water, by the introduction of an additional amphiphilic component or components. Solubilization can be determined by measuring the concentration of the chemical that can be incorporated in a surfactant solution while remaining isotropic as a function of its concentration. At concentrations below the cmc, the amount of chemical that can be incorporated in the solution increases slightly above its solubility in water. However, just above the cmc, the concentration of the chemical that can be incorporated in the micellar solution increases rapidly with further increase in surfactant concentration. This rapid increase, just above the cmc, is usually described as the onset of solubilization. One may differentiate three different locations of the substrate in the micelles. The most common location is in the hydrocarbon core of the micelle. This is particularly the case for a lipophilic nonpolar molecule, as is the case with most agrochemicals. Alternatively, the substrate may be incorporated between the surfactant chains of the micelle, i.e. by co-micellization. This is sometimes referred to as penetration in the palisade layer, in which one may distinguish between deep and short penetration. The third way of incorporation is by simple adsorption on the surface of the micelle. This is particularly the case with polar compounds.

Several factors affect solubilization, of which the structures of the surfactant and solubilizate, temperature and addition of electrolyte are probably the most important. Generalizations about the manner in which the structural characteristics of the surfactant affect its solubilizing capacity are complicated by the existence of different solubilization sites within the micelles. For deep penetration within the hydrocarbon core of the micelle, solubilization increases with increase in the alkyl chain length of the surfactant. On the other hand, if solubilization occurs in the hydrophilic portion of the surfactant molecules, e.g. its poly(ethylene oxide) chain, then the capacity increases with increase in the hydrophilic chain length. The solubilizate structure can also play a major role. For example, polarity and polarizability, chain branching, molecular size and shape and structure have been shown to have various effects. The temperature also has an effect on the extent of micellar solubilization which is dependent on the structures of the solubilizate and of the surfactant. In most cases, solubilization increases with increase in temperature. This is usually due to the increase in the solubility of the solubilizate and increase in the micellar size with nonionic ethoxylated surfactants. Addition of electrolytes to ionic surfactants usually causes an increase in the micelle size and a decrease in the cmc and hence an increase in the solubilization capacity. Non-electrolytes which are capable of being incorporated in the micelle, e.g. alcohols, lead to an increase in the micelle size and hence to an increase in solubilization.

As discussed in Chapter 10, microemulsions, which may be considered as swollen micelles, are more effective in the solubilization of many agrochemicals. Oil-in-water microemulsions contain a larger hydrocarbon core than surfactant micelles and hence they have a larger capacity for solubilizing lipophilic molecules such as agrochemicals. However, with polar compounds, O/W microemulsions may not be as effective as micelles of ethoxylated surfactants in solubilizing the chemical. Therefore, one has to be careful in applying microemulsions without knowledge of the interaction between the agrochemical and the various components of the microemulsion system.

The presence of micelles or microemulsions will have significant effects on the biological efficacy of an insoluble pesticide. In the first instance, surfactants will affect the rate of solution of the chemical. Below the cmc, surfactant adsorption can aid wetting of the particles and consequently increase the rate of dissolution of the particles or agglomerates. Above the cmc, the rate of dissolution is affected as a result of solubilization. According to the Noyes–Whitney relation [8], the rate of dissolution is directly related to the surface area of the particles, A, and the saturation solubility, C_s:

$$\frac{dC}{dt} = kA(C_s - C) \tag{30}$$

where C is the concentration of the solute.

Higuchi [9] assumed that an equilibrium exists between the solute and solution at the solid/solution interface and that the rate of movement of the solute into the bulk is governed by the diffusion of the free and solubilized solute across a stagnant layer. Hence the effect of surfactant on the dissolution rate will be related to the dependence of that rate on the diffusion coefficient of the diffusing species and not on their solubilities as suggested by equation. However, experimental results have not confirmed this hypothesis and it was concluded that the effect of solute solubilization involves more steps than a simple effect on the diffusion coefficient. For example, it has been argued that the presence of surfactants may facilitate the transfer of solute molecules from the crystal surface into solution, since the activation energy of this process was found to be lower in the presence of a surfactant than its absence in water. On the other hand, Chan *et al.* [10] considered a multi-stage process in which surfactant micelles diffuse to the surface of the crystal, become adsorbed (as hemimicelles) and form mixed micelles with the solubilizate. The latter is dissolved and it diffuses away into bulk solution, removing the solute from the crystal surface. This multi-stage process, which directly involves surfactant micelles, will probably enhance the dissolution rate.

Apart from the above effect on dissolution rate, surfactant micelles also affect the membrane permeability of the solute. Solubilization can, under certain circumstances, aid the transport of an insoluble chemical across a membrane. The driving force for transporting the substance through an aqueous system is always the difference in its chemical potential (or to a first approximation to the difference in its relative saturation) between the starting point and its destination. The principal steps involved are dissolution, diffusion or convection in bulk liquid and crossing of a membrane. As mentioned above, solubilization will enhance the diffusion rate by affecting transport away from the boundary layer adjacent to the crystal. It should be mentioned, however, that to enhance transport the solution should remain saturated, i.e. excess solid particles must be present since an unsaturated solution has a lower activity.

Diffusion in bulk liquid obeys Fick's first law, i.e.

$$J_\mathrm{D} = D\left(\frac{\partial C}{\partial x}\right) \tag{31}$$

where J_D is the flux of solute (amount of solute crossing a unit cross-section in unit time), D is the diffusion coefficient and $\partial C/\partial x$ is the concentration gradient. The presence of the chemical in a micelle will lower D since the radius of a micelle is obviously greater than that of a single molecule. Since the diffusion coefficient is inversely proportional to the radius of the diffusing particle, D is generally reduced when the molecule is transported by a micelle. Assuming that the volume of the micelle is about 100 times greater than that of a single molecule, the radius of the micelle will only be about 10 times larger than that of a single molecule. Hence D will be reduced by a factor of about 10 when the molecule diffuses within a micelle compared with that of a free molecule. However, the presence of micelles increases the concentration gradient in direct proportionality to the increase in incorporation of the chemical by the micelle. This is because Fick's law involves the absolute concentration gradient, which is necessarily small provided that the solubility is small, and not its relative rate. If the saturation is represented by S, Fick's law may be written as

$$J_\mathrm{D} = D \times 100S\left(\frac{\partial \% S}{\partial x}\right) \tag{32}$$

where $\partial\%S/\partial x$ is the gradient in relative value of S. Equation 32 shows that for the same gradient of relative saturation, the flux caused by diffusion is directly proportional to saturation. Hence solubilization will in general increase transport by diffusion, since it can increase the saturation value by many orders of magnitude (which outweighs the decrease in D).

Solubilization also increases transport by convection since the flux of this process, J_C, is directly proportional to the velocity of the moving liquid and the concentration of the solute, C. Moreover, one would expect that solubilization enhances transport through a membrane by an indirect mechanism. Since solubilization reduces the steps involving diffusion and convection in bulk liquid, it permits the application of a greater fraction of the total driving force to transport through the membrane. In this way, solubilization accelerates the transport through the membrane, even if the resistance to this step remains unchanged. It should also be mentioned that enhancement of transport as a result of solubilization does not necessarily involve transport of any micelles. The latter are generally too large to pass through membranes.

The above discussion clearly demonstrates the role of surfactant micelles in the transport of agrochemicals. Since the droplets applied to foliage undergo rapid evaporation, the concentration of the surfactant in the spray deposits can reach very high values which allow considerable solubilization of the agrochemical. This will certainly enhance transport, as discussed above. Since the lifetime of a micelle is relatively short, usually less than 1 ms (see Chapter 2), such units break up quickly, releasing their contents near the site of action, and produce a large flux by increasing the concentration gradient. However, there have been few systematic investigations to study this effect in more detail and this should certainly be a topic of research in the future.

13.8 Interaction Between Surfactant, Agrochemical and Target Species

For the selection of adjuvants that can be used for the enhancement of biological efficacy, one has to consider the specific interactions that may take place between the surfactant, agrochemical and target species. This is usually described in terms of an activation process for uptake of the chemical into the plant. This mechanism is particularly important for systemic agrochemicals.

Several key factors may be identified in the uptake activation process: (1) in the spray droplet; (2) in the deposit formed on the leaf surface; (3) in the cuticle before or during penetration; and (4) in tissues underlying the site of application. Four main sites were considered by Stock and Holloway [11] for the increase in uptake of the agrochemical into a leaf: (1) on the surface of the cuticle; (2) within the cuticle itself; (3) in the outer epidermal wall underneath the cuticle; and (4) at the cell membrane of internal tissues.

The activator surfactant is initially deposited together with the agrochemical and it can penetrate the cuticle, reaching other sites of action, hence the role of surfactant in the activation process can be very complex. The net effect of surfactant interactions at any of the sites of action is to enhance the mass transfer of an agrochemical from a solid or liquid phase on the outside of the cuticle to the aqueous phase of the internal tissues of the treated leaf. As discussed above, solubilization can play a major role in activating the transport of the agrochemical molecules. With many nonpolar systemic fungicides, which are mostly applied as suspension concentrates, the presence of micelles can enhance the rate of dissolution of the chemical and this results in increased availability of the molecules. It also leads to an increase in the flux, as discussed above.

It has been suggested that cuticular wax can be solubilized by surfactant micelles (by the same mechanism as solubilization of the agrochemical). However, no evidence could be presented (for example, using scanning electron microscopy) to show the wax disruption by the micelles. Schönherr [12] suggested that the surfactants interact with the waxes of the cuticle and thus increase the fluidity of this barrier. This hypothesis is sometimes referred to as wax 'plasticization' (similar to the phenomenon of glass transition temperature reduction of polymers by addition of plasticizers). Some measurements of uptake using surfactants with various molecular weights and HLB numbers offered some support for this hypothesis.

Several other mechanisms were suggested by Stock and Holloway [11] for uptake activation: (1) prevention of crystal formation in deposits; it is often assumed that the foliar uptake of an agrochemical from a crystalline deposit will be less that from an amorphous deposit; (2) retention of moisture in deposits by humectant action; the humectant theory has arisen mainly from the observation that the uptake of highly soluble chemicals was promoted by high-EO surfactants such as Tween 20; and (3) promotion of uptake of solutions via stomatal infiltration; this hypothesis stemmed from the observation of rapid uptake of agrochemicals (within the first 10 min) when using superwetters such as Silwett L-77, which is capable of reducing the surface tension of water to values as low as 20 mN m^{-1}.

References

1 R.T. Brunskill, in *Proceedings of the Third Weed Conference*, Association of British Manufacturers, p. 593 (1956).

2 G.S. Hartley and R.G. Brunskill, *Surface Phenomena in Chemistry and Biology*, Pergamon Press, Oxford, pp. 214–223 (1958).

3 J.J. Bikerman, *Ind. Eng. Chem.*, *13*, 443 (1941).

4 C.G.L. Furmidge, *J. Colloid Interface Sci.*, *17*, 309 (1962).

5 A.D.B. Cassie and S. Baxter, *Trans. Faraday Soc.*, *40*, 546 (1944).

6 B.V. Deryaguin, *C. R. Acad. Sci. USSR*, *51*, 361 (1946).

7 W.A. Zisman, *Adv. Chem. Ser.*, *43* (1964).

8 D. Attwood and A.T. Florence, *Surfactant Systems*, Chapman and Hall, London (1983).

9 W.I. Higuchi, *J. Pharm. Sci.*, *53*, 532 (1964); *56*, 315 (1967).

10 A.F. Chan, D.F. Evans and E.L. Cussler, *AIChE J.*, *22*, 1006 (1976).

11 D. Stock and P.J. Holloway, *Pestic. Sci.*, *37*, 233 (1993).

12 J. Schönherr, *Pestic. Sci.*, *39*, 213 (1993).

Subject Index

Colloids and Interface Science Series, Vol. 5
Colloids in Agrochemicals. Tharwat F. Tadros

ISBN: 978-3-527-31465-2

t